T0257780

Biodegradation: Science and Technology

Biodegradation: Science and Technology

Edited by **William Chang**

New York

Published by Callisto Reference,
106 Park Avenue, Suite 200,
New York, NY 10016, USA
www.callistoreference.com

Biodegradation: Science and Technology
Edited by William Chang

International Standard Book Number: 978-1-63239-090-5 (Hardback)

Printed in the United States of America.

Contents

Preface

This book aims to highlight the current researches and provides a platform to further the scope of innovations in this area. This book is a product of the combined efforts of many researchers and scientists, after going through thorough studies and analysis from different parts of the world. The objective of this book is to provide the readers with the latest information of the field.

This book presents an array of various research works discussing different technologies that have been used for the escalation of biodegradation process. The book deals with various factors and aspects of biodegradation. These include hydrocarbons biodegradation, and biodegradation and anaerobic digestion.

I would like to express my sincere thanks to the authors for their dedicated efforts in the completion of this book. I acknowledge the efforts of the publisher for providing constant support. Lastly, I would like to thank my family for their support in all academic endeavors.

Editor

Biodegradation of Hydrocarbons

Biodegradability of Water from Crude Oil Production

Edixon Gutiérrez and Yaxcelys Caldera

Additional information is available at the end of the chapter

1. Introduction

According to Gutiérrez *et al.* (2007) the waters of formation (WOF), are those that are naturally in the rocks and are present before the perforation of the well. Their composition depends on the origin of the water and the modification that could happen as soon as they enter in contact with the environment of the subsoil. WOF must be obtained from the bottom of the well; nevertheless, for costs reason the samples are taken at the surface level, in the head of the well. As they rise in the column from the well up to the surface, their characteristics change due to the changes of pressure, temperature and composition of the gases. For this reason the name adapted for these samples of waters is water associated with crude oil production. Other researchers name these waters as water from petroleum, water from oil field production, oily waters, effluent from the extraction of oil, water from petroleum. In this work they are named waters associated with crude oil production (WCP).

Among the characteristics of WCP are their high content of free and emulsified crude oil and hydrocarbons, suspended solid, H₂S and mercaptans (Gutiérrez *et al.*, 2002), aromatic, poliaromatic and phenols compounds (Rincón *et al.*, 2008), high temperature and high salinity (Guerrero *et al.*, 2005; Li *et al.*, 2005), saturated, aromatics, resins and asphaltenes compounds (SARA) (Díaz *et al.*, 2007), and metal traces Na, Ca, Mg, Fe, Sr, Cr, As and Hg (Gutiérrez *et al.*, 2009). According to García *et al.* (2004) among the pollutants with a major potential impact related to the petroleum industry are polycyclic aromatic hydrocarbons (PAH), voltaic organic compounds (VOC) and total hydrocarbons of the oil (THO). The first ones have high carcinogenic, mutagenic and teratogenic potential in aquatic organisms; the second ones contribute to the greenhouse effect and are involved in the direct ozone formation on the soil level and indirectly on the acid rain, besides some individual compounds are toxic, carcinogenic, mutagenic or bioaccumulative, and the last ones present diverse effects on the flora an fauna.

Gives that the WCP volumes generated in the Ulé tank farm, on the east cost of Maracaibo Lake, Venezuela, belonging to the petroleum industry in Venezuela, would exceed the needs for secondary recovery and the systems of reinjection would be rapidly saturated, different research works were realized to present alternatives to the petroleum industry, to diminish the potential pollutant of WCP.

In this aspect, some proposals for the treatment of WCP are aerobic and anaerobic biological processes, physicochemical treatment and some new technologies as constructed wetlands. Among the anaerobic processes are the batch reactors (BR) and the upflow anaerobic sludge blanket reactors (UASB).

The biological mesophilical and thermophilical anaerobic systems have been successful in the treatment of complex waters, with low, moderate and high organic load (Lettinga, 2001). In the case of UASB, these reactors are outlined by their capacity to retain biomass, to form granular sludge with high properties of sedimentation, to handle high organic load to short hydraulic retention time (HRT), produce biogas and remove high concentration of biodegradable organic matter (Lepisto and Rintala, 1990; Lettinga, 2005).

On the other hand, the aerobic systems have been efficient for the treatment of wastewater containing chemical compounds resistant to be biodegraded. Among these systems are the sequential biological reactors (SBR), which have showed excellent results in the degradation of toxic compounds present in industry effluents (Díaz et al., 2005a; González et al., 2007). As well as, the rotating biological contactor reactors (RBC), which produce good quality effluents including total nitrification, low costs and ease of operation and maintenance (Behling et al., 2003).

Among the physicochemical treatment applied to reduce the pollutants in wastewater are the dissolve air flotation (DAF) and the coagulation. The most applied products to treat natural water and wastewater by coagulation and flocculation are iron and aluminium salts. However, the cationic polymers have demonstrated their efficiency in the removal of oils and phenols from industrial wastewater (Renault et al., 2009; Ahmad et al., 2006).

In this investigation was reviewed a several papers from studies conducted at the Universidad del Zulia during 2002 to 2012, to analyze the efficiency of biological and physicochemical systems BR, UASB, SBR and RBC, and the physicochemical treatment as coagulation and flotation (DAF), which have been evaluated to remove COD, hydrocarbons, SARA and phenols, present in the WCP.

The instrument used was a matrix register of the treatment, considering criteria like WCP type, system of treatments, operation conditions, organic load, retention times, temperature, pollutant contents and dose of coagulant. The efficiency of the treatments was compared considering the parameters COD, phenols, hydrocarbons and SARA.

2. Results

2.1. Origin and composition of the waters associated with the crude oil production

The WCP samples were obtained from the Ulé tank farm, located on the east coast of Maracaibo Lake, Tía Juana, Zulia state, Venezuela (Figure 1). The water samples come from the segregations: Tía Juana light (TJL), Urdaneta heavy (UH), Tía Juana medium (TJM), and the dehydrations of the Punta Gorda tank farm (Rosa medium-RM), Shell Ulé (F-6/h-7) and lacustrine terminal of La Salina (LTLS). These waters were obtained from the separation of the water associated with the extraction of light crude oil (>31.8°API) WCPL, from the water associated with the extraction of medium crude oil (22°API-29.9°API) WCPM, from the water associated with the extraction of heavy crude oil (10°API-21.9°API) WCPH, classified according to the American Petroleum Institute. Also, water samples were taken from the converged point of the three cuts (WCPC).

The Tables 1, 2, 3 and 4 present the principal characteristics of WCPL, WCPM, WCPH and WCPC, respectively. In general, it is observed that the physicochemical characteristics of the WCP are different depending on the contact of these waters with the crude oil associated. They are waters with high pollutant contents and they do not comply with the Venezuelan environmental regulations to be discharged into water bodies (Gaceta Oficial, 1995). On the other hand, the differences in the characteristics reported by the researchers, might be related to the changes that have been given in the productive processes of the petroleum industry in the last years.

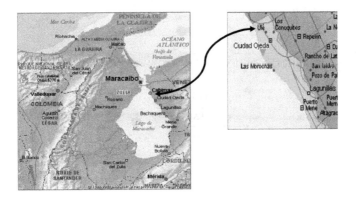

Figure 1. Geographical location of the Ulé tank farm, Tía Juana Zulia state, Venezuela.

Parameters	Díaz et al. (2005a)	Díaz et al. (2005b)	Gutiérrez et al. (2012)	González et al. (2007)	Rincón et al. (2008)
pH	7.9	8.0	8.3	7.99	NR
Alkalinity (mg CaCO₃/L)	2933	2215	2670	2412	NR
COD soluble (mg/L)	1065.2	799	1400	1105	106.2
Phenols (mg/L)	19.36	1.73	NR	16.8	NR
Nitrogen NTK (mg/L)	23.82	28.8	20	21.2	23.82
Phosphorous (mg/L)	1.07	1.0	7.7	1.57	1.07
Hydrocarbons (mg/L)	NR	91	224.2	78.0	NR
Chlorines (mg/L)	NR	NR	NR	NR	NR
TSS (mg/L)	NR	NR	104	NR	NR
VSS (mg/L)	NR	NR	54	NR	NR
O&G (mg/L)	NR	NR	66	100.7	NR
Saturated (mg/L)	NR	NR	76.6*	NR	1.24
Aromatics (mg/L)	NR	NR	7.04*	NR	17.64
Resins (mg/L)	NR	NR	6.34*	NR	8.51
Asphaltenes (mg/L)	NR	NR	7.73*	NR	7.49

*Values in (%), NR: No register

Table 1. Physicochemical parameters of WCPL from tank farm of Ulé

Parameters	Díaz et al. (2005a)	Gutiérrez et al. (2012)	Rincón et al. (2008)	Castro et al. (2008)
pH	8.0	8.5	NR	8.04
Alkalinity (mg CaCO₃/L)	3440	2800	NR	2906
COD soluble (mg/L)	782.6	933	782.6	880
Phenols (mg/L)	1.40	NR	NR	NR
Nitrogen NTK (mg/L)	39.20	15.1	39.20	NR
Phosphorous (mg/L)	1.05	3.5	1.05	NR
Hydrocarbons (mg/L)	NR	148.7	NR	NR
Chlorines (mg/L)	NR	NR	NR	NR
TSS (mg/L)	NR	NR	NR	82.57
VSS (mg/L)	NR	NR	NR	69.71
Saturated (mg/L)	NR	25.32*	5.73	0.24

Parameters	Díaz et al. (2005a)	Gutiérrez et al. (2012)	Rincón et al. (2008)	Castro et al. (2008)
Aromatics (mg/L)	NR	5.86*	9.77	50.34
Resins (mg/L)	NR	6.49*	5.30	33.22
Asphaltenes (mg/L)	NR	5.99*	5.30	16.10

*Values in (%), NR: No register

Table 2. Physicochemical parameters of WCPM from tank farm of Ulé

Parameters	Díaz et al. (2005a)	Gutiérrez et al. (2012)	González et al. (2007)	Gutiérrez et al. (2009)	Caldera et al. (2011)
pH	8.0	8.2	8.3	7.08	8.41
Alkalinity (mg CaCO$_3$/L)	885	1000	885	NR	803.33
COD soluble (mg/L)	307	864	320	1029	259.6
Phenols (mg/L)	2.70	NR	2.5	NR	0.83
Nitrogen NTK (mg/L)	10.61	15.7	9.2	8.26	5.60
Phosphorous (mg/L)	2.68	2.0	9.8	0.013	3.01
Hydrocarbons (mg/L)	NR	52.7	78	35.0	123.21
Chlorines (mg/L)	NR	NR	NR	NR	1101.21
TSS (mg/L)	NR	NR	NR	NR	573.33
VSS (mg/L)	NR	NR	NR	NR	220.00
Color (CU)	NR	NR	NR	NR	718.80
Turbidity (NTU)	NR	NR	NR	NR	140.00
Chrome (mg/L)	NR	NR	NR	4.75	NR
Lead (mg/L)	NR	NR	NR	4.35	0.0
Sodium (mg/L)	NR	NR	NR	89.94	NR
Zinc (mg/L)	NR	NR	NR	2.50	0.30
O&G (mg/L)	NR	NR	113.3	NR	NR
Saturated (mg/L)	NR	23.97*	NR	NR	NR
Aromatic (mg/L)	NR	6.15*	NR	NR	NR
Resins (mg/L)	NR	64.7*	NR	NR	NR
Asphaltenes (mg/L)	NR	5.14*	NR	NR	NR

*Values in (%). NR: No register

Table 3. Physicochemical parameters of WCPH from tank farm of Ulé

Parameters	Behling et al. (2003)[a]	Rincón et al. (2004)[a]	Rojas et al. (2008)[b]	Blanco et al. (2008)[c]
pH	7.72	8	7.74	8.03
Alkalinity (mg CaCO₃/L)	2460	2238	2477	2635
COD soluble (mg/L)	823	NR	NR	1391.85
COD total (mg/L)	NR	700	NR	NR
Phenols (mg/L)	NR	5	NR	2.14
Nitrogen NTK (mg/L)	12.92	NR	NR	17.55
Phosphorous (mg/L)	1.40	NR	NR	3.67
Hydrocarbons (mg/L)	NR	100	NR	276.68
Chlorine (mg/L)	NR	NR	1802	1404.87
TSS (mg/L)	170	NR	122	550
VSS (mg/L)	50	NR	NR	82.35
Sulfides (mg/L)	NR	NR	NR	7.32
Turbidity (NTU)	NR	NR	480	NR
Chrome (mg/L)	NR	NR	NR	0.31
Lead (mg/L)	NR	NR	NR	0.17
Sodium (mg/L)	NR	NR	NR	8880.32
Nickel (mg/L)	NR	NR	NR	0.20
Zinc (mg/L)	NR	NR	NR	0.32
Copper (mg/L)	NR	NR	NR	0.19
O&G (mg/L)	NR	181	737	NR

[a] Combination of light, medium and heavy crude oil, and exit of the clarifier

[b] Combination of medium and heavy crude oil, API 5.

[c] Combination of light, medium and heavy crude oil, and in of the clarifier

NR: No register

Table 4. Physicochemical parameters of WCPC from tank farm of Ulé

2.2. Treatment of the waters associated with crude oil production

The Tables 5, 6, 7 and 8 show a summary of the methodology used by each researcher, showing the operational conditions for each system. On the other hand, Table 9 and Table 10 compare the different treatments: physicochemical treatments, aerobic and anaerobic biological treatment, and combined treatments.

2.3. Biological treatment applied to the waters associated with crude oil production

The Tables 5 and 6 show a resume of the aerobic and anaerobic biological treatments applied to WCP, and Table 8 shows the operation conditions of the combined system aerobic-anaerobic applied to WCP. Among the aerobic biological systems are the rotating biological contactor reactors (RBC), the sequential biological reactors (SBR) and the continuous flow reactors (CR); and among the anaerobic biological treatments are the batch reactors (BR) and the upflow anaerobic sludge blanket reactors (UASB), working under mesophilic and thermophilic conditions. Likewise, Table 9and Table 10 present a summary of the results of applying these treatments to WCP.

Researcher, year	Kind of WCP	Treatment systems	Characteristics of the experimental equipment	Operation conditions	Parameters evaluated
Behling et al. (2003)	WCPC (WCPL, WCPM and WCPH)	RBC	RBC of 9.5 L, with 50 circular disc of PVC, 0.8 cm separation, supported in an axis of carbon steel 3/8 " diameter, rotation speed of 2.5 rpm. The discs were immersed 40 % in the effluent. The area of contact was 2.44 m². The water volume was 7.5 L	The RBC worked under mesophilic condition. The organic load average applied was 2.04 ± 0.7 g COD/m²d and 5.2 mL/min, TRH of 24 h, temperature 27-32ºC.	pH COD TSS VSS Total alkalinity
Díaz et al. (2005a)	WCPL, WCPM and WCPH	SBR	The SBR of 4 L were constructed in material of plastic and cylindrical form, with a volume of operation of 2 L, in which 600 mL sludge and 1.4 L of WCP. At the bottom of the reactors were located air diffusers connected to a compressor.	After acclimated and stabilized, they worked with HRT of 16 hours with sequence of 15 hours of ventilation, 30 minutes of sedimentation and 30 minutes for capture of sample and recharges of the reactor. The temperature was mesophilic (37 ºC). The SBR-1, SBR-2, SBR-3 operated with organic charges of 1.6; 1.17 and 0.46 kg/m³d for the WCPL, WCPM and WCPH, respectively.	pH Alkalinity COD Phenols
Díaz et al. (2005b)	WCPM	SBR	The SBR of 2 L was constructed in material of plastic, with 600 mL of sludge and 1.4 L of WCPM. They gave oxygen to the reactor by means of a compressor.	After acclimated and stabilized, they were operated at the first stage of 15 hours the HRT and time of cellular retention of 15-20 days with sequence of 14 hours for mixed, ½ hour of rest and ½ hour for discharge and load. Whereas in the second stage the HRT was 24 hours with sequence of 23 hours for mixed and ventilation and one hour of discharge and load. The temperature was 37 ºC. The	COD Hydrocarbons Phenols

Researcher, year	Kind of WCP	Treatment systems	Characteristics of the experimental equipment	Operation conditions	Parameters evaluated
				organic load applied was between 0.89 and 0.51 kg/m³d	
González et al. (2007)	WCPL and WCPH	SBR	The SBR of 2 L was constructed in material of plastic, in cylindrical form, in which they added 600 mL of sludge and 1.4 L of WCP. They gave oxygen to the reactor by a compressor.	HRT of 8 hours and time of cellular retention of 20 days. Nutrients were added. The COD in the inflow was 1105 and 320 mg/L for WCPL and WCPH, respectively.	COD Hydrocarbons Phenols
Castro et al. (2008)	WCPM	Batch reactor	The reactor was a receptacle adjusted as Plexiglas of 3 L, provided with a porous circular stone and a hose connected to the tubes for the supply of compressed air. As effective volume of 0.3 L of bacterial suspension and 0.7 L of WCPM.	They used several functional groups and consortiums of bacteria. The systems were operated under mesophilic conditions (27 °C) and HRT of 144 h. The COD of feeding was 880 mg/L.	pH COD TSS VSS Alkalinity

Table 5. Methodology for aerobic treatment of WCPM

Researcher, year	Kind of WCP	Treatment systems	Characteristics of the experimental equipment	Operation conditions	Parameters evaluated
Gutiérrez et al. (2007)	WCPL WCPM and WCPH	Batch rectors	They placed four (4) reactors of 500 mL each one, containing 20 % of the useful volume of mesophilic granular sludge proceeding from a beer industry, and 80 % of effluent to treat. The reactors were immersed in a thermal bath that allowed controlling the temperature. The produced biogas was meter by water displacement.	Initially the reactors were loaded, for ten days, with D +glucose on an equivalent concentration in COD of 1500 mg/L and solution of nutrients, for a retention time (RT) of 24 hours. Later they added to three reactors WCPL, WCPM and WCPH with concentrations of 1200-1300 mgCOD/L, 857-960 mgCOD/L and 860-870 mgCOD/L, respectively. The fourth reactor worked with glucose (D+ glucose). To reach the thermophilic conditions (55°C ± 1°C) the temperature was increased from the mesophilic conditions (37°C ± 1°C) at the reason of 1°C/day. The RT in all the cases was 24 hours.	pH COD TSS and VSS Alkalinity VFA Methane

Researcher, year	Kind of WCP	Treatment systems	Characteristics of the experimental equipment	Operation conditions	Parameters evaluated
Gutiérrez et al. (2009)	WCPM and WCPH	Batch reactors	They placed three (3) reactors of 500 mL each one, containing 20 % of the useful volume mesophilic granular sludge proceeding from a beer industry, and 80 % of effluent to treat. The reactors were immersed in a thermal bath that allowed controlling the temperature. The produced biogas was meter by water displacement.	Initially the reactors were loaded, for ten days, with D +glucose on an equivalent concentration in COD of 1500 mg/L and solution of nutrients, for a retention time (RT) of 24 hours. Later they added to two reactors WCPM and WCPH with concentrations of 1876.9 and 1029.0 mgCOD/L, respectively. The third reactor worked with glucose (D+glucose). To reach the thermophilic conditions (55ºC ± 1ºC) the temperature was increased from the mesophilic conditions (37ºC ± 1ºC) at the reason of 1ºC/day. The RT in all the cases was 24 hours.	pH COD TSS and VSS Alkalinity VFA Methane
Rincón et al. (2002)	WCPL	UASB reactors	There were employed at a UASB reactor of 4 L, 0.098 m of diameter, 0.67 m high and 0.53 m high of water, inoculated with 30 % of granular sludge from a UASB reactor that treats residual waters of a brewery of the locality.	Initially, the reactor was fed with residual synthetic water that was containing glucose as the only source of carbon (1 g/L) and nutrients. Later, it was operated for 275 days with HRT from 38 to 5 h. The reactors were evaluated for organic loads of 0.78; 1.20; 1.46; 1.64; 1.90; 3.17 and 4.70 kg COD/m³d for HRT of 36, 24, 21, 17, 11, 8 and 6 hours, respectively. They worked under mesophilic conditions (37ºC ± 1ºC).	pH Alkalinity COD Phenols
Díaz et al. (2005a)	WCPL WCPM WCPH	UASB reactors	They worked with 3 UASB reactors of 4 L, inoculated with 1.2 L of granular sludge from an UASB reactor treating residual waters of a brewery of the locality.	Initially, the reactor was fed with residual synthetic water that was containing glucose as the only source of carbon (850 mg/L) and nutrients. Later, the reactors UASB-1, UASB-2 and UASB-3 were fed by WCPL, WCPM and WCPH why organic loads of 1.06; 0.78 and 0.31 kg COD/m³d, respectively. They worked under mesophilic conditions (37ºC ± 1ºC) during 1 month with HRT of 24 h.	pH Alkalinity COD SARA

Researcher, year	Kind of WCP	Treatment systems	Characteristics of the experimental equipment	Operation conditions	Parameters evaluated
Gutiérrez et al (2006)	WCPL	UASB reactors	They used two UASB reactors constructed in Plexiglas with volumes of 1.7 and 2.5 L, operating under temperatures of 37 ± 1°C and 55 ± 1°C, respectively. The reactors were provided with a jacket, supporting the temperature for recirculation of warm water. Both reactors were inoculated with mesophilic anaerobic granular sludge (20 % of the useful volume) proceeding from a beer industry.	Initially the reactors were load, for two days, with D+glucose on an equivalent concentration in COD of 1500 mg/L and solution of nutrients and TRH of 24 h; then WCPL was added. Later to reach the thermophilic conditions (55°C ± 1°C) in the thermophilic reactor, the temperature was increased from the mesophilic condition (37°C ± 1°C) to a rate of 1°C/day. The reactors were evaluated for organic loads of 1.4, 1.9, 2.8 and 5.6 kg COD/m³d and RTH of 24, 18, 12 and 6 hours, respectively.	pH Alkalinity COD VFA Methane Enzymes
Caldera et al. (2007)	WCPL	UASB reactor	They used a UASB reactor constructed in Plexiglas with volume of 2.5 L, inoculated with anaerobic mesophlic granular sludge (30 % of the useful volume) proceeding from a beer industry. The reactor was provided with a jacket, supporting the temperature of 55 ± 1°C for recirculation of warm water.	Initially the reactor was loaded, for two days, with D+glucose on an equivalent concentration in COD of 1500 mg/L and solution of nutrients, and HRT of 24 h; then WCPL was added. Later to reach the thermophilic condition (55°C ± 1°C), in the thermophilic reactor, the temperature was increased from the mesophilic condition (37°C ± 1°C) to a rate of 1°C/day. The reactor was evaluated for 42 days, with HRT of 24 and 12 and organic loads of 1.4 and 2.8 kg COD/m³d, respectively.	pH Alkalinity COD VFA Methane
Rincón et al. (2008)	WCPL WCPM	UASB reactors	They used two UASB reactors of 2.5 L, inoculated with 0.75 L of granular sludge from an UASB reactor treating residual waters of a brewery of the locality.	Initially, the reactors were fed with residual synthetic water that was containing glucose as the only source of carbon (850 mg/L) and nutrients. Later, the reactors UASB-1 and UASB-2 were fed with WCPL and WCPM APPL and organic load of 1.06 and 0.78 kg COD/m³d respectively. They worked mesophilic conditions (37°C ± 1°C) during 1 month with HRT of 24 h.	pH Alkalinity COD SARA

Table 6. Methodology for anaerobic treatment of WCP

Researcher, year	Kind of WCP	Treatment systems	Characteristics of the experimental equipment	Operation conditions	Parameters evaluated
Rojas et al. (2008)	WCPC (WCPM and WCPH)	Coagulation and DAF	The DAF, consisted of a pressurization cell or saturation camera, constructed in material of transparent plastic of 90 mm of external diameter and 270 mm high. Inside the camera was finding a manual agitator of stainless steel and a filter that worked as diffuser; in addition, a series of connections and valves of the distribution and pressure of the water and air.	They worked with pressures of 30, 40 and 50 psi and recycle of 30%, 40% and 50%, and temperature of 25°C. They evaluated a cationic flocculants of high molecular weight, in concentration of 0.006 % in volume (3.54 mg/L)	TSS Turbidity O&G
Caldera et al. (2009)	WCPH	Coagulation-flocculation	They used a Jar Test model JLT6; adding 1 L of WCP, to each of six precipitation jar of 1000 mL, taking one of these as a control.	They simulated coagulation, flocculation, and sedimentation processes to 100 rpm for rapid agitation for 1 minute and 30 rpm for slow agitation by 20 minutes. The sedimentation was 30 minutes. The initial turbidity of the water was 140 NTU. They used as coagulant commercial chitosan (CCH) (Sigma Chemical Co.) and chitosan obtained in the laboratory (LCH) to 100 °C dissolved in acetic acid 0.10 M, preparing solutions of 0.6 %. They worked with concentrations of 24, 30, 36, 42 and 48 mg/L of solution of LCH and CCH, respectively.	pH COD TSS VSS Turbidity Color O&G Hydrocarbons
Caldera et al. (2011)	WCPH	Coagulation-flocculation	They used a Jar Test model JLT6; adding 1 L of WCP, to each of six precipitation jar of 1000 mL, taking one of these as a control.	They simulated coagulation, flocculation and sedimentation processes to 100 rpm for rapid agitation for 2 minutes, and 100 rpm for slow agitation for 30 minutes. The sedimentation was 30 minutes. The turbidity initial was 52 NTU. As coagulant agent was used commercial chitosan (CCH) dissolved in acetic acid 0.10 M, preparing solutions of 1.0%. The concentrations evaluated were 40,	pH COD TSS VSS Turbidity Color O&G Hydrocarbons

Researcher, year	Kind of WCP	Treatment systems	Characteristics of the experimental equipment	Operation conditions	Parameters evaluated
				42, 44, 46 and 48 mg/L of CCH solution.	

Table 7. Methodology for physicochemical treatment of WCP

Researcher year	Kind of WCP	Treatment systems	Characteristics of the experimental equipment	Operation conditions	Parameters evaluated
Rincón et al. (2004)	WCPL WCPC	UASB-SBR system	They used two types of reactors placed in series, a reactor UASB of 2.5 L of useful volume and a SBR. The reactor UASB was inoculated by sludge from an UASB reactor treating residual waters of a brewery. While the SBR reactor was inoculated with aerobic sludge from a wastewater treatment plant.	The system worked 195 days, in two stages. The first was feeding with WCPL from 1100 to 1230 mg COD/L (133 days) and the second one with WCPC of 176 and 264 mg COD/L (66 days). The effluent treated in the UASB was fed in the SBR. The HRT was 24 hours and the temperatures were UASB 37°C and SBR 28 °C.	pH Alkalinity COD Hydrocarbons Phenols
Paz et al. (2012)	WCPC	Superficial constructed wetlands (SCWFF)	They used two superficial constructed wetlands of free flow (SCWFF) to pilot scale. The support material was gravel and soil, and aquatic emergent plants that counted of support of gravel and soil, and aquatic emergent plants (Cyperus luzulae y Cyperus ligulari – SCWFF I, y Cyperuz feraz, Paspalum sp. y Typha dominguesis – SCWFF II), and a control (C) without plants.	They placed 30 plants for each species. The depth of the support was 0.25 cm with 7 % of gravel and 93 % of soil, and a water layer of 0.05 m of water. The flow fed was 8 mL/min, with a HRT of 7 days and organic load of 23.5 g COD/m²d. The samples were collected weekly for 80 days.	COD pH Sulphide Phenols TSS VSS DO
Blanco et al. (2008)	WCPC	Sub-superficial constructed wetlands (SSCW)	The system SSCW consisted of three polyethelene tray of 1.28 m long for 0.45 m wide and 0.45 m high, one that of them as control (without plants) and the others two with emergent aquatic plants Cyperus luzulae, Cyperus feraz L.C, Cyperus ligularis L. y Typha dominguensis (SSCW I y SSCW II). The beds of the tray were constituted by 86400 cm³ of gravel as support and a water level of 1.5 L to simulate a natural system of wetland.	The systems worked to continue flow, without recirculation of the effluent with an organic load of 29.42 g/m²d, a flow of 10 mL/min and HRT of 7 days.	pH Alkalinity COD VSS Hydrocarbons Phenols

Table 8. Methodology for combined treatment of WCP

Researcher year	Treatment systems / WCP	COD (%)	TSS (%)	VSS (%)	Hydro-carbons (%)	O&G (%)	Phenols (%)	Turbidity (%)	pH	Alkalinity (mgCaCO$_3$/L)
Rojas et al. (2008)	Coagulation and DAF WCPC	—	77	—	—	90	—	69	—	—
Caldera et al. (2009)	Coagulation-flocculation WCPH	50.7	—	—	70.1	—	—	90.7	8.0-8.2	—
Caldera et al. (2011)	Coagulation-flocculation WCPH	12.5	55-61	41-63	70-90	39-59	—	76-78	7.9	—
Behling et al. (2003)	RBC WCPC	76.1	<4	<3	—	—	—	—	8.9	2343
Díaz et al. (2005a)	SBR WCPL, WCPM and WCPH	88.8 65.2 62.9	—	—	—	—	96.8 89.2 82.8	—	9.0-9.9 9.0-9.6 8.9-9.4	—
Díaz et al. (2005b)	SBR WCPM	65.1[a] 60.9[b]	—	—	76.8 79.5	55.5 62.4	87.5 92	—	—	—
González et al. (2008)	SBR WCPL and WCPH	88 66	—	—	84.4 73.8	—	95.6 79.4	—	—	—
Castro et al. (2008)	Batch reactor WCPM	62.4-89.8	63.3-9.5	—	—	—	—	—	7.4-6.6	—

a and b: different HRT

Table 9. Results of the treatment of WCP

Researcher year	Treatment systems / WCP	COD (%)	VSS (%)	Hydro-carbons (%)	Phenols (%)	pH	Alkalinity (mgCaCO$_3$/L)	SARA (%)	Methane content (%)
Gutiérrez et al. (2007)	Batch reactors WCPL, WCPM and WCPH	70.7 59.9 62.1	—	—	—	7.6 7.6 7.2	2673.7 2620.0 936.7	—	73.1 51.9 54.1
Gutiérrez et al. (2009)	Batch reactors WCPM and WCPH	68.2-69.2 55.9-50.4	—	—	—	8.2 7.5	—	—	—

Rincón et al. (2002)	UASB WCPL	23.8-86.1	—	—	10-59	7.6-8.0	2500-2800		24-95
Díaz et al. (2005a)	UASB	81.7	—	—	55.1	7.3-8.6	—	—	—
	WCPL, WCPM	23.5			74.7	7.1-8.5			
	and WCPH	35.7			92.5	7.2-8.4			
Gutiérrez et al. (2006)	UASB	40-80[M]	42-73	—	—	7.4-8.5	1960-2633	—	53-79
	WCPL	67-84[T]	52-67			7.9-8.0	2190-2454		54-80
Caldera et al. (2007)	UASB	78[a]	—	—	—	8.0	2413		87
	WCPL	77[b]				8.2	2046		77
Rincón et al. (2008)	UASB	93	—	—	—	7.5	1955	84	—
	WCPL and WCPM	26				7.8	2520	54	
Rincón et al. (2004)	UASB-SBR	95	—	74	99.9	9	2468	—	—
	WCPL-WCPC	79		82	90	9	2405		
Paz et al. (2012)	Superficial constructed wetlands WCPC	-	-		64.3	9.13-10.5	-	-	-
					61.3	8.84-9.93			
Blanco et al. (2008)	Sub-superficial constructed wetlands WCPC	31.4-65.7	45.2-91.9	77.5	94.7	8.9	2508	—	—

M: Mesophilic T: Thermophilic; a and b : different HRT

Table 10. Results of the treatment of WCP

2.4. Biological treatment of the waters associated with light crude oil production

The waters associated with the production of light crude oil (WCPL) are biodegradable in aerobic biological treatments, in anaerobic biological treatments and in a combination of these treatments. Díaz *et al.* (2005a) report that the COD removal in SBR was 88.8%, and the removal of phenols was 96.8%.

Likewise, the WCPL showed be biodegradable in anaerobic conditions in batch and continuous systems, under mesophilic conditions (37ºC) and thermophilic conditions (55ºC). In batch systems the COD removal reached 70.7% under mesophilic conditions (Gutiérrez *et al.*, 2007), while in UASB reactors under both temperature conditions, the efficiency of COD removal reached over 75%.

In UASB reactors the HRT influenced in the COD removal; so, Rincón *et al.* (2002) reported that under mesophilic conditions the optimal HRT was between 15 and 10 hours, with COD removal above 80%, but for HRT under 10 hours the system did not allow the methanogenic microorganisms to be able to transform volatile fatty acid (VFA), provoking the inhibition of the system. On the other hand, Gutiérrez *et al.* (2006) indicated that for the same temperature conditions, the HRT optimal was 18 hours with COD removal of 80%; they indicate also that for thermophilic conditions, the optimal HRT was 18 hours with COD removal of 84%, maintaining good COD efficient removal for HRT of 6 hours (67%).

When the efficiency of COD removal of WCPL in UASB reactors under mesophilic and thermophilic conditions were compared, major percentages of COD removal under thermophilic conditions for HRT under at 15 hours were observed. This removal of COD can be associated to high temperature accelerate the enzymatic biological systems. Nevertheless, there were not significant differences (p>0.05) between the values obtained for mesophilic and thermophilic temperature conditions, for the HRT from 12 to 24 hours.

When combined systems were used, the COD removal of the system was higher than those obtained in each separated system (Rincón *et al.*, 2004).

The maximum COD removal reached for the systems applied to WCPL were between 67% and 95%. In this aspect, the petroleum industry has alternatives to treat the WCPL; however, the final decision will be an economic decision between the temperature, the size of the reactor and energy costs. In the case of thermophilical route, it is necessary to considerate the cost of raising the temperature of the water, because the WCPL is at atmospheric conditions. For aerobic processes, the costs of the energy associated must be considered.

2.5. Biological treatment of the waters associated with medium crude oil production

It is observed in the Table 9 and Table 10 that the WCPM presented lower biodegradability than WCPL, for both aerobic and anaerobic systems. In discontinuous batch aerobic systems, Castro *et al.* (2008) report that the COD removal was between 62.4% and 89.8%; while for SBR reactors was between 60.9% and 65.2% (Díaz *et al.*, 2005 a; Díaz *et al.*, 2005b). On the other hand, in batch anaerobic reactors under thermophilic conditions, the COD removal was between 59.9% and 69.2% (Gutiérrez *et al.* 2007; Gutiérrez *et al.*, 2009). In UASB reactors, the COD removal was between 23.5% (Díaz *et al.*,2005a) and 26% (Rincón *et al.*, 2008).

In the different treatment systems it is observed that the COD removal for WCPM was between 23.5% and 89.8%.

2.6. Biological treatment of the waters associated with heavy crude oil production

In the case of the waters associated with heavy crude oil production (WCPH), the behavior was similar to WCPM. In SBR systems the COD removal was between 62.9% (Díaz *et al.*, 2005a) and 66% (González *et al.*, 2007). While in anaerobic batch reactor systems under thermophilic conditions, the COD removal was between 50.4% and 62.1% (Gutiérrez *et al.*, 2007; Gutiérrez *et al.*, 2009). On the other hand, in UASB reactors under mesophilic conditions, the COD removals were lower than 40% (Díaz *et al.*, 2005a; Rincón *et al.*, 2008).

2.7. Biological treatment of the combination of waters associated with crude oil production

The WCPC represent the combination of the waters in contact with different fractions of crude oil, whether produced in plant or by the researchers. The biodegradability of these waters has been studied in RBC and combined systems UASB-SBR (Table 8).

Behling *et al.* (2003) commented that the COD removal in RBC system used to treat WCPC was 76.1%, while Rincón *et al.* (2004) studied a UASB-SBR system and reported that the COD removal reached 79%, indicating that was important removals of phenols and hydrocarbons were obtained.

3. Discussion

Comparing the biodegradability of WCPL, WCPM and WCPH, it is observed that the WCPL present the major biodegradability in the different treatment systems and operating conditions studied.

The biodegradability of the WCP has been associated to diverse factors as SARA composition, phenols concentration, alkalinity, organic load, metals concentration and temperature.

Some researchers (Rincón *et al.*, 2002; Gutiérrez *et al.*, 2006; Gutiérrez *et al.*, 20007) argue that the WCPL biodegradability is good in anaerobic systems under mesophilic conditions and under thermophilic conditions. The final decision between the temperature used and size of the reactor will be economical, because the WCPL are at atmospheric temperature and the termophilic route implicates to consider the costs associated of warming the water. In the cases of WCPM and WCPH the studies realized up to the moment are not conclusive.

Other researchers (Gutiérrez *et al.*, 2007; Gutiérrez *et al.*, 2001; Gutiérrez *et al.*, 2012) share that the biodegradability of WCP is associated to the SARA composition present in these waters, as product of the contact with the crude oil associated. The difference of composition of these fractions confer characteristics that influence in their biodegradability, because the SARA fractions change in relation to the crude oil that is in contact with the WCP, being the WCPL the waters with the biggest percentages of saturated, considered more biodegradable than WCPM and WCPH.

When the organic fractions present in the WCP are compared, it is observed that WCPM and WCPH present a similar content of organic fractions (p>0.05). The opposite case was observed with the WCPL, which organic fractions are different in saturated, aromatics and resins, in comparing to WCPM and WCPH (p>0.05).

On the other hand, there is a tendency to increase the saturated fractions in WCP (r=0.871) with the increase of the API gravity of the crude oil with the water associated, following the order WCPL>WCPM>WCPH. In relation to the resins, it was observed that it increases with regard to the decrease of the API gravity of the crude oil with the WCP were associated following the order WCPL<WCPM<WCPH.

A study realized by Díaz *et al.* (2007) with WCPM from other tank farm of the Venezuelan petroleum industry, indicated that the SARA fractions can be removed from the WCP using UASB reactors. They obtained removals of 72% of saturated, 91% of resins and 71% of asphaltenes, and did not obtain removals of aromatics. They associated these results with the increases of the aromatic fractions for degradation of the fractions like resins and asphaltenes to aromatics.

Also the researchers have presented biodegradability percentages of different types of WCP under anaerobic conditions. They report values for mesophilic and thermophilic anaerobic systems of 80% and 78%, 45% and 86%, and 20% and 0%, for WCPL, WCPM and WCPP respectively in batch reactors (Gutiérrez and Caldera, 2011; Gutiérrez *et al.*, 2007; Rincón *et al.*, 2006).

In regard to phenols concentration, the studies mention that the consortium of microorganisms developed in mesophilic UASB reactors were influenced by the initial phenols concentrations, indicating that the phenols removal might be associated with the presence of different phenols compounds in the different types of WCP, with varied resistance to degradation and metabolism (aerobic/anaerobic).

Additionally, the studies indicate that the alkalinity values in the WCP were between 900 and 3000 mg $CaCO_3$/L. It has been commented that the difference of COD removal might be due to the acidity-basicity conditions in the WCP. The WCPH presented lower values of alkalinity (642.9-580.4 mg $CaCO_3$/L) and lower COD removal than WCPM. The alkalinity of WCPM was superior to 2000 mg $CaCO_3$/L. As for the pH, the WCP presented basic pH (7-10) for the different treatments.

In other cases, it is mentioned that the presence of metals in the WCP makes the treatment more complex. However, the metals K, Na, Fe, Cr, Pb and Zn can be used by thermophilic microorganisms or can be removed from the WCP and reach to the sludge by diverse mechanisms (Gutiérrez *et al.*, 2009).

In relation to degrading microorganisms present in the WCP, some have been isolated, and identified the genus *Aeromonas, Klebsielle, Xanthomona, Bacteroides* and *Acinetobacter*, as well as a consortium of them, that resulted to be effective in COD decrease (Castro *et al.*, 2008).

The Table 7 shows that WCP has been treated by coagulation-flocculation at laboratory level using chitosane as a coagulating agent in concentrations of 24 to 38 mg/L of solution of commercial chitosane (CCH), and by dissolved air flotation (DAF) using a cationic flocculants of high molecular weight.

Rojas *et al.* (2008) reported that the TSS removal and the turbidity in the WCPC were 77% and 69% respectively. On the other hand, Caldera *et al.* (2009, 2011) commented that the turbidity removal in the WCPH was 90.7%, accompanied of COD removal of 50.7%. In any case, the hydrocarbons removal and oils removal by physicochemical methods were between 70% and 90%, concluding that the cationic polymers represent an alternative to remove oily compounds in the WCP.

Table 8 shows other alternatives applied to treat WCP. In constructed sub-superficial wetlands COD removal of WCPC was between 31.4% and 65.7%, while in constructed superficial wetlands there was no COD removal. Both systems showed efficiency to remove more than 60% of the hydrocarbons present in the WCPC (Paz *et al.*, 2012; Blanco *et al.*, 2008).

The application of ozone also has been proposed to increase the biodegradability of the WCP. According to Gutiérrez *et al.* (2002), the application of ozone improves considerably the biodegradability of the WCP, with an increase of up to 87%. They concluded that the application of doses of ozone to WCP in the order of 30 mg/L of ozone, would affect favorably in the later biological processes applied.

4. Conclusions

The WCP from the different cuts: light (WCPL), medium (WCPM), heavy (WCPH) and combinations of them (WCPC), have different characteristics and their biodegradability or treatment are associated on the SARA compositions, organic matters concentration, hydro carbons and phenols concentrations, and the operation conditions (HRT and temperature).

The biodegradability of the WCP followed the order WCPL>WCPM>WCPH.

The COD removal in biological systems changed between 67%-95%, 23.5%-89.8% and 35%-66% for WCPL, WCPM and WCPH, respectively.

The physicochemical treatment DAF and coagulation, removed hydrocarbons and oils between 70% and 90%.

Other parameters like phenols, hydrocarbons and SARA fractions, can be removed from the WCP by biological treatments.

It is necessary to analyze other parameters and operating conditions, as well as to conduct an economic evaluation before the treatment selection.

Author details

Edixon Gutiérrez[1] and Yaxcelys Caldera[2]

*Address all correspondence to: egutierr12@gmail.com; yaxcelysc@gmail.com

1 Centro de Investigación del Agua. Facultad de Ingeniería, Universidad del Zulia, Maracaibo, estado Zulia, Venezuela

2 Laboratorio de Investigaciones Ambientales, Núcleo Costa Oriental del Lago, Universidad del Zulia. Cabimas, estado Zulia, Venezuela

References

[1] Ahmad, A, Sumathi, S, & Hameed, B. (2006). Coagulation of residue oil and suspended solid in palm oil milk effluent by chitosan, alum and PAC. Chemical Engineering Journal, 118 (1-2), 99-105.

[2] Behling, E, Marín, J, Gutiérrez, E, & Fernández, N. (2003). Aerobic treatment of two industrial effluents using a rotating biological contactor reactor. Multiciencias, 3 (2), 126-135.

[3] Blanco, E, Gutiérrez, E, Caldera, Y, Núñez, M, & Paz, N. (2008). Tratamiento de aguas de producción a través de humedales construidos de flujo subsuperficial a escala piloto. Memorias del XXXI Congreso Interamericano de Ingeniería Sanitaria y Ambiental, AIDIS, 12 al 15 de octubre, Santiago de Chile, Chile.

[4] Caldera, Y, Rodriguez, Y, Oñate, H, Prato, J, & Gutiérrez, E. (2011). Efficiency of chitosan as coagulant during treatment of low turbidity water associated crude oil production. Revista Tecnocientífica URU, 1 (1), 54-60.

[5] Caldera, Y, Clavel, N, Briceño, D, Nava, A, Gutiérrez, E, & Mármol, Z. (2009). Chitosan as a coagulant during treatment of water from crude oil production. Boletín del Centro de Investigaciones Biológicas, 43 (4), 541-555.

[6] Caldera, Y, Gutierrez, E, Madueño, P, Griborio, A, & Fernández, N. (2007). Anaerobic biodegradability of industrial effluents in an UASB reactor. Impacto Científico, 2 (1), 11-23.

[7] Castro, F, Fernández, N, & Chávez, M. (2008). Diminution of the COD in formation waters using bacterial stocks. Rev. Téc. Ing. Univ. Zulia, 31 (3), 256-265.

[8] Díaz, A, Rincón, N, López, F, Fernández, N, Chacín, E, & Debellefontaine, H. (2005b). The biological treatment in sequencing batch reactor (SBR) of effluents from the extraction of medium oil production. Multiciencias, 5 (2), 150-156.

[9] Díaz, A, Rincón, N, Marín, J, Behling, E, Chacín, E, & Fernández, N. (2005a). Degradation of total phenols during biological treatment of oilfields produced water. Ciencia, 13 (3), 281-291.

[10] Díaz, V, Bauza, R, Cepeda, N, Behling, E, Díaz, A, Fernández, N, & Rincón, N. (2007). Development of micro-SARA method for organic fractions of crude oil determination on the oil extraction production waters with anaerobic treatment. Ciencia, 15, (1), 95-104.

[11] Gaceta Oficial 5021 de la República de Venezuela ((1995). Caracas 18 de diciembre de 1995. 5021, Extraordinaria. Decreto 883. Normas para la clasificación y el control de las aguas de los cuerpos de agua y vertidos o efluentes líquidos.

[12] García, A, Arreguín, F, Hernández, S, & Lluch, D. (2004). Impacto ecológico de la industria petrolera en sonda de Campeche, México, tras tres décadas de actividad: Una revisión. Interciencia, 29(6), 311-319.

[13] González, Y, Rincón, N, López, F, & Díaz, A. (2007). Organic matter removal from the petroleum effluents by a sequencing batch reactor (SBR). Rev. Téc. Ing. Univ. Zulia, 30 (Edición Especial), 82-89.

[14] Guerrero, C, Escobar, S, & Ramírez, D. (2005). Manejo de la salinidad en aguas aso-
ciadas de producción de la industria petrolera. Investigación e Ingeniería, 25 (3),
27-33.

[15] Gutiérrez, E, & Caldera, Y. (2011). Tratamiento de aguas provenientes de la extrac-
ción de crudo en sistemas biológicos anaerobios. Intellectus, 1 (1), 11-21.

[16] Gutiérrez, E, Caldera, E, Ruesga, L, Villegas, C, Gutiérrez, R, Paz, N, Blanco, N, &
Mármol, Z. (2012). Organic fractions in water from crude oil production. Revista Tecnocientífica URU, 2 (2), 31-37.

[17] Gutiérrez, E, Caldera, Y, Contreras, K, Blanco, E, & Paz, N. (2006). Anaerobic mesophilic and thermophilic degradation of waters from light crude oil production. Boletín del Centro de Investigaciones Biológicas, 40 (3), 242-256.

[18] Gutiérrez, E, Caldera, Y, Fernández, N, Blanco, E, Paz, N, & Mármol, Z. (2007). Thermophilic anaerobic biodegradability of water from crude oil production in batch reactors. Rev. Téc. Ing. Univ. Zulia, 30 (2), 111- 117.

[19] Gutiérrez, E, Caldera, Y, Perez, F, Blanco, E, & Paz, N. (2009). Behavior of metals in
water from crude oil production during thermophilic anaerobic treatment. Boletín
del Centro de Investigaciones Biológicas, 43 (1), 145-160.

[20] Gutiérrez, E, Fernández, N, Herrera, L, Sepúlveda, J, & Mármol, Z. (2002). The effect
of ozone applications on biodegradability in water used for oil production. Multiciencias, 2 (1), 50-54

[21] Lepisto, R, & Rintala, J. (1999). Extreme thermophilic (70ºC), VAF-FED UASB reactor;
performance, temperature response, load potential and comparation with 35 and
55ºC UASB reactor. Water Research, 33 (14), 3162-3170.

[22] Lettinga, G. (2005). A good life environment for all through conceptual, technological
and social innovations. Memorias del VIII Taller y Simposio Latinoamericano sobre
Digestión Anaerobia. Punta del Este, Uruguay, 1-15.

[23] Lettinga, G. (2001). Digestion and degradation, air for life. Water Science Tech., 44
(8), 157-176.

[24] Li, Q, Kang, C, & Zhang, C. (2005). Wastewater produced from an oilfield and continuous treatment with an oil-degrading bacterium. Process Biochemistry, 40 (2),
873-877.

[25] Paz, N, Blanco, E, Gutiérrez, E, Núñez, M, & Caldera, Y. (2012). Pilot scale superficial
flow constructed wetlands for sulfide and phenols removals from oil field producer
water. Rev. Téc. Ing. Univ. Zulia, 35 (1), 71-79.

[26] Renault, F, Sancey, B, Bodot, P, & Crini, G. (2009). Chitosan for coagulation/floccula-
tion process- An eco-friendly approach. European Polymer Journal, 45 (5), 1337-1348.

[27] Rincón, N, Behling, E, & Cepeda, N. (2004). Combinación de tratamientos anaerobio-
aerobio de aguas provenientes de la industria petrolera venezolana. Memorias del

XXXI Congreso Interamericano de Ingeniería Sanitaria y Ambiental, AIDIS, 22 al 27 de agosto, San Juan, Puerto Rico.

[28] Rincón, N, Cepeda, N, Díaz, A, Behling, E, Marín, J, & Bauza, R. (2008). Behavior of organic fraction in water separated from extrated crude oil with anaerobic digestion. Rev. Téc. Ing. Univ. Zulia, 31 (2), 169-176.

[29] Rincón, N, Chacín, E, Marín, J, Moscoso, L, Fernández, L, Torrijos, M, Moletta, R, & Fernández, N. (2002). Optimum time of hydraulic retention for the anaerobic treatment of light oil production wastewater. Rev. Téc. Ing. Univ. Zulia, 25 (2), 90-99.

[30] Rincón, N, Torrijos, M, Colina, G, Behling, E, Chacín, E, Marín, J, & Fernández, N. (2006). Method to determine anaerobic biodegradability of oil production waters. Impacto Científico, 1 (1), 9-20.

[31] Rojas, C, Rincón, N, Díaz, A, Colina, G, Behling, E, Chacín, E, & Fernández, N. (2008). Evaluation of dissolved air flotation unit for oil produced water. Rev. Téc. Ing. Univ. Zulia, 31 (1), 50-57.

Chapter 2

Crude Oil Biodegradation in the Marine Environments

Mehdi Hassanshahian and Simone Cappello

Additional information is available at the end of the chapter

1. Introduction

Petroleum is a viscous liquid mixture that contains thousands of compounds mainly consisting of carbon and hydrogen. Oil fields are not uniformly distributed around the globe, but being in limited areas such as the Persian Gulf region. The world production of crude oil is more than three billion tons per year, and about the half of this is transported by sea. Consequently, the international transport of petroleum by tankers is frequent. All tankers take on ballast water which contaminates the marine environment when it is subsequently discharged. More importantly, tanker accidents exemplified by that of the Exxon Valdez in Prince William Sound, Alaska, severely affect the local marine environment. Off-shore drilling is now common to explore new oil resources and this constitutes another source of petroleum pollution. However, the largest source of marine contamination by petroleum seems to be the runoff from land. Annually, more than two million tons of petroleum is estimated to end up in the sea. Fortunately, petroleum introduced to the sea seems to be degraded either biologically or abiotically.

2. The composition of crude oil

Petroleum has been known for several years to occur in the surface seepage and was first obtained in pre-Christian times by the Chinese. The modern petroleum industry had its beginning in Romania and in a well-sunk in Pennsylvania by Colonel E. A. Drake in 1859 [1]. The principal early use of the product of the petroleum industry was for the replacement of expensive whale oil for lighting. Today, its consumption as a fuel and its dominance in the world market as a source of chemicals has diversified tremendously.

Petroleum is defined as any mixture of natural gas, condensate, and crude oil. Crude oil which is a heterogeneous liquid consisting of hydrocarbons comprised almost entirely of the elements hydrogen and carbon in the ratio of about 2 hydrogen atoms to 1 carbon atom. It also contains elements such as nitrogen, sulfur and oxygen, all of which constitute less than 3% (v/v).

There are also trace constituents, comprising less than1% (v/v), including phosphorus and heavy metals such as vanadium and nickel. Crude oils could be classified according to their respective distillation residues as paraffins, naphthenes or aromatics and based on the relative proportions of the heavy molecular weight constituents as light, medium or heavy. Also, the composition of crudes may vary with the location and age of an oil field, and may even be depth dependent within an individual well. About 85% of the components of all types of crude oil can be classified as either asphalt base, paraffin base or mixed base. Asphalt base contain little paraffin wax and an asphaltic residue [2].The sulfur, oxygen and nitrogen contents are often relatively higher in comparison with paraffin base crudes, which contain little or no asphaltic materials. Mixed crude oil contains considerable amount of oxides of nitrogen and asphalt [2].

Crude oil is perhaps the most complex mixture of organic compounds that occurs on earth. Recent advances in ultra-high-resolution mass spectrometry have allowed the identification of more than 17,000 distinct chemical components, and the term petroleomics has been coined to express this newly uncovered complexity [3]. Furthermore, crude oil is not a homogeneous mat erial, and different crude oils have a range of chemical and physical properties that affect their susceptibility to biodegradation and their environmental fate. Within this complexity, however, crude oil can be classified into four main operationally defined groups of chemicals: the saturated hydrocarbons and the aromatic hydrocarbons, and the more polar, non-hydro-carbon components the resins and the asphaltenes. Light oils are typically high in saturated and aromatic hydrocarbons, with a smaller proportion of resins and asphaltenes. Heavy oils, which result from the biodegradation of crude oil under anoxic conditions *in situ* in petroleum reservoirs, have a much lower content of saturated and aromatic hydrocarbons and a higher proportion of the more polar chemicals, the resins and asphaltenes [4] (figure 1). Biodegrada-tion of crude oil in surface environments results in similar changes in crude oil composition and the loss of saturated and aromatic hydrocarbons, together with an increase in the relative abundance of the polar fractions (which are more resistant to biodegradation), is a character-istic signature of crude-oil biodegradation. Because saturated hydrocarbons constitute the largest fraction of crude oil by mass, the biodegradation of saturated hydrocarbons is quanti-tatively the most important process in the removal of crude oil from the environment. Nevertheless, the aromatic hydrocarbons and polar fractions, which are more toxic and persistent, could be of greater long-term environmental significance [5].

3. Oil pollution as an environmental problem

It is no exaggeration that oil fuels the world's economy, and it is used on a staggering scale. World production was some 80 Mbbl (11 Mt/day) by the end of 2000, and this is expected to increase by 1.9% year in the next decade [6, 7]. Approximately 40% of the world's oil travels by water at some time between its production and final consumption, and again the volumes

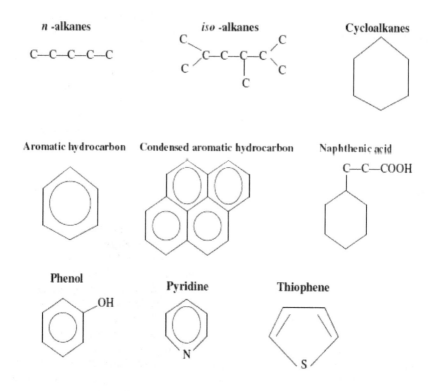

Figure 1. Structural classification of some crude oil components [1].

are staggering. For example, the US imported 350 000 t of oil per day from the Middle East alone in 1999 [7]. Unfortunately, despite the best efforts of the major part of the petroleum industry, a small amount is inevitably spilled. Fortunately this is only a tiny fraction of that transported, and there has been a general improvement in oil spill statistics in the last two decades [7, 8]. Massive releases from pipelines, wells and tankers receive the most public attention, but in fact these account for only a relatively small proportion of the total petroleum entering the environment. The National Research Council has recently updated its classic oil in the sea [7] and now estimates that the total input of petroleum into the sea from all sources is approximately 1.3 Mt/year. Almost 50% comes from natural seeps, and less than 9% emanates from catastrophic releases. Consumption, principally due to non-tanker operational discharges and urban run-off, is responsible for almost 40% of the input (figure 2) skimmers and adsorbents is generally the first priority of responders, but this is neither rarely easy, nor very effective after a large spill. There is therefore a continuing search for alternative and additional responses. Amongst the most promising are those that aim to stimulate the natural process of oil biodegradation [9].

The marine environment is subject to contamination by organic pollutants from a variety of sources. Organic contamination results from uncontrolled releases from manufacturing and refining installations, spillages during transportation, direct discharge from effluent treatment plants and run-off from terrestrial sources.

In quantitative terms, crude oil is one of the most important organic pollutants in marine environments and it has been estimated that worldwide some where between 1.7 and 8.8 ′ 106 tons of petroleum hydrocarbons impact marine waters and estuaries annually [7]. Large oil spills, such as the Exxon Valdez and Sea Empress incidents, invariably capture media attention but such events are relatively rare; however, a substantial number of smaller releases of petroleum hydrocarbons occur regularly in coastal waters. Around the coast of the UK alone, between the years of 1986 and 1996, 6,845 oil spills were reported. Of these, 1,497 occurred in environmentally sensitive areas or were of sufficient magnitude to require clean-up (23). As a consequence of the importance of oil spills relative to other sources of organic contaminants in the marine environment, there is a large body of research on oil-spill bioremediation. Furthermore, studies of oiled shorelines have been far more numerous than open water studies, which have often been equivocal [11, 12].

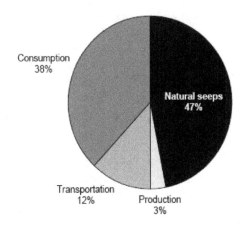

Figure 2. Sources of oil into the sea.

4. The fate of oil in the marine environment

The fate of petroleum in marine ecosystems has been intensively studied [5]. Crude oil and petroleum distillate products introduced to the marine environment are immediately subject to a variety of physical and chemical, as well as biological, changes (figure 3) [13].

Abiological weathering processes include evaporation, dissolution, dispersion, photochemical oxidation, water-in-oil emulsification, adsorption onto suspended particulate material, sinking, and sedimentation. Biological processes include ingestion by organisms as well as microbial degradation [l]. These processes occur simultaneously and cause important changes in the chemical composition and physical properties of the original pollutant, which in turn may affect the rate or effectiveness of biodegradation. The most important weathering process during the first 48 hours of a spill is usually evaporation, the process by which low to medium-weight crude oil components with low boiling points volatilize into the atmosphere. Evaporation can be responsible for the loss of one to two-thirds of an oil spill's mass during this period, with the loss rate decreasing rapidly over time [13].

Roughly one-third of the oil spilled from the Amoco Cadiz, for example, evaporated within the frost 3 days. Evaporative loss is controlled by the composition of the oil, its surface area and physical properties, wind velocity, air and sea temperatures, sea state, and the intensity of solar radiation [14]. The material left behind is richer in metals (mainly nickel and vanadium), waxes, and asphaltenes than the original oil [15]. With evaporation, the specific gravity and viscosity of the original oil also increase. For instance, after several days, spilled crude oil may begin to resemble Bunker C (heavy) oil in composition.

None of the other abiological weathering processes accounts for as significant a proportion of the losses from a spill. For example, the dissolving, or dissolution, of oil in the water column is a much less important process than evaporation from the perspective of mass lost from a spill; dissolution of even a few percent of a spill's mass is unlikely. Dissolution is important, however, because some water soluble fractions of crude oil (e.g., the light aromatic compounds) are acutely toxic to various marine organisms (including microorganisms that may be able to degrade other fractions of oil), and their impact on the marine environment is greater than mass balance considerations might imply [14, 15].

Dispersion, the breakup of oil and its transport as small particles from the surface to the water column extremely important process in the disappearance of a surface slick [15]. Dispersion is controlled largely by sea surface turbulence: the more turbulence, the more dispersion. Chemical dispersants have been formulated to enhance this process. Such dispersants are intended as a first-line defense against oil spills that threaten beaches and sensitive habitats such as salt marshes and mangrove swamps although used widely in other countries, dispersants have had trouble being accepted in the United States. The National Research Council has generally approved their use, but effectiveness and, to a lesser degree, toxicity remain concerns. Dispersed oil particles are more susceptible to biological attack than undispersed ones because they have a greater exposed surface area. Hence, dispersants may enhance the rate of natural biodegradation Water-in-oil emulsions, often termed "mousses are formed when seawater, through heavy wave action, becomes entrained with the insoluble components of oil. Such emulsions can form quickly in turbulent conditions and may contain 30 to 80 percent water [16].

Heavier or weathered crudes with high viscosities form the most stable mousses. Mousse will eventually disperse in the water column and/or be biodegraded, but may first sink or become stranded on beaches. A water-in-oil emulsion is more difficult for microorganisms to degrade than oil alone [17].

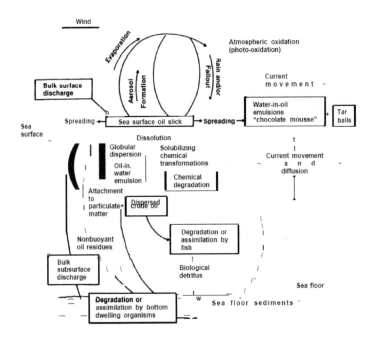

Figure 3. the fate of oil in the marine environment [7].

Mousse formation, for example, has been suggested as a major limiting factor in petroleum biodegradation of the Ixtoc I and Metula spills, probably because of the low surface area of the mousse and the low flux of oxygen and mineral nutrients to the oil-degrading microorganisms within it [17]. Natural biodegradation is ultimately one of the most important means by which oil is removed from the marine environment, especially the nonvolatile components of crude or refined petroleum.

In general, it is the process whereby microorganisms (especially bacteria, but yeasts, fungi, and some other organisms as well) chemically transform compounds such as petroleum hydrocarbons into simpler products. Although some products can actually be more complex, ideally hydrocarbons would be converted to carbon dioxide (i.e., mineralized), nontoxic water-soluble products, and new microbial biomass. The mere disappearance of oil (e.g., through emulsification by living cells) technically is not biodegradation if the oil has not actually been chemically transformed by microbes [17].

The ideal may be difficult to reach, particularly in a reasonably short time, given the recalcitrance of some petroleum fractions to biodegradation (discussed below) and the many variables that affect its rate and extent. Man-made bioremediation technologies are intended to improve the effectiveness of natural biodegradation [17].

5. Response of marine microbial community to oil pollution

Hydrocarbon-degrading microorganisms usually exist in very low abundance in marine environments. Pollution by petroleum hydrocarbons, however, may stimulate the growth of such organisms and cause changes in the structure of microbial communities in the contaminated area [18]. For example Hassanshahian et al (2010) show that oil contamination can induce major changes in marine microbial communities at Persian Gulf and Caspian Sea, that when the pollution occur the number of crude oil degrading bacteria increased and also inhibit some catalytic enzymes [19].

Identification of the key organisms that play roles in pollutant biodegradation is important for understanding, evaluating and developing in situ bioremediation strategies. For this reason, many efforts have been made to characterize bacterial communities, to identify responsible degraders, and to elucidate the catalytic potential of these degraders. In a natural marine environment, the amounts of nutrients, especially those of nitrogen and phosphorus, are insufficient to support the microbial requirements for growth, especially after a sudden increase in the hydrocarbon level associated with an oil spill. Therefore, nitrogen and phosphorus nutrients are added to a contaminated environment to stimulate the growth of hydrocarbon degrading microorganisms and, hence, to increase the rate of biodegradation of the polluting hydrocarbons [20, 21].

6. Crude oil degrading microorganisms

Hydrocarbon-degrading bacteria were first isolated almost a century ago [22] and a recent review lists 79 bacterial genera that can use hydrocarbons as a sole source of carbon and energy, as well as 9 cyanobacterial genera, 103 fungal genera and 14 algal genera that are known to degrade or transform hydrocarbons (Table 1) [23, 24].

Despite the difficulty of degrading certain fractions, some hydrocarbons are among the most easily biodegradable naturally occurring compounds. Many more as-yet-unidentified strains are likely to occur in nature [25]. Moreover, these genera are distributed worldwide. All marine and freshwater ecosystems contain some oil-degrading bacteria. No one species of microorganism, however, is capable of degrading all the components of given oil. Hence, many different species are usually required for significant overall degradation. Both the quantity and the diversity of microbes are greater in chronically polluted areas. In waters that have not been polluted by hydrocarbons, hydrocarbon-degrading bacteria typically make up less than 1 percent of the bacterial population, whereas in most chronically polluted systems (harbors, for example) they constitute 10 percent or more of the total population [26].

Hydrocarbon degrading bacteria and fungi are widely distributed in marine, freshwater, and soil habitats. Similarly, hydrocarbon degrading cyanobacteria have been reported [27, 28] although contrasting reports indicated that growth of mats built by cyanobacteria in the Saudi coast led to preservation of oil residues [29]. Typical bacterial groups already known for their capacity to degrade hydrocarbons include *Pseudomonas, Marinobacter, Alcanivorax, Microbulbifer, Sphingomonas,Micrococcus, Cellulomonas, Dietzia,* and *Gordonia* groups [30]. Molds belonging to the genera *Aspergillus, Penicillium, Fusarium, Amorphoteca, Neosartorya, Paecilomyces, Talaromyces, Graphium* and the yeasts *Candida, Yarrowia* and *Pichia* have been implicated in hydrocarbon degradation [27, 31]. However, reports in literature on the actual numbers of hydrocarbon utilizes are at variance with one another because of the methodological differences used to enumerate petroleum-degrading microorganisms.

Diverse petroleum-degrading bacteria inhabit marine environments. They have often been isolated as degraders of alkanes or of such aromatic compounds as toluene, naphthalene and phenanthrene. Several marine bacteria capable of degrading petroleum hydrocarbons have been newly isolated. These are bacteria of the genera *Alcanivorax* [32], *Cycloclasticus* [33], *Marinobacter* [34], *Neptunomonas* [25], *Oleiphilus* [35] and *Oleispira* [36] within the γ-Proteobacteria, and of the genus *Planococcus* within Gram-positive bacteria [37]. These bacteria, with the possible exception of *Marinobacter* and *Neptunomonas*, use limited carbon sources with a preference for petroleum hydrocarbons and are thus 'professional hydrocarbonoclastic' bacteria. For example, *Alcanivorax* strains grow on n-alkanes and branched alkanes, but cannot use any sugars or amino acids as carbon sources. Similarly, *Cycloclasticus* strains grow on the aromatic hydrocarbons, naphthalene, phenanthrene and anthracene, whereas *Oleiphilus* and *Oleispira* strains grow on the aliphatic hydrocarbons, alkanoles and alkanoates. Many 'nonprofessional' hydrocarbonoclastic bacteria have been isolated: for example, *Vibrio, Pseudoalteromonas, Marinomonas* and *Halomonas* have been isolated as marine bacteria capable of degrading phenanthrene or chrysene [38].

Some hydrocarbon-degrading bacteria isolated from marine environments have been classified into several genera that include terrestrial hydrocarbon degrading bacteria: namely, naphthalene-degrading *Staphylococcus* and *Micrococcus* [39], 2-methylphenanthrene-degrading *Sphingomonas* [40] and alkane-degrading *Geobacillus* [41]. Although some *Cycloclasticus* strains have been isolated using the extinction culturing method, other strains were isolated by conventional enrichment techniques with petroleum hydrocarbons used as the sources of carbon and energy. Therefore, a greater variety of hydrocarbon-degrading marine bacteria are likely to be isolated if hydrocarbon enrichment is done in combination with the specific resuscitation techniques already described.

7. Pathway for biodegradation of some compartment of crude oil

7.1. Fundamental reactions of aerobic degradation

The fundamental reactions of the aerobic hydrocarbon decomposition have been well known for several decades. Suitable surveys are contained in the books of [42, 43]. Even though many details have been published since, such as the degradation of aliphatic alkenes [44], the fundamental steps are still valid and enable us to understand the dependence of the processes on environmental conditions (Figures 4 and 5). Experiments on the laboratory scale as well as

Bacteria	Yeast	Fungi
Achromobacter	Candida	Aspergillus
Acinetobacter	Cryptococcus	Cladosporium
Alcanivorax	Debaryomyces	Corollasporium
Alcaligenes	Hamsenula	Cunninghamella
Bacillus	Pichia	Dendryphiella
Brevibacterium	Rhodotorula	Fusarium
Burkholderia	Saccharomyces	Gliocladium
Corynebacterium	Sporobolomyces	Luhworthia
Flavobacterium	Torulopsis	Penicillium
Mycobscterium	Trichosporon	Varicospora
Nocardia	Yarrowia	Verticillium
Pseudomonas		
Rhodococcus		
Sphingomonas		
Streptomyces		

Table 1. Crude-oil degrading microorganisms

observation of polluted sites have made it possible to estimate the impact of oil degradation on sediment.

The key step of hydrocarbon degradation is the addition of one oxygen atom, in some cases, two oxygen atoms, to the hydrocarbon molecule, which is then converted to an alkanol (in the case of aliphatic hydrocarbons) or to a phenol (in the case of aromatic molecules). In some species, an epoxide is the first intermediate. This activation makes the hydrocarbon more soluble in water, marks a reactive site, and introduces a reactive site for the next reactions. The reaction requires energy, which is typically generated via the oxidation of a reduced biological intermediate such as NADH, which itself is reoxidized by an electron acceptor. For the degradation of alkanes, different enzyme systems are known which carry out the primary attack. An omega-hydroxylase system consisting of three proteins (the rubredoxin reductase, a rubredoxin and an omega-hydroxylase) was isolated and characterized from *Pseudomonas* [45]. In some bacterial or fungal species as well as in mammalian cells, there are enzyme systems which depend on cytochrome P450 acting as a terminal oxidase. The main intermediates of the alkane degradation are fatty acids, which are produced from the alkanols via aldehydes. These acids can be further decomposed by the pathway typical of physiologica carboxylic acid degradation, in which the molecule is shortened stepwise. However, fatty acids can also be excreted by the cells and accumulate in the environment.

Once released, they can produce ambiguous effects. On the one hand, fatty acids can serve as a carbon source for bacteria of a community, thus enhancing the hydrocarbon degradation. On the other hand, fatty acids (chain length 14 C) can inhibit growth and hydrocarbon metabolism because they interfere with the cell membrane [47]. This provokes a toxic effect and reduces growth. Different degradative pathways have been demonstrated for aromatic substrates. The choice of the pathway depends on the type of the organism and/or on the type of the aromatic molecule, especially on its substituents and (in the case of polyaromatic molecules, PAH) on the number of rings [48]. For an overview of the fundamental possibilities of PAH biodegradation, three different metabolic routes considered to be the main pathways are summarized here.

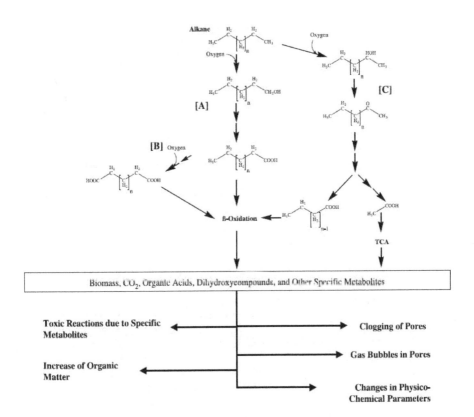

Figure 4. Aerobic degradation of crude oil hydrocarbons with its environmental impact. Biodegradation of n-alkanes: metabolism begins with the activity of a monooxygenase which introduces a hydroxyl group into the aliphatic chain. [A]-monoterminal oxidation, [B]-biterminal oxidation, [C]- subterminal oxidation); TCA-tricarboxylic acid cycle [44]

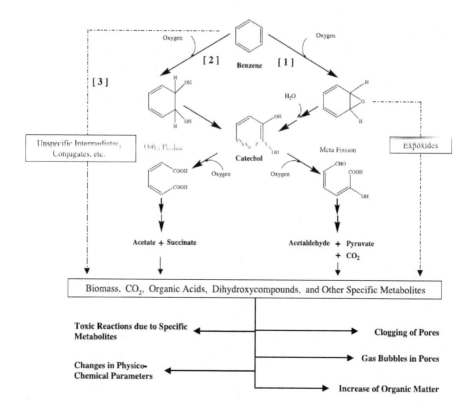

Figure 5. Biodegradation of aromatic hydrocarbons: metabolism begins with the activity of a monooxygenase [1] or a dioxygenase [2] which introduce one or two atoms of oxygen; it can also begin with unspecific reactions [3] [48].

7.2. Complete mineralization or the dioxygenase pathway

This pathway is taken mainly by bacteria. The monoaromatic molecule or one ring of the polyaromatic system is attacked by a dioxygenase, and the molecule is oxidized stepwise via formation of a diol and subsequent ring cleavage. Pyruvate is one of the main intermediates of the pathway. The main products are biomass and carbon dioxide. An accumulation of dead-end products is rare and occurs mostly when cells are deficient in their degradation pathway. The disadvantage of this pathway is that only ring systems of up to four rings are mineralized. Systems with a higher number of rings seem to be recalcitrant [49].

7.3. Cometabolic transformation or the monooxygenase pathway

This pathway has been mainly demonstrated for yeasts and fungi, but it also occurs in bacteria and in some algae. The respective PAH-degrading species can only perform the degradation if a compound is available which can serve as a source of carbon and energy. The characteristic enzymes which perform ring cleavage are monooxygenases (e.g., Cyt P450). The monooxygenase activity results in the formation of an epoxide which is highly reactive, resulting in toxic or mutagenic activity. Epoxides may also be transformed to trans-dihydrodiols. The latter have not been metabolized further in pure cultures in the laboratory and have to be regarded as dead-end products. However, no such metabolites have been detected in soil or in sediment [50].

7.4. Unspecific oxidation via radical reactions

The wood-destroying white rot fungi, e.g., have been shown to destroy the structure of lignin via the activity of extracellular peroxidases and phenol oxidases. They attack the phenolic molecule structure by a nonspecific action, thus also attacking other aromatic structures such as PAH. The type of cleavage product is not predictable. Frequent metabolites of PAHs are quinones, quinoles, and ring systems with a ring number lower than that of the original substance. These compounds may be incorporated into sediments and alter the sediment structure [51].

7.5. Anaerobic hydrocarbon degradation

For many decades, it was assumed that hydrocarbons undergo biodegradation only in the presence of molecular oxygen. However, in 1988 Evans and Fuchs [50] published a review paper on the anaerobic degradation of aromatic compounds, and Aeckersberg et al. (1991) [52] reported on a sulphate-reducing bacterium able to anaerobically mineralize hexadecane. Since that time, a great deal of work has been done on the anaerobic degradation of aliphatic and aromatic hydrocarbons. It has been demonstrated that anaerobic hydrocarbon degradation is not uncommon in nature although, in most cases it is considerably slower than aerobic degradation. Denitrifying, sulfate-reducing, and iron (III)-reducing strains collected at different sites (terrestrial, aquifers, fresh-water and marine systems) are able to anaerobically metabolize hydrocarbons. The same has been demonstrated for the phototrophic bacterium *Blastochloris sulfoviridis* strain ToP1, which uses light as an energy source [53]. Even methanogenic consortia have been shown to degrade hydrocarbons [54, 55]. The metabolic routes of alkane degradation seem to function differently and are not completely understood yet. Several authors have discussed a terminal or sub terminal addition of a one-carbon moiety or a fumarate molecule to the alkane as an activation mechanism [56, 57] (Figure 6). For aromatic molecules, it has been demonstrated that alkyl benzenes which have a methyl group as a side chain undergo an enzymes addition of fumarate, most likely via a radical mechanism. This was demonstrated for toluene. Alkyl benzenes with side chains of two or more carbon atoms are activated by dehydrogenation of the side chain.

This has been shown for ethyl- and propylbenzene [53]. A scheme of the anaerobic degradation is shown in Figure (7).

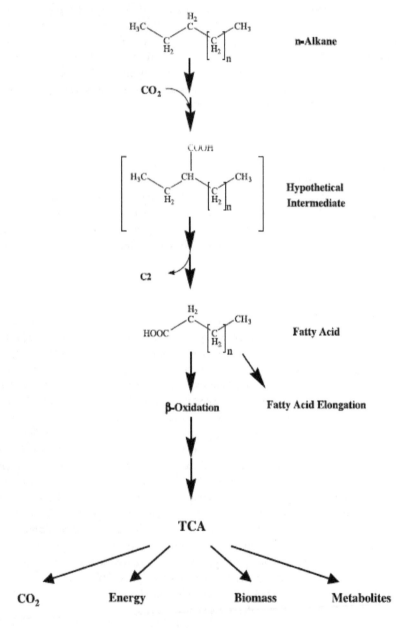

Figure 6. Proposed pathway for anaerobic degradation of n-alkanes; activation via addition of a C1-moiety (subterminal carboxylation at C3). Pathway according to So et al. (2003); TCA tricarboxylic acid cycle [55].

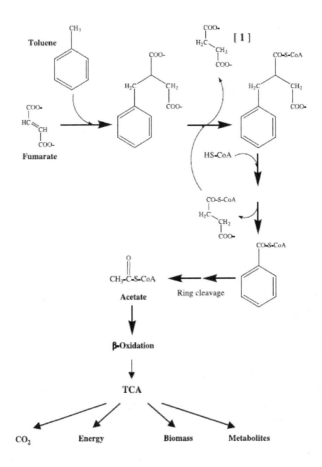

Figure 7. Proposed pathways of anaerobic degradation of aromatic hydrocarbons; activation via addition of fumarate, [1]—succinate. Pathways according to Spormann and Widdel (2000), and Wilkes et al. (2002); TCA—tricarboxylic acid cycle [55].

7.6. Competing processes

The ideal preconditions for biodegradation cited above occur only rarely, e.g., in the case of a rough and nutrient-rich sea or on energy-rich tidal flats. Mostly, however, the reality of oil spills is very different. The ideal steps are rendered difficult, slowed down, or made impossible by competing processes. Such influences are exemplified by the case studies. Heavy oils or heavy oil products such as heavy fuel oil or bunker oil C behave very differently from the light oils described above. Heavy oils incorporate suspended matter, debris, biomass, and even garbage, which increases their viscosity and decreases their biodegradability. Due to their viscosity, the energy needed to emulsify heavy oils is very great. Solar irradiation causes the evaporation of the light components and photodecomposition, resulting in unpredictable compounds. Oil carpets are formed. Where they meet the coastline, beaches are covered. Their removal by natural forces is very slow or even impossible, and technical purification is expensive and troublesome. Biological degradation is extremely slow because the low oil surface to volume ratio limits the bioavailability of the oil.

Oil biodegradation works well on the open sea but proceeds differently on beaches. Vast areas of tidal beaches can be covered by oil when there is wind onshore during the ebb tide. If this oil cover is subjected to strong sun irradiation, the oil does not float up during the next flood because the light components have evaporated. The sediment is soaked with the sticky oil. Tides and wind add further sediment, and the initially liquid, later viscous, pollutant becomes more and more solidified [58].

This solidified material is only slowly attacked by waves, hampering biodegradation because the available surface is too small. Irradiation and the catalyzing capacity of particle surfaces help to convert a part of the original mixture of small molecules into high molecular mass material of low solubility, forming tar and finally asphalt. Such products appear as geological rather than organic matter. Experience has shown that it is difficult for organisms to settle on oil layers. The Persian Gulf spill presented a new experience in so far as thick and vital cyanobacterial mats developed on oil covers within a few months, introducing biomass as well as Aeolian and hydrodynamic sediments fixed by the growing mats. This observation was welcomed initially [59] but then turned out to be disappointing because biodegradation was not favored [60]. In some cases, colonization opened the oil crusts; in other cases, it formed stable covers which prevented the access of oxygen to deeper layers, helping to preserve the pollution. The latter clearly transformed the polluted system, resulting in geo-biological matter that had never been present before. This geo-biological matter dominated the sites after the spill. In the upper eulittoral and the lower supra tidal zone, calcareous incrustation and solid salt supported the conversion of the oil into rock-like matter with a life span of 10 or more years [61].

8. Factors affected crude oil biodegradation in marine environment

Environmental variables can also greatly influence the rate and extent of biodegradation. Variables such as oxygen and nutrient availability can often be manipulated at spill sites to enhance natural biodegradation (i.e., using bioremediation). Other variables, such as salinity, are not usually controllable. The great extent to which a given environment can influence biodegradation accounts for some of difficulty in accurately predicting the success of biore-mediation efforts. Lack of sufficient knowledge about the effect of various environmental factors on the rate and extent of biodegradation is another source of uncertainty [19, 62].

8.1. Oxygen

Oxygen is one of the most important requirements for microbial degradation of hydrocarbons. However, its availability is rarely a rate-limiting factor in the biodegradation of marine oil spills. Microorganisms employ oxygen-incorporating enzymes to initiate attack on hydrocarbons. Anaerobic degradation of certain hydrocarbons (i.e., degradation in the absence of oxygen) also occurs, but usually at negligible rates. Such degradation follows different chemical paths, and its ecological significance is generally considered minor. For example, studies of sediments impacted by the Amoco Cadiz spill found that, at best, anaerobic biodegradation is several orders of magnitude slower than aerobic biodegradation. Oxygen is generally necessary for the initial breakdown of hydrocarbons, and subsequent reactions may also require direct incorporation of oxygen. Requirements can be substantial; 3 to 4 parts of dissolved oxygen are necessary to completely oxidize 1 part of hydrocarbon into carbon dioxide and water. Oxygen is usually not a factor limiting the rate of biodegradation on or near the surface of the ocean, where it is plentiful and where oil can spread out to provide a large, exposed surface area. Oxygen is also generally plentiful on and just below the surface of beaches where wave and tide action constantly assist aeration. When oxygen is less available, however, the rates of biodegradation decrease. Thus, oil that has sunk to the sea floor and been covered by sediment takes much longer to degrade. Oxygen availability there is determined by depth in the sediment, height of the water column, and turbulence (some oxygen may also become available as the burrowing of bottom-dwelling organisms helps aeration) [63, 64]. Low-energy beaches and fine-grained sediments may also be depleted in oxygen; thus, the rate of biodegradation may be limited in these areas. Pools of oil are a problem because oxygen is less available below their surfaces. Thus, it may be preferable to remove large pools of oil on beaches, as was done in Alaska, before attempting bioremediation [18, 65].

8.2. Nutrients

Nutrients such as nitrogen, phosphorus, and iron play a much more critical role than oxygen in limiting the rate of biodegradation in marine waters. Several studies have shown that an inadequate supply of these nutrients may result in a slow rate of biodegradation [52]. Although petroleum is rich in the carbon required by microorganisms, it is deficient in the mineral nutrients necessary to support microbial growth [53]. Marine and other ecosystems are often deficient in these substances because non-oil degrading microorganisms (including phytoplankton) consume them in competition with the oil degrading species. Also, phosphorus precipitates as calcium phosphate at the pH of seawater. Lack of nitrogen and phosphorus is most likely to limit biodegradation, but lack of iron or other trace minerals may sometimes be important. Iron, for instance, is more limited in clear offshore waters than in sediment-rich coastal waters Scientists have attempted to adjust nutrient levels (e.g., by adding nitrogen- and phosphorus-rich fertilizers) to stimulate biodegradation of petroleum hydrocarbons. This is the experimental bioremediation approach used recently on about 110 miles of beaches in Prince William Sound, Alaska. Researchers have also experimented with alternative methods of applying nutrients. Given the necessity of keeping nutrients in contact with oil, the method of application is itself likely to be an important factor in the success of bioremediation [65, 66].

8.3. Temperature

The temperature of most seawater is between –2 and 35 °C (55). Biodegradation has been observed in this entire temperature range, and thus in water temperatures as different as those of Prince William Sound and the Persian Gulf. The rates of biodegradation are fastest at the higher end of this range and usually decrease—sometimes dramatically in very cold climates-with decreasing temperature. One experiment showed that a temperature drop from 25 to 5 °C caused a tenfold decrease in response [56]. At low temperature, the rate of hydrocarbon metabolism by microorganisms decreases [57]. Also, lighter fractions of petroleum become less volatile, thereby leaving the petroleum constituents that are toxic to microbes in the water for a longer time and depressing microbial activity. Petroleum also becomes more viscous at low temperature. Hence, less spreading occurs and less surface area is available for colonization by microorganisms. In temperate regions, seasonal changes in water temperature affect the rate of biodegradation, but the process continues year-round.

8.4. Other factors

Several variables, including pressure, salinity, and pH may also have important effects on biodegradation rates. Increasing pressure has been correlated with decreasing rates of biodegradation; therefore, pressure may be very important in the deep ocean [67,68]. Oil reaching great ocean depths degrades very slowly and, although probably of little concern, is likely to persist for a long time [59]. Microorganisms are typically well adapted to cope with the range of salinities common in the world's oceans. Estuaries may present a special case because salinity values, as well as oxygen and nutrient levels, are quite different from those in coastal or ocean areas. However, there is little evidence to suggest that microorganisms are adversely affected by other than hyper saline environments. Extremes in pH affect a microbe's ability to degrade hydrocarbons. However, like salinity, pH does not fluctuate much in the oceans it remains between 7.6 and 8.1 and does not appear to have an important effect on biodegradation rates in most marine environments. In salt marshes, however, the pH maybe as low as 5.0, and thus may slow the rate of biodegradation in these habitats [69, 70].

9. Biodegradation strategy for crude oil removal from marine environment (biostimulation and bioaugmentation)

Bioremediation technologies for responding to marine oil spills may be divided into three discrete categories:

1. Nutrient enrichment (Biostimulation)

2. Seeding with naturally occurring microorganisms (Bioaugmentation)

3. Seeding with genetically engineered microorganisms (Bioaugmentation with GEMs)

9.1. Nutrient enrichment (biostimulation)

Of all the factors that potentially limit the rate of petroleum biodegradation in marine environments, lack of an adequate supply of nutrients, such as nitrogen and phosphorus, is probably the most important and perhaps the most easily modified. Nutrient enrichment (sometimes called nutrition) also has been more thoroughly studied than the other two approaches, especially now that EPA, Exxon, and the State of Alaska have carried out extensive nutrient enrichment testing on beaches polluted by oil from the Exxon Valdez [71]. In part for these reasons, many scientists currently view nutrient enrichment as the most promising of the three approaches for those oil spill situations in which bioremediation could be appropriate. This approach involves the addition of those nutrients that limit biodegradation rates (but not any additional microorganisms) to a spill site and conceptually is not much different than fertilizing a lawn [71]. The rationale behind the approach is that oil-degrading microorganisms are usually plentiful in marine environments and well adapted to resisting local environmental stresses. However, when oil is released in large quantities, microorganisms are limited in their ability to degrade petroleum by the lack of sufficient nutrients. The addition of nitrogen, phosphorus, and other nutrients is intended to overcome these deficits and allow petroleum biodegradation to proceed at the optimal rate. Experiments dating to at least 1973 have demonstrated the potential of this approach. Researchers, for example, have tested nutrient enrichment in near shore areas off the coast of New Jersey, in Prudhoe Bay, and in several ponds near Barrow, Alaska. In each case, the addition of fertilizer was found to stimulate biodegradation by naturally occurring microbial populations. The recent nutrient enrichment experiments in Alaska provided a wealth of experimental data about bioremediation in an open environment (box B) [72]. Since previous research findings had already demonstrated the general value of this approach, the experiments were intended to determine for one type of environment how much enhancement of natural biodegradation could be expected and to evaluate the most effective methods of application. The results provided additional evidence that application of nutrients could significantly enhance the natural rate of biodegradation on and below the surface of some beaches. As a result, Exxon was authorized by the Coast Guard on-scene coordinator, in concurrence with the Alaska Regional Response Team, to apply fertilizers to the oiled beaches in Prince William Sound [73]. To date, about 110 miles of shoreline have been treated with nutrients, and a monitoring program has been established. Without additional research, however, it is premature to conclude that nutrient enrichment will be effective under all conditions or that it will always be more effective than other bioremediation approaches, other oil spill response technologies, or merely the operation of natural processes. The results of the Alaska experiments were influenced by the beach characteristics (mostly rocky beaches, well-washed by wave and tide action), the water temperature (cold), the kind of oil (Prudhoe Bay crude), and the type and quantity of indigenous microorganisms in Prince William Sound. Few detailed analyses or performance data are yet available for different sets of circumstances. One smaller-scale test using the same fertilizer as in Alaska was recently conducted on beaches in Madeira polluted by the Spanish tanker Aragon. Results in this very different setting and with a different type of oil were not especially encouraging. Researchers speculated that the unsatisfactory results could have been due to differences in the type of oil, the concentration of fertilizer used the lower initial bacterial activity, and/or different climatic conditions. At the same time, Exxon recently used what it learned in Alaska to help degrade subsurface no. 2 heating oil spilled in a wildlife refuge bordering the Arthur Kill at Prall's Island, New Jersey. An innovative aspect of this application was the use of two trenches parallel to the beach in which to distribute fertilizer. Nutrients were dissolved with the incoming tide and pulled down the beach with the ebb tide, enabling a more even distribution than point sources of fertilizer. Exxon reports those 3 months after applying fertilizers, the oil in the treated zone had been reduced substantially relative to that in an untreated control zone [74, 75, 76].

9.2. Seeding with naturally occurring microorganisms (bioaugmentation)

Seeding (also called inoculation) is the addition of microorganisms to a polluted environment to promote increased rates of biodegradation. The inoculums maybe a blend of non indigenous microbes from various polluted environments, specially selected and cultivated for their oil-degrading characteristics, or it may be a mix of oil-degrading microbes selected from the site to be remediate and mass-cultured in the laboratory or in on-site bioreactors. Nutrients would usually also accompany the seed culture. The rationale for adding microorganisms to a spill site is that indigenous microbial populations may not include the diversity or density of oil-degraders needed to efficiently degrade the many components of a spill. Some companies that advocate seeding with microorganisms also claim that commercial bacterial blends can be custom-tailored for different types of oil in advance of a spill, that the nutritional needs and limitations of seed cultures are well understood, that microbes can easily be produced in large quantities for emergency situations, and that seed cultures can be stored, ready for use, for up to 3 years.

The value of introducing nonindigenous microorganisms to marine environments is still being evaluated. With some exceptions, the scientilc community has not been encouraging about the promise of seeding marine oil spills. Controlled studies have not been conducted in such settings, so no data are available to evaluate the effectiveness of this approach. Many scientists question the necessity of adding microbes to a spill site because most locales have sufficient indigenous oil-degrading microbes, and in most environments biodegradation is limited more by lack of nutrients than by lack of microbes [72]. At many spill sites, a very low level of oil is often present as "chronic" input, inducing oil-degrading capability in naturally occurring microorganisms. Moreover, the requirements for successful seeding are more demanding than those for nutrient enrichment. Not only would introduced microbes have to degrade petroleum hydrocarbons better than indigenous microbes, they would also have to compete for survival against a mixed population of indigenous organisms well adapted to their environment. They would have to cope with physical conditions (such as local water temperature, chemistry, and salinity) and predation by other species, factors to which the native organisms are likely to be well adapted [77].

The time required for introduced microbes to begin metabolizing hydrocarbons is also important. If a seed culture can stimulate the rapid onset of biodegradation, it would have an advantage over relying on indigenous microbes that may take time to adapt. Despite some claims, seed cultures have not yet demonstrated such an advantage over indigenous microbial communities. Seed cultures are typically freeze-dried (and therefore dormant) and require time before they become active [73].

Seed cultures also must be genetically stable, must not be pathogenic, and must not produce toxic metabelites. Some laboratory and small-scale experiments in controlled environments have demonstrated that seeding can promote biodegradation [75].

However, it is exceedingly difficult to extrapolate the results of such tests to open water where many more variables enter the picture. Results of experimental seeding of oil spills in the field have thus far been inconclusive. As noted in box B, recent EPA tests of two commercial products applied to contaminated beaches in Alaska concluded that, during the period of testing, there was no advantage from their use [77]. In a well-publicized attempt to demonstrate seeding at sea, one company applied microorganisms to oil from the 1990 Mega Borg spill in the Gulf of Mexico [78]. Although the experiment aroused some interest, the results were inconclusive and illustrated the difficulty of conducting a controlled bioremediation experiment at sea and measuring the results. Although there were changes observed in the seeded oil, in the absence of controls the experiment could not tell whether they were due to biodegradation or bioemulsification (the process in which microbes assist the dispersal of surface oil), or were unrelated to the seeding. (Even if bioemulstilcation rather than biodegradation was the process at work in this experiment, it may be of potential interest for oil spill response and could be investigated further.) An attempt has been made to apply a seed culture to a polluted salt marsh [78, 79]. In July 1990 the Greek tanker Shinoussa collided with three barges in the Houston Ship Channel, resulting in a spill of about 700,000 gallons of catalytic feed stock, partially refined oil. Some of this oil impacted neighboring Marrow Marsh. Microbes were applied to experimental areas within the marsh, and control areas were established. Visual observations made by the scientific support coordinator who monitored the application for the National Oceanic and Atmospheric Administration (NOAA) indicated that treated oil changed color within a few minutes to a few hours after treatment, but that after several days there were no significant visual differences between treated and untreated plots.

More importantly, chemical analyses indicated "no apparent chemical differences in petroleum hydrocarbon patterns between treated and untreated plots several days after treatment [70, 80]. Not all of the monitoring data have been analyzed yet, so a final determination of effectiveness has not been made. Seed cultures may be most appropriate for situations in which native organisms are either present as slow growers or unable to degrade a particular hydrocarbon. Especially difficult-to-degrade petroleum components, such as polynuclear aromatic hydrocarbons, might be appropriate candidates for seeding [80]. In other cases, if a time advantage can be realized; there may be some utility in seeding with a culture consisting of indigenous organisms [81]. Thus, the potential environmental adaptation problems of nonindigenous organisms might be avoided. In many cases, fertilizers would also have to be added. Seeding may offer promise in environments where conditions can be more or less controlled. In such cases one would have to consider the proper choice of bacteria, a suitable method of application, and suitable site engineering. Arrangements would have to be made for keeping cells moist and in contact with the oil; for protecting them from excess ultraviolet light; for providing adequate nutrients; and for controlling temperature, pH, and salinity.

However, before claims about the utility of seeding marine oil spills can be proved (or disproved) additional research [82, 83, 84].

9.3. Seeding with genetically engineered microorganisms (bioaugmentation with GEMs)

Although it was not demonstrably superior to indigenous organisms and has never been tested in the field, the frost organism ever patented was a microorganism genetically engineered to degrade oil [82]. The rationale for creating such organisms is that they might possibly be designed either to be more efficient than naturally occurring species or to have the ability to degrade fractions of petroleum not degradable by naturally occurring species. To be effective, such microorganisms would have to overcome all of the problems related to seeding a spill with nonindigenous microbes. EPA has not yet conducted any GEM product reviews for commercial applications, although at least two companies are considering using genetically engineered products for remediating hazardous waste [84, 85]. Since the development and use of GEMs are still limited by scientific, economic, regulatory, and public perception obstacles, the imminent use of bioengineered microorganisms for environmental cleanup is unlikely. Lack of a strong research infrastructure, the predominance of small companies in the biore-mediation field, lack of data sharing, and regulatory hurdles are all barriers to the commercial use of genetically engineered organisms [83]. The development of GEMs for application to marine oil spills does not have high priority. Many individuals, including EPA officials, believe that we are so far away from realizing the potential of naturally occurring microorganisms to degrade marine oil spills that the increased problems associated with GEMs render them unnecessary at this time [86, 87, 88].

10. Field evaluation of marine oil spill

There have been several oil spill incidents in which bioremediation products have been used in an attempt to enhance oil biodegradation. In some cases, the response authorities have allowed products to be used for experimental purposes [89]. However, in general, it is difficult to draw valid conclusions from many of these efforts because of the time constraints in planning experiments with appropriate controls after a major spill. Moreover, many of the results are reported second hand with little reliable quantitative information. Despite these limitations, some of these spills have been given as examples of bioremediation success and therefore qualify for scientific appraisal [90, 91]. One notable exception is the work carried out in the aftermath of the Exxon Valdez spill. The assessments of bioremediation products and techniques are based on experiments carried out with considerable scientific rigor, and the work after the Exxon Valdez incident is therefore given prominence in this section. The scientific results of this research have been only recently published in primary publications and conference proceedings. A majority of the papers were not peer reviewed prior to publication in the scientific literature (a fact that applies to much work conducted after oil spill incidents), and thus the results from these studies should be assessed with caution [92, 93].

Also, it is important to emphasize that even in this case, there were significant limitations in the scope of the work. For example, the studies concentrated on North Slope crude oil on cobble shorelines in a high-latitude environment. During the early 1990s, there was an increase in bioremediation field trials associated with accidental spills, largely as a result of the perceived success of the bioremediation program following the Exxon Valdez incident [93]. These are mentioned herein, but many are characterized by having been carried out over a short period and, in some cases, with products in the early stage of development [94].

10.1. Amoco Cadiz

On 16 March 1978, the tanker Amoco Cadiz containing 223,000 tones of Arabian Light and Iranian Light crude oil was wrecked off the coast of France. Rough sea conditions resulted in rapid emulsification of the spilled oil, resulting in an increase in the volume of pollutant. Despite efforts to treat the oil at sea, extensive contamination of the shoreline occurred. Most of the beach cleanup effort focused on pumping and mechanical recovery, particularly during the first few weeks of the operation when there was a thick emulsion on the sand and rocks and in the crevices between the rocks. These operations caused some oil to penetrate the sand. In some places, oily sand was overlaid with clean sand deposited as a result of natural coastal processes. Repeated ploughing and harrowing were used to clean the intertidal zone, and four different products were tested to assess the possibility of promoting the biodegradation of oil trapped in sand [95]: (i) a commercial cleaning compound containing nutrients especially adapted to restore oiled soils; (ii) a mixture of lyophilized adapted bacteria, dispersant, and nutrient; (iii) a chemical fertilizer used in agriculture; and (iv) a talc treated with 0.1% of surfactant. The approaching tourist season seems to have prevented extended experimentation, and other techniques were used to complete the cleanup operations. Hence, the limited results were inconclusive [95, 96]. Some changes in oil content were found in these experiments, but it was not clear if the removal was physically or biologically mediated.

10.2. Apex barge

On 28 July 1990, the Greek tanker Shinoussa collided with two Apex tank barges in the Houston Ship Channel, Galveston Bay, Tex., causing a release of approximately 3,000 m^3 of partially refined catalytic feedstock oil over 2 days, which spread onto the surrounding coastline. Alpha BioSea (Alpha Environmental, Houston, Tex.), a product composed of a lyophilized bacterial mixture and inorganic phosphorus and nitrogen nutrients, was applied 8 days after the spill in selected areas of Pelican Island and Marrow Marsh [97, 98]. Two plots on the beach were treated, and two were left untreated as controls. The 15-m diameter experimental plots (separated by 45 to 75 m) were sampled on a routine basis [99]. The results of the detailed chemical analysis showed that there were no significant differences between pre- and post treatment samples after 96 h of treatment with any of the selected methods. Although visual signs indicated that the condition of the marsh areas improved after treatment [100], there was no conclusive evidence to show significant degradation of the oil within the 4-day monitoring period. Numerous compromises in the experimental design of this study have been identified [99]. For example, the separation of treated and untreated plots and the booming methods used to isolate them may not have prevented mixing and cross-contamination Furthermore, our knowledge from previous laboratory studies and field trials suggests that the 96-h duration of the experiment was insufficient for a definitive test of bioremediation. Unfortunately, no attempt was made to establish which factor (if any) was limiting biodegradation and what the most appropriate bioremediation strategy might be.

10.3. Mega Borg

On 8 June 1990, the Norwegian tanker Mega Borg was carrying out a lightering operation with the Italian tanker Fraqmura about 57 mile off the Texas coast. Following an explosion and fire, the Fraqmura carried out an emergency breakaway operation from the Mega Borg, which resulted in the release of approximately 45 m3 of Angolan Palanca crude oil [101]. The next day, further oil was lost before the situation was controlled. While it was initially predicted that no oil would reach the shoreline, the Louisiana coast was littered with tiny tar balls 16 days after the accident (55). In terms of bioremediation strategies, the On-Scene Coordinator granted permission to conduct a field trial 1 day after the accident occurred. Two portions of the slick were treated with a product containing Alpha BioSea [102]. A 16-hectare patch of slick located about 5 km from the *Mega Borg* was treated 7 days after the accident with 50 kg of microbial agent (Alpha BioSea) which had been rehydrated with seawater. The product was applied with the standard shipboard fire-hose system. The equipment and treatment preparation time of approximately 1 h (105) indicates that very little rehydration time was given to the product. Four traverses of the treatment area were made over a 30-min period. Following large-scale application of the product at sea, visual observations indicated that the treated oil changed from a continuous film of brown oil and sheen to discrete areas of mottled brown and yellow material and sheen. An aerial reconnaissance 16 h after treatment was not able to detect oil in the area. However, there is considerable uncertainty about the fate of the treated oil [102]. The measurements on water samples from the treated slick showed no evidence of acute toxicity to marine life or significantly elevated levels of nutrients or total hydrocarbons. Attempts to assess the effect of the microbial agent from measurements of oil content in the emulsion samples were unsuccessful because of sample variability. By 8 h after treatment, the slick had largely broken up and dissipated. Although little change was observed in the control area, conclusive evidence of bioremediation effectiveness was not achieved because of limitations in the sampling strategy and the chemical evidence obtained. This study demonstrated the potential problems with the application of bioremediation products at sea, including difficulties with uniform product application, representative sampling, and uncertainties about the ultimate fate of the oil. The short periods over which monitoring are often possible may not be sufficient to validate the presence and activity of oil-degrading bacteria or the effectiveness of bioremediation treatments. The observed visual effects may well have been caused by physical or chemical processes such as surfactant action associated with the treatment [102].

10.4. Prall's Island

In January 1990, fuel oil from a pipeline failure spilled into the Arthur Kill waterway in New Jersey and contaminated a gravel beach on the Prall's Island bird sanctuary. Mechanical methods were used to remove the bulk of the oil. Cleanup was suspended in March 1990 to minimize possible adverse effects on migrating birds. However, Exxon was granted permission to carry out a bioremediation experiment on part of a contaminated beach. Two shallow trenches were dug in the intertidal zone to bury bags of beach substrate containing known concentrations of oil and to help overcome possible problems of variable distribution of oil on the beach. A slow-release fertilizer (Customblen, Sierra Chemicals) was placed in the trenches to encourage biodegradation. Over a 92-day period, sub samples were periodically taken from the oiled bags, together with beach samples and water samples for analysis of total petroleum hydrocarbons, GC-MS detection of hydrocarbons, microbial counts, and water quality (nitrogen, phosphorus, ammonia, and dissolved oxygen) determination. No clear trends of increased biodegradation from the fertilized plots could be identified during the experiment, and there was high variability in the levels of total petroleum hydrocarbons, which may have masked any effects of the treatment [89].

10.5. Seal Beach

On 31 October 1990, a well blowout off Seal Beach, Calif., resulted in the release of approximately 2 m^3 of crude oil that contaminated 8,000 to 12,000 m^2 of marsh grassland in the Seal Beach National Wildlife Refuge. One week after the incident, the marsh was hand sprayed with a combination of a microbial product used in sewage treatment plants (INOC 8162) and a commercial fertilizer (Miracle Gro 30-6-6). Two weeks later, the fertilizer alone was applied. Oiled, oiled and treated, and unoiled samples were collected and analyzed for oil content by GC-MS [89].

Measurements were also made of the microbial mineralization of the phenanthrene, microbial respiration, and biomass. The results of a 35-day monitoring effort showed no differences between the treated and untreated oil plots. Subsequently, laboratory tests were carried out with the microbial product and Prudhoe Bay crude oil to compare the performance of the microbial product with nutrient-only controls. After 16 days of incubation, little or no difference was found between treated and control flasks. It was concluded that the microbial product was not effective in accelerating biodegradation of oil under controlled laboratory conditions [89].

Moreover, the salt marsh environment may be difficult to bioremediate simply by adding sources of nitrogen and phosphorus. Oxygen depletion may have been a significant factor in the inhibition of oil biodegradation [103].

10.6. Exxon Valdez

The tanker Exxon Valdez ran aground on Bligh Reef in the Gulf of Alaska on 24 March 1989, spilling approximately 41,000 m^3 of Alaskan North Slope crude oil (primarily Prudhoe Bay crude oil). A major response effort was mounted at sea to recover the oil, but the prevailing conditions and circumstances resulted in the contamination of about 2,090 km of coastline [104]. Some beaches were heavily oiled, particularly those on islands in Prince William Sound that were directly in the path of the slick. Many techniques were adopted in a massive effort to clean up the shoreline of the Sound (72% rock face, 24% mixed boulder and cobble, 3.5% mixed cobble and pebble, and 0.5% fine-grain sand/mud or marsh). These included cold- and warm-water washing, steam cleaning, and manual oil recovery techniques. Initially, the main aim was to remove the heaviest concentrations of oil to minimize the impact on wildlife and fisheries [104, 105]. A bioremediation option based on nutrient enrichment was proposed shortly after the spill. However, it was thought necessary to carry out some research first to establish the potential for effective and safe use of this technique. The limited success of the initial field tests led to the approval of full-scale application in August 1989, and 119 km of shoreline was subsequently treated that year. By 1990, the previous cleanup efforts and winter storms had greatly reduced the extent of shoreline oiling [106] and natural recovery processes were already well advanced [7, 8]. The National Oceanic and Atmospheric Administration applied the concept of net environmental benefit analysis in an evaluation of the main alternative to bioremediation at this time, namely, excavation and rock-washing treatment [107].

It was concluded that this technique would be particularly damaging to the environment. Bioremediation was therefore adopted as a prime cleanup strategy. In 1990 and 1991, bioremediation was used in combination with storm berm relocation, tilling, and manual pickup. On 12 June 1992, the U.S. Coast Guard and the State of Alaska declared the cleanup officially concluded on the basis that there would be no further net environmental benefit from continuing the effort.

Shortly after the Exxon Valdez spill, it was suggested that bioremediation may be able to enhance the rates of oil removal from the contaminated beaches [108]. As a preliminary step, the number of oil degrading microorganisms on oiled beaches in comparison with untreated controls was determined. Pritchard et al. (2005) reported that the hydrocarbon-degrading microorganisms on oiled shorelines had increased by as much as 10,000 times to an average level of 106 cells per g of beach material. Once it was clear that hydrocarbon degraders were present in abundance, it was necessary to establish which factors were likely to limit biodegradation and which specific hydrocarbon components were biodegradable. The research was conducted in the laboratory with Prudhoe Bay crude oil weathered by distillation to remove the volatile fraction. Biodegradation by indigenous microorganisms was monitored by noting changes in the concentration of components of the oil by GC-MS, by monitoring carbon dioxide evolution and oxygen consumption by the microorganisms, and by determining the evolution of radioactive $14CO_2$ from specific 14C-labeled oil components such as phenanthrene [109].

The experiments demonstrated unequivocally that the microbial population in Prince William Sound could rapidly biodegrade the aliphatic and aromatic fractions of Prudhoe Bay crude in the presence of suitable nitrogen and phosphorus sources. The microbial community decomposed C1 dibenzothiphene, C_2 fluorenes, C_3 naphthalenes, phenanthrene, and anthracene among others (113). Studies of CO_2 production suggested that the oil was not just being biotransformed but that it was being completely mineralized to CO_2 and H_2O. For example, over 30% of [U-14C]phenanthrene could be mineralized to $14CO_2$ within 4.5 days when incubated with oil-contaminated beach material from Prince William Sound [83].

The highest mineralization rates were noted in the test systems treated with the highest concentration of nitrogen. From these results, it is clear that the main factor limiting the biodegradation of oil on the beaches in Prince William Sound was the concentration of nutrients, particularly nitrogen. A substantial microbial biomass had already developed in the contaminated areas of Prince William Sound which was able to decompose many components within the contaminant oil. Hence, addition of nutrients, and not seeding, was thought to be the most appropriate bioremediation strategy [17, 80].

11. Conclusion and future prospects

Despite the growing acceptance of bioremediation as a means to treat spilled oil in marine environments the mechanisms that promote the process under field conditions remain poorly constrained. Although general statements can be made regarding the enhancement of biodegradation by nutrient amendment, there is no consensus on how to best optimize nutrient additions. Subsequently, oil spill treatment strategies are largely developed empirically from previous experience and/or from laboratory feasibility studies. Introduction of a theoretical framework to explain observations from primarily empirical studies of oil-spill bioremediation would be a fundamental step towards the development of more objective spill management practices. Resource ratio theory has recently been put forward as a theoretical basis to explain some of the effects of bioremediation and many of the observations made in bioremediation studies are consistent with the theory's predictions. Although the introduction of this theory may simply augment current empirical approaches, in the longer term it has the potential to form the basis of more predictable bioremediation strategies, and the introduction of theory to the field of bioremediation is an important progression. To further test the applicability of resource-ratio theory it will be necessary to conduct systematic studies on the effect of different nutrient amendments on bacterial populations and concomitant alterations in biodegradation rates, to identify patterns of microbial diversity associated with optimum contaminant removal. Until recently, such an approach would not have been possible due to the limitations of the methods available to characterize the composition of microbial communities. With the introduction of molecular methods to study indigenous microorganisms, this limitation has been alleviated to some extent. Integrated studies combining careful field evaluation of crude oil biodegradation with molecular approaches to study microbial populations involved in degradation of spilled oil have already begun and promise to reveal much regarding the relationship between microbial population structure and the progress of bioremediation. Anaerobic hydrocarbon degradation in marine environments has only recently been widely accepted and there is a need to determine both how widespread an occurrence this is and in what circumstances it will have a significant impact on the dissipation of crude oil contamination. The environmental factors that promote the process must also be identified if it is to be exploited for the treatment of spilled oil.

Author details

Mehdi Hassanshahian[1*] and Simone Cappello[2,3*]

*Address all correspondence to: mshahi@uk.ac.ir

1 Department of Biology - Faculty of Science- Shahid Bahonar University of Kerman - Kerman, Iran

2 Istituto per l'Ambiente Marino Costiero (IAMC) - C.N.R. U.O.S. di Messina, Italy

3 Istituto Sperimentale Talassografico (IST) di Messina, Italy

References

[1] Alloway BJ, Ayres DC. Organic Pollutants. In: Chemical Principles of Environmental Pollution. 1st edition. Chapman and Hall, India; 1993.

[2] Atlas RM. Microbial degradation of petroleum hydrocarbons: an environmental perspective. Microbiology Review 1981; 45 180-209.

[3] Marshall AG, Rogers RP. Petroleomics: the next grand challenge for chemical analysis. Analytical Chemistry Research 2003; 37 53–59.

[4] Head IM, Jones DM Larter S R. Biological activity in the deep subsurface and the origin of heavy oil. Nature 2003; 426344–352.

[5] Texas B. In Proceedings of the Fifteenth Arctic and Marine Oil Spill Program Technical Seminar. Canada Ottawa Canada, 2001.

[6] Cappello S, Denaro R, Genovese M, Giuliano L, Yakimov MM. Predominant growth of Alcanivorax during experiments on oil spill bioremediation in mesocosms. Microbiology Research 2006; 162 185-190.

[7] National Research Council. Oil in the Sea III: Inputs, Fates and Effects. National Academy of Sciences Washington DC; 2002.

[8] Mahon A, Labelle RP. Update of comparative occurrence rates for offshore oil spills. Spill Science Technology Bulletin 2000; 6 303-321.

[9] Cappello S, Caruso G, Zampino D, Monticelli LS, Maimone G, Denaro R, Tripodo B, Troussellier M, Yakimov MM, Giuliano L. Microbial community dynamics during assays of harbour oil spill bioremediation: a microscale simulation study. Journal of Applied Microbiology 2007. 102 (1), 184-194.

[10] Department of the Environment. Transport and the Regions: Digest of Environmental Statistics. London HMSO; 1998.

[11] Prince RC. Bioremediation of oil spills. Trends Biotechnology 1997, 15158-160.

[12] Swannell RPJ, Lee K, McDonagh M. Field evaluations of marine oil spill bioremediation. Microbiology Review 1996; 60 342-365.

[13] Cappello S, Gabriella G, Vivia B. Crude oil-induced structural shift of coastal bacterial communities of rod bay and characterization of cultured cold-adapted hydrocarbonoclastic bacteria. FEMS Microbiology Ecology 2004; 49 419–432.

[14] Cappello S, Crisari A, Hassanshahian M, Genovese M, Santisi S, Yakimov MM. "Effect of a Bioemulsificant Exopolysaccharide (EPS2003) on Abundance and Vitality of Marine Bacteria." Water Air Soil Pollution 2012; DOI 10.1007/s11270-012-1159-8.

[15] U.S. Environmental Protection Agency. Interim report. Oil Spill Bioremediation Project. Office of Research and Development, U.S. Environmental Protection Agency 1990; Gulf Breeze, Fla.

[16] Cappello S, Genovese M, Torre CD, Crisari A, Hassanshahian M, Santisi S, Calogero R, Yakimov MM. Effect of bioemulsificant exopolysaccharide (EPS2003) on microbial community dynamics during assays of oil spill bioremediation: A microcosm study. Marine Pollution Bulletin 2012; http://dx.doi.org/10.1016/j.marpolbul.2012.07.046.

[17] Cappello S, Santisi S, Calogero R, Hassanshahian M, Yakimov MM. Characterization of Oil-Degrading Bacteria Isolated from Bilge Water. Water Air Soil Pollution 2012; 223 3219-3226.

[18] Caruso G, Denaro R, Genovese M, Giuliano L, Mancuso M, Yakimov MM. New methodological strategies for detecting bacterial indicators. Chemistry and Ecology 2004; 20 (3) 167–181.

[19] Kohno T, Sugimoto Y, Sei K, Mori K. Design of PCR Primers and gene probes for general detection alkane-degrading bacteria. Microbes and Environment 2002; 17 (3) 114-212.

[20] Hassanshahian, M., Emtiazi, G., Kermanshahi, R., Cappello, S. 2010. Comparison of oil degrading microbial communities in sediments from the Persian Gulf and Caspian Sea. Soil and Sediment Contamination. 19 (3), 277-291.

[21] Das K, Ashis K, Mukherjee F. Crude petroleum-oil biodegradation efficiency of Bacillus subtilis and Pseudomonas aeruginosa strains isolated from a petroleum-oil contaminated soil from North-East India. Bioresource Technology 2006; 98 1339-1345.

[22] Naughton SJ, Stephen JR, Venosa AD, Davis GA, Chang Y-J, White DC. Microbial population changes during bioremediation of an experimental oil spill. Applied Environmental Microbiology 1999, 65 3566-3574.

[23] Söhngen NL. Benzin Petroleum, Paraffinöl und Paraffin als Kohlenstoff- und Energiequelle für Mikroben. Zentralbl. Bakteriol 1913; 2 37 595–609 (in German).

[24] Prince RC. Petroleum Microbiology. American Society for Microbiology Press Washington DC; 2005.

[25] Emtiazi G, Hassanshahian M, Golbang N. Development of a microtiter plate method for determination of phenol utilization, biofilm formation and respiratory activity by environmental bacterial isolates. International Biodeterioration & Biodegradation 2005; 56 231-235.

[26] Hedlund BP, Geiselbrecht AD, Bair TJ, Staley JT. Polycyclic aromatic hydrocarbon degradation by a new marine bacterium, Neptunomonas naphthovorans gen. nov., sp. nov. Applied Environmental Microbiology 1999; 65 251-259.

[27] Emtiazi G, Saleh T, Hassanshahian M. The effect of bacterial glutathione S-transferase on morpholine degradation. Biotechnology Journal 2009 ; 4, 202–205.

[28] Chaillana F, Flècheb A, Burya E, Phantavonga Y, Saliot A, Oudot J. Identification and biodegradation potential of tropical aerobic hydrocarbon-degrading microorganisms. Research Microbiology 2004; 155(7) 587-595.

[29] Lliros M, Munill X, Sole A, Martinez-Alonso M, Diestra E, Esteve I. Analysis of cyanobacteria biodiversity in pristine and polluted microbial mats in microcosms by confocal laser scanning microscopy (CLSM). Science Technology and Education of Microscopy 2003; 52 483–499.

[30] Barth HJ. The influence of cyanobacteria on oil polluted intertidal soils at the Saudi Persian Gulf shores. Marine Pollution Bulletin 2003; 46 1245-52.

[31] Hassanshahian M, Emtiazi G, Cappello S. Isolation and characterization of crude-oil-degrading bacteria from the Persian Gulf and the Caspian Sea. Marine Pollution Bulletin 2012; 64 7–12.

[32] Ghanavati H, Emtiazi G, Hassanshahian M. Synergism effects of phenol degrading yeast and Ammonia Oxidizing Bacteria for nitrification in coke wastewater of Esfahan Steel Company. Waste Management & Research 2008; 26(2) 203-208.

[33] Yakimov MM, Golyshin PN, Lang S, Moore ER, Abraham WR, Lunsdorf H, Timmis KN. Alcanivorax borkumensis gen. nov., sp. nov., a new hydrocarbon-degrading and surfactant producing marine bacterium. International Journal Systematic Bacteriology 1998; 48 339-348.

[34] Dyksterhouse SE, Gray JP, Herwig RP, Lara JC, Staley JT. Cycloclasticus pugetii gen. nov., sp. nov., an aromatic hydrocarbon-degrading bacterium from marine sediments. International Journal Systematic Bacteriology 1995; 45 116-123.

[35] Gauthier MJ, Lafay B, Christen R, Fernandez L, Acquaviva M, Bonin P, Bertrand JC: Marinobacter hydrocarbonoclasticus gen. nov., sp. nov., a new extremely halotolerant, hydrocarbondegrading marine bacterium. Int J Syst Bacteriol 1992, 42:568-576.

[36] Golyshin PN, Chernikova TN, Abraham WR, Lunsdorf H, Timmis KN, Yakimov MM. Oleiphilaceae fam. nov., to include Oleiphilus messinensis gen. nov., sp. nov., a

novel marine bacterium that obligatory utilizes hydrocarbons. International Journal Systematic Bacteriology 2002; 52 901-911.

[37] Yakimov MM, Giuliano L, Gentile G, Crisafi E, Chernikova TN, Abraham WR, Lunsdorf H, Timmis KN, Golyshin PN. Oleispira antarctica gen. nov., sp. nov., a novel hydrocarbonoclastic marine bacterium isolated from Antarctic coastal sea water. International Journal Systematic Evolutionary Microbiology 2003; 53 779-785.

[38] Engelhardt MA, Daly K, Swannell RP, Head IM. Isolation and characterization of a novel hydrocarbon-degrading, Gram positive bacterium, isolated from intertidal beach sediment, and description of Planococcus alkanoclasticus sp. nov. Journal Applied Microbiology 2001; 90 237-247.

[39] Melcher RJ, Apitz SE, Hemmingsen BB. Impact of irradiation and polycyclic aromatic hydrocarbon spiking on microbial populations in marine sediment for future aging and biodegradability studies. Applied Environment Microbiology 2002; 68 2858-2868.

[40] Zhuang WQ, Tay JH, Maszenan AM, Tay ST. Isolation of naphthalene-degrading bacteria from tropical marine sediments. Water Science Technology 2003; 47 303-308.

[41] Gilewicz M, Nimatuzahroh T, Nadalig H, Budzinski P, Doumenq V, Michotey JC, Bertrand JC. Isolation and characterization of a marine bacterium capable of utilizing 2-methylphenanthrene. Applied Microbiology Biotechnology 1997; 48 528-533.

[42] Button DK, Schut F, Quang P, Martin R, Robertson BR. Viability and isolation of marine bacteria by dilution culture: theory, procedures, and initial results. Applied Environmental Microbiology 1993; 59 881-891.

[43] Atlas RM. Petroleum Microbiology. Macmillan Publishing New York; 1984.

[44] Gibson DT. Microbial Degradation of Organic Compounds. Dekker New York; 1984.

[45] Ensign SA. Microbial metabolism of aliphatic alkenes. Biochemistry 2001; 40 5845–5853.

[46] Van Beilen JB, Wubbolts MG, Witholt B. Genetics of alkane oxidation by Pseudomonas oleovorans. Biodegradation 1994; 5 161– 174.

[47] Atlas RM, Bartha R. Inhibition by fatty acids of the biodegradation of petroleum. Antonie van Leeuwenhoek 1973; 39 257– 271.

[48] Cerniglia CE. Biodegradation of polycyclic aromatic hydrocarbons. Biodegradation 1992; 3 351– 368.

[49] Maneerat S, Kulnaree P. Isolation of biosurfactant-producing marine bacteria and characteristics of selected biosurfactant. Applied Microbiology 2007; 29 783-791.

[50] Mathew M. Obbard JP. Optimization of the dehydrogenase assay for measurement of indigenous microbial activity in beach sediments contaminated with petroleum. Biotechnology Letters 2001; 23 227–230.

[51]   Evans WC, Fuchs G. Anaerobic degradation of aromatic compounds. Annual Review Microbiology 1988; 42 289– 317.

[52]   Pruthi V, Cameotra SS. Rapid identification of biosurfactant-producing bacterial strains using a cell surface hydrophobicity technique. Biotechnology Techniques 1997; 11 671-674.

[53]   Aeckersberg F, Bak F, Widdel F. Anaerobic oxidation of saturated hydrocarbons to carbon dioxide by a new type of sulfate-reducing bacterium. Archive Microbiology 1991; 156 5 – 14.

[54]   Spormann AM, Widdel F. Metabolism of alkylbenzenes, alkanes, and other hydrocarbons in anaerobic bacteria. Biodegradation 2000; 11 85– 105.

[55]   Anderson RT, Lovley DR. Hexadecane decay by methanogenesis. Nature 2000; 404, 722-723.

[56]   So CM, Phelps CD, Young LY. Anaerobic transformation of alkanes to fatty acids by a sulfate-reducing bacterium, strain Hxd3. Applied Environmental Microbiology 2003; 69 3892– 3900.

[57]   Edwards EA, Grbic-Galic D. Anaerobic degradation of toluene and o-xylene by a methanogenic consortium. Applied Environmental Microbiology 1994; 60 313– 322.

[58]   Wilkes H, Rabus R, Fischer T, Armstroff A, Behrends A, Widdel F. Anaerobic degradation of n-hexane in a denitrifying bacterium: further degradation of the initial intermediate (1-methylpentyl) succinate via C-skeleton rearrangement. Achieve Microbiology 2002; 177 235– 243.

[59]   Tannenbaum E, Starinsky A, Aizenshtat Z. Light-oils transformation to heavy oils and asphalts—assessment of the amounts of hydrocarbons removed and the hydro-logical-geological control of the process. Exploration for Heavy Crude Oil and Natural Bitumen. The American Association of Petroleum Geologists 1987, Tulsa, Oklahoma

[60]   Sorkhoh NA, Ghannoum MA, Ibrahim AS, Stretton RJ, Radwan S. Crude oil and hydrocarbon-degrading strains of Rhodococcus rhodochrous isolated from soil and marine environments in Kuwait. Environmental Pollution 1990; 65 1 – 17.

[61]   Hfpner T, Felzmann H, Struck H, van Bernem KH. The nature and extent of oil contamination on Saudi Persian Gulf beaches: examinations of beaches of Dawhat ad Dafi and Dawhat ad Musallamiya in summer 1991 and winter 1991/92. Arab Journal Science Engineering 1993; 18 243-255.

[62]   Hasanshahian M, Emtiazi G. Investigation of alkane biodegradation using the micro-titer plate method and correlation between biofilm formation, biosurfactant production and crude oil biodegradation. International Biodeterioration & Biodegradation 2008; 62 170-178.

[63] Hassanshahian M, Tebyanian H, Cappello S. Isolation and characterization of two crude-oil degrading yeast strains, Yarrowia lipolytica PG-20 and PG-32 from Persian Gulf. Marine Pollution Bulletin 2012; 64 1386-1391.

[64] Kubota M, Nodate M, Yasomoto H, Taku U, Osamu K, Misawa R. Isolation and functional analysis of cytochrom P450 genes from various environment. Biosience Biotechnology and Biochemistry 2005; 69 (12) 2421-2430.

[65] Lebaron P, Servais P, Trousellier M. Changes in bacterial community structure in seawater mesocosms differing in their nutrient status. Acquatic Microbial Ecology 1999; 19 255-267.

[66] Kloos J, Charles M, Schloter M. New method for the detection of alkane-monooxygenase homologous genes (alkB) in soils based on PCR-hybridization. Journal of Microbiological Methods 1999 ; 66 486–496.

[67] Lee M, Hwang G, Hung J, Young K, Kyung H. Physical structure and expression of alkb encoding alkane hydroxylase and rubredoxin reductase from Pseudomonas maltophilia. Biochemical and Biophysical Research Communications 1996; 218 17–21.

[68] Pukall R, Pauker O, Buntefu BD, Ilichs G, Lebaron P, Bernard, L, Guindulain T, Vives-Rego J, Stackebrandt E. High sequences diversity of Alteromonas macleodii-related cloned and cellular 16S rDNAs from Mediterranean seawater mesocosm experiment. FEMS Microbiology Ecology 1999; 28 335-344.

[69] Radwan SS, Al-Hasan RH, Salamah A, Khanafer M. Oil-consuming microbial consortia floating in the Persian Gulf. International Biodeterioration & Biodegradation 2005; 56 28-33.

[70] Li ZY, Kravchenko I, Xu H, Zhang C. Dynamic changes in microbial activity and community structure during biodegradation of petroleum compounds: A laboratory experiment. Journal of Environmental Science 2007; 19 1003–1013.

[71] Liu C, Zongze S. Alcanivorax dieselolei sp. nov., a novel alkane-degrading bacterium isolated from sea water and deep-sea sediment. International Journal of Systematic and Evolutionary Microbiology 2005; 55 1181–1186.

[72] Maa FB, Jing B, Guo L, Zhao C, Chein-chi C, Di C. Application of bioaugmentation to improve the activated sludge system into the contact oxidation system treating petrochemical wastewater. Bioresource Technology 2009; 100 597–602.

[73] Macnaughton S J, Stephen JR, Venosa AD, Davis GA, Chang YJ, White DC. Microbial population changes during bioremediation of an experimental oil spill. Applied and Environmental Microbiology 1999; 65 3566-3574.

[74] Rahman KSM, Thahira-Rahman J, Lakshmanaperumalsamy P, Banat IM. Towards efficient crude oil degradation by a mixed bacterial consortium. Bioresource Technology 2004; 85 257–261.

[75] Makrarn T. Suidan Effects of nitrogen source on crude oil biodegradation. Journal of Industrial Microbiology 1994; 13 279-286.

[76] Malatova A, Sbirova P, Rastosdia R. Isolation and characterization of hydrocarbon degrading bacteria from enviromental habitats in western New York State. Journal of Applied Microbiology 2005; 65 780-790.

[77] Manee P, Prayad P, Edward S, Upatham A, Ladda T. Biodegradation of crude oil by soil microorganisms in the tropic. Biodegradation 1998; 9 83–90.

[78] Margesin R, Labbe D, Schinner FC, Gmur W, Whyte LG, Characterization of hydro-carbon-degrading microbial populations in contaminated and pristine alpine soils. Applied and Environmental Microbiology 2003; 69 3085–3092.

[79] Margesin R, Feller G, Hämmerle M, Stegner U, Schinner F. colorimetric method for the determination of lipase activity in soil. Biotechnology Letters 2002; 24 27–33.

[80] Marquez MC, Ventosa A. Marinobacter hydrocarbonoclasticus Gauthier et al. 1992 and Marinobacter aquaeolei Nguyen et al. 1999 are heterotypic synonyms. International Journal of Systematic and Evolutionary Microbiology 2005; 55 1349–1351.

[81] Mckew B, Coulon F, Yakimov MM, Denaro R, Genovese M, Smith J, Osborn M, Timmis KN, Mcgenity TJ. Efficacy of intervention strategies for bioremediation of crude oil in marine systems and effects on indigenous hydrocarbonoclastic bacteria. Environmental Microbiology 2007; 9 (6) 1562–1571.

[82] Muratova AY, Turkovskaya V. Degradation of petroleum oils by a selected microbial association. Applied Biochemistry and Microbiology 2001; 37 155–159.

[83] Muyzer G, Kornelia S. Application of denaturing gradient gel electrophoresis (DGGE) and temperature gradient gel electrophoresis (TGGE) in microbial ecology. Antonie van Leeuwenhoek 1998; 73 127–141.

[84] Muyzer G, Waal EC, Uitterlinden AG. Profiling of complex microbial populations by denaturing gradient gel electrophoresis analysis of polymerase chain reaction-amplified genes encoding for 16S rRNA. Applied and Environmental Microbiology 1993; 59, 695-700.

[85] Nakamuraa S, Sakamotoa Y, Ishiyamaa M, Tanakaa M, Kuniib K, Kubob C, Sato P. Characterization of two oil-degrading bacterial groups in the Nakhodka oil spil. International Biodeterioration & Biodegradation 2007; 60 202–207.

[86] Narhi LO, Wen LP, Fulco AJ. Characterization of the protein expressed in Escherichia coli by a recombinant plasmid containing the Bacillus megaterim cytochrome P-450 BM-3 gene. Molecular Cell Biochemistry 1988; 79 63-71.

[87] Nodate M, Mitsutoshi K, Norihiko M. Functional expression system for cytochrome P450 genes using the reductase domain of self-sufficient P450RhF from Rhodococcus sp. NCIMB 9784. Applied Microbiology Biotechnology 2006; 71 455–462.

[88] Odum EP. The mesocosm. Bioscience 1984; 34 558-562.

[89] Okerentugba PO, Ezeronye OU. Petroleum degrading potentials of single and mixed microbial cultures isolated from rivers and refinery effluent in Nigeria. African Journal of Biotechnology 2003; 2 (9) 288-292.

[90] Hoff R. A summary of bioremediation applications observed at marine oil spills. Report HMRB 91-2. Hazardous Materials Response Branch, National Oceanic and Atmospheric Administration 1991; Washington DC.

[91] Petersen JE, Cornwell JC, Kemp WM. Implicit scaling in the design of experimental aquatic ecosystems. Oikos 1999; 85 3-18.

[92] Powell SM, Ferguson SH, Bowman P, Snape I. Using real-time PCR to assess changes in the hydrocarbon-degrading microbial community in antarctic soil during bioremediation. Microbial Ecology 2006; 52 523–532.

[93] National Academy of Sciences. Oil in the Sea: Inputs, Fates and Effects. Washington DC: National Academy Press; 1985.

[94] Hoff R. Bioremediation: an overview of its development and use for oil spill clean-up. Marine Pollution Bulletin 1993; 26 476–481.

[95] Atlas RM. Microbial hydrocarbon degradation—bioremediation of oil spills. Journal Chemistry Technology Biotechnology 1991; 52 149–156.

[96] Bocard CP, Renault J, Croquette S. Cleaning products used in operations after the Amoco Cadiz disaster. In Proceedings of the International Oil Spill Conference. American Petroleum Institute 1979; Washington D.C.

[97] Pruthi V, Cameotra S S. Rapid identification of biosurfactant-producing bacterial strains using a cell surface hydrophobicity technique. Biotechnology Techniques 1997; 11 671-674.

[98] Mearns A J. 1991. Observations of an oil spill bioremediation activity in Galveston Bay, Texas. NOAA Technical Memorandum NOS OMA 57. National Oceanic and Atmospheric Administration 1997; Washington D.C.

[99] Texas General Land Office. Combating oil spills along the Texas coast: a report on the effect of bioremediation. Texas General Land Office 1989; Austin.

[100] Nadeau R, Singhvi J, Ryabik YH, Lin J, Syslo C. Bioremediation efficacy in Marrow Marsh following the Apex Oil Spill, Galveston 1992.

[101] Greene TC. The Apex Barges spill, Galveston Bay, July 1990, In Proceedings of the 1991 Oil Spill Conference. American Petroleum Institute 1991; Washington, D.C.

[102] Leveille TP. The Mega Borg fire and oil spill: a case study. In Proceedings of the 1991 International Oil Spill Conference. American Petroleum Institute 1991; Washington, D.C.

[103] Texas General Land Office. Mega Borg oil spill off the Texas coast: an open water bi-
      oremediation test. Texas General Land Office 1990; Texas Water Commission Austin
      273–278.

[104] Lee K, Levy EM. Bioremediation: waxy crude oils stranded on low-energy shorelines,
      In Proceedings of the 1991 Oil Spill Conference (Prevention, Behaviour, Control,
      Cleanup). American Petroleum Institute 1991; Washington, D.C 541–547.

[105] Bragg JR, Prince JB, Wilkinson RM, Atlas D. Biore- 362 SWANNELL ET AL. MICRO-
      BIOL. REV. mediation for shoreline cleanup following the 1989 Alaskan oil spill. Fi
      tion Co. USA 1992) Houston.

[106] Owens EH. Changes in shoreline oiling conditions 1 1/2 years after the 1989 Prince
      William Sound spill. Technical Report 1991; Woodward-Clyde Consultants, Coastal
      Science and Engineering Center, Seattle.

[107] Jahns HO, Bragg LC, Dash EH, Owens. Natural cleaning of shorelines following the
      Exxon Valdez oil spill. In Proceedings of the 1991 Oil Spill Conference. American Pe-
      troleum Institute 1991; Washington, D.C., 167–176.

[108] National Oceanic and Atmospheric Administration. Excavation and rock washing
      treatment technology—net environmental benefit analysis. Hazardous Materials Re-
      sponse Branch, National Oceanic and Atmospheric Administration 1990; Seattle.

[109] Pritchard PH, Costa C. EPA's Alaska oil spill bioremediation project. Environment.
      Science. Technology 1991; 25 372–379.

Chapter 3

Microbial Hydrocarbon Degradation: Efforts to Understand Biodegradation in Petroleum Reservoirs

Isabel Natalia Sierra-Garcia and
Valéria Maia de Oliveira

Additional information is available at the end of the chapter

1. Introduction

The understanding of the phylogenetic diversity, metabolic capabilities, ecological roles, and community dynamics taking place in oil reservoir microbial communities is far from complete. The interest in studying microbial diversity and metabolism in petroleum reservoirs lies mainly but not only on providing a better comprehension of biodegradation of crude oils, since it represents a worldwide problem for petroleum industry. Generally, biodegradation of oil affects physical and chemical properties of the petroleum, resulting in a decrease of its hydrocarbon content and an increase in oil density, sulphur content, acidity and viscosity, leading to a negative economic consequence for oil production and refining operations [1,2]. Another important point for studying biodegradation lies on its important role in the global carbon cycle and the direct impact on bioremediation of polluted ecosystems. Furthermore, many of the enzymes involved in the degradation pathways are considered key catalysts in industrial biotechnology [3].

Despite these motivations and long recognition of petroleum as a the most important "primary energy" source, at present, microorganisms and factors involved in biodegradation of crude oil hydrocarbons in petroleum reservoirs are still not fully understood. The inaccessibility and complex microbiological sampling of petroleum reservoirs as well as the inherent limitations of the traditional culturing methods conventionally employed can explain this fact. Culture-based techniques have traditionally been the primary tools utilized for studying the microbiology of terrestrial and subsurface environments [4], which allowed the recovery and documentation of a large collection of bacteria capable of hydrocarbon utilization. Studies of numerous aerobic and anaerobic bacterial isolates have revealed mechanisms, which allow them to degrade specific classes of the highly diverse range of hydrocarbon compounds.

Therefore, all we know about the degradation of petroleum compounds has come from studying isolated microorganisms. Here, we provide an overview of what is currently known about the mechanisms of aerobic and anaerobic degradation of hydrocarbons, as a result from biochemical and genomic approaches, we give a perspective of the petroleum microbial diversity unraveled so far, and finally we discuss the common oil reservoir characteristics that can be used to predict the most probable mechanism of degradation into deep petroleum reservoirs.

It is well known that microbial diversity in environment is several orders of magnitude higher than the one assumed based on previous cultivation methods [6]. A particularly large number of novel techniques have been developed, which now allow the determination of the *in situ* microbial diversity and activity on a particular site, screening for a particular gene or activity of interest, gene quantification, and DNA and mRNA sequencing and analysis from total communities. This book chapter will address how the implementation of such culture-independent molecular methods allow the access to the microbial diversity and metabolic potential of microorganisms and bring novel information about microbial diversity and new pathways involved in biodegradation processes taking place in petroleum reservoirs. This information will certainly contribute to a broader perspective of the biodegradation processes and corroborate with previous findings that degradation of pollutants in many cases is carried out by microbial consortia rather than a single species [6], where key species and catabolic genes are often not identical to those that have been isolated and described in the laboratory [7, 8].

2. Microbial diversity in oil reservoirs

Recognition of indigenous microbiota harbored by oil reservoirs has been discussed for a long time. Actually, determining the nature of isolated microorganisms from oil reservoirs (indigenous or nonindigenous) is a difficult issue concerning petroleum microbiologists. The reasons for this controversy rely mainly on the difficulty of aseptic sampling in deep oil reservoirs. This means that microorganisms observed in oil field fluids conceivably could be contaminants introduced during drilling operations and/or during sample retrieval, or could be material sloughed from biofilms growing in installed pipes. Another reason for skepticism is the commonplace practice of "water-flooding" (injection of surface waters or re-injection of natural formation waters to maintain reservoir pressure for oil production); since in this case microbes would be introduced during injection and therefore would not necessarily represent indigenous species [9].

In addition to this controversy, there is the fact that petroleum reservoirs are considered extreme environments where *in situ* conditions, like high pressure, temperature, salinity and anaerobic conditions, are considered as inhospitable to microbial activity. In fact, perception of deep subsurface as a sterile environment has only changed during the past two decades with the increasing awareness of the ability of microbes to colonize extreme environments. Actually, with the use of more sophisticated and appropriate sampling and cultivation techniques, as well as the application of molecular biological techniques to oil field fluids, the dogma of the sterile deep subsurface has been dispelled [9]. Rather, it has become clear that many oil reservoirs do harbor indigenous microbes (*e.g.* the genera *Geotoga* and *Petrotoga* isolated only from oil reservoirs) [10]. Nowadays it is clear that worldwide petroleum reserves are dominated by deposits that have been microbially degraded over geological time and biodegraded petroleum reservoirs represent the most dramatic manifestation of the deep biosphere [11]

In spite of the polemics on which micro-organisms would actually be native and which would be contaminants in oil reservoirs, a wide range of microbial taxonomic groups have been identified in oil reservoirs geographically distant using traditional techniques adapted to *in situ* conditions, as described by L'Haridon et al. [12], Grassia et al. [13] and reviewed by Magot et al [14], or combined with cultivation-independent molecular methods, as reported by Orphan et al. [15]. Table 1 summarizes the various physiological and taxonomical groups and species that have been isolated from oil reservoirs.

3. Aspects from oil reservoir determining microbial degradation

For a long time, the mechanism considered to be prevalent for oil degradation in petroleum reservoirs was the well documented aerobic microbial metabolism and it has long been thought that the flow of oxygen through meteoric waters was necessary for in-reservoir petroleum biodegradation [16]. This mechanism has been widely accepted despite the fact that oxygen would likely be consumed by oxidation of organic matter in near surface sediments and therefore, would be very unlikely for oxygen to reach deep petroleum reservoirs [11].

Recently, the discovery of the ability of microorganisms to degrade anaerobically hydrocarbon oil components and the detection of metabolites characteristic of anaerobic hydrocarbon degradation in oil samples from biodegraded reservoirs, but not in non-degraded reservoirs or aerobically degraded oils [11], have provided valuable information to determine the processes involved in the degradation of oil reservoirs. Nowadays, evidences of such degradation through anaerobic rather than aerobic processes are becoming more substantial and compelling [17].

It is known that microorganisms in anaerobic conditions can use a variety of final electron acceptors, including nitrate, iron, sulfate, manganese and, more recently, chlorate. Anaerobic degradation has also been coupled to methanogenesis, fermentation and phototrophic metabolism but growth of these microorganisms and, therefore, biodegradation rates are significantly lower compared to aerobic degraders. These anaerobic processes have been demonstrated in surface sediments and pure cultures or enrichments in laboratories [18] and all of them potentially play a role in oil biodegradation in anoxic petroleum reservoirs [11]. However, nitrate, like oxygen, is highly reactive and would likely be completely consumed before it could reach the oil reservoir [17]. In deep reservoirs, the supply of large amounts of Fe(III) or manganese(IV) via meteoric water influx are unlikely due to poor solubility and slow water recharge rates in subterranean cycles. Therefore, iron and manganese, which could be used as electro acceptors for oil oxidation, are unlikely to be responsible for significant compositional changes in the oil, considering their limited availability in the reservoir. Accordingly, oil degradation linked to sulfate reduction and methanogenic would therefore explain the consistent hydrocarbon compositional patterns seen in degraded oils worldwide [17]. Sulfate arises from geological sources, such as evaporitic sediments and limestone, or from the injection of seawater for pressure stabilization, and may lead to significant oil degradation and increased residual-oil sulfur content. Methanogenic oil degradation, on the other hand, does not require external electron acceptors and leads to less overall souring of the oil reservoir. Several studies have described in vitro methanogenic degradation of crude oil related compounds [19, 20] Jones et al., 2008), including n-alkanes [21, 20] and aromatic hydrocarbons [17].

Organism	Taxonomical group	Metabolism	Origin	Reference
Thermodesulforhabdus norvegicus	Deltaproteobacteria	Sulfate-reducer	Oil field in Norway	[22]
Desulfacinum infernum	Deltaproteobacteria	Sulfate-reducer	North see petroleum reservoir near Scotland	[23]
Desulfomicrobium norvegicum	Deltaproteobacteria	Sulfate reducer	Petroleum reservoir in Canada	[24]
Desulfovibrio sp.	Deltaproteobacteria	Sulfate reducer	Petroleum reservoir in Canada	[24]
Dethiosulfovibrio peptidovorans	Bacteria, Synergistetes	Sulfate reducer	Oil well in the Emeraude oilfield in Congo, Central Africa,	[25]
Desulfotomaculum thermocisternum	Bacteria, Firmicutes	Sulfate reducer	Oil reservoir in the North sea	[26]
Deferribacter sp.	Bacteria, Deferribacteres	Sulfate reducer	California oil fields	[15]
Halanaerobium congolense	Bacteria, Firmicutes	Thiosulfate- and sulfur-reducing bacterium	African oil field	[27]
Thauera phenylacetica	Betaproteobacteria	Nitrate reducer	Petroleum reservoir in Canada	[24]
Pseudomonas stutzeri	Gammaproteobacteria	Nitrate reducer	Petroleum reservoir in Canada	[24]
Garciella nitratireducens	Bacteria, Firmicutes	Nitrate reducer	Oil field in Tabasco, Gulf of Mexico	[28]
Geobacillus subterraneus, Geobacillus uzenensis	Bacteria, Firmicutes	Nitrate reducer	Petroleum reservoir in China	[29]

Organism	Taxonomical group	Metabolism	Origin	Reference
Lactosphaera pasteurii	Bacteria, Firmicutes	Fermentative	Petroleum reservoir in Canada	[24]
Propionicimonas paludicola	Bacteria, Firmicutes	Fermentative	Petroleum reservoir in Canada	[24]
Anaerobaculum	Bacteria, Synergistetes	Fermentative	California oil fields	[15]
Thermococcus sp.	Archaea, Euryarchaeota	Fermentative	California oil fields	[15]
Thermococcus sibericus	Archaea, Euryarchaeota	Fermentative	Petroleum reservoir in Western Siberia	[30]
Petrotoga sp.	Bacteria, Thermotogae	Fermentative	California oil fields	[15]
Petrotoga olearia; P. siberica	Bacteria, Thermotogae	Fermentative	Petroleum reservoir in Western Siberia	[12]
Thermoanaerobacter	Bacteria, Firmicutes	Fermentative	California oil fields	[15]
Thermotoga sp.	Bacteria, Thermotogae	Fermentative	California oil fields	[15]
Thermosipho geolei	Bacteria, Thermotogae	Fermentative	Petroleum reservoir in Western Siberia	[12]
Anaerobaculum thermoterrenum	Bacteria, Synergistetes	Fermentative	Oil well in Utah	[23]
Fusibacter paucivorans	Bacteria, Firmicutes	Fermentative	Oil well in the Emeraude oilfield in Congo, Central Africa	[31]
Thermovirga lienii	Bacteria, Synergistetes	Fermentative	Oil reservoir in the North sea	[32]
Methanococcus	Archaea, Euryarchaeota	Methanogen	California oil fields	[15]
Methanococcus thermolithotrophicus	Archaea, Euryarchaeota	Methanogen	North sea old field in Norway	[33]
Methanoculleus	Archaea, Euryarchaeota	Methanogen	California oil fields	[15]
Methanobacterium	Archaea, Euryarchaeota	Methanogen	California oil fields	[15]

Table 1. Summary of bacteria isolated from oil reservoirs worldwide.

Deep subsurface environments such as petroleum reservoirs are logistically much more difficult to study than contaminated shallow subsurface environments [17]. Since in many biodegraded petroleum reservoirs most biodegradation occurs close to the oil water transition zone, it has been proposed that the oil–water transition zone (OWTZ) provides suitable physical and chemical conditions for microbial activity [17].

There are other physical and chemical parameters influencing *in situ* biodegradation. Temperature is one of the main factors which limits oil degradation in reservoir, and, empirically, it has been repeatedly observed that biodegradation does not occur in oil reservoirs with *in situ* temperatures >80-90°C [34]. Salinity is another factor that affects in-reservoir oil biodegradation, especially in combination with temperature [13]. Typically, reservoirs with highly saline waters show limited oil biodegradation [11]. This is consistent with the observations that it has not been possible to cultivate microorganisms from reservoir waters with salinity greater than 100 g/L [13]. Pressure seems to be a less limiting factor, except that it may select for certain physiological types and influences the pH of pore waters by increasing dissolution of CO_2 [9]. The availability of electron donors and acceptors governs the type of bacterial metabolic activities within oil field environments [14]. The potential electron donors include CO_2, hydrocarbons, H_2 and numerous organic molecules. Availability of fixed nitrogen is unlikely to limit microbial activity in reservoirs. However, the availability of water-soluble nutrients, like phosphorus and/ or oxidants (terminal electron acceptors such as ferrous iron, sulfate or CO2), is more likely to limit *in situ* microbial activity [9]. Nonetheless, physiological characteristics of microorganisms indigenous to petroleum reservoirs shed light on the conditions under which petroleum degradation may occur and the potential degradation mechanisms.

4. Hydrocarbon degradation

Hydrocarbons are understood as the compounds that consist exclusively of carbon and hydrogen. Because of the lack of functional groups, hydrocarbons are largely apolar and exhibit low chemical reactivity at room temperature. Differences in their reactivities are primarily determined by the occurrence, type and arrangement of unsaturated bonds. Therefore, in this chapter, we will use the common way to classify hydrocarbons according to their bonding features: i) aliphatic group, which includes straight-chain (n-alkanes), branched-chain and cyclic compounds and ii) aromatic group which includes mono or polycyclic hydrocarbons an many important compounds which also contain aliphatic hydrocarbon chains (*e. g.*, alkylbenzenes).

Already a century ago, bacterial isolates had been reported to use aliphatic and aromatic hydrocarbons as sole carbon and energy sources [35]. Since then, numerous aerobic, and also anaerobic, bacterial isolates have been studied in order to understand the mechanisms which allow them to degrade specific members of the highly diverse aliphatic and aromatic compounds. Degradation by such isolates has been investigated thoroughly and results have revealed that they can completely degrade most classes of hydrocarbons, including alkanes, alkenes, alkynes and aromatic compounds. Such degradation can occur aerobically, with oxygen, or anaerobically, with nitrate, ferric iron, sulfate or other electron acceptors [36].

Efforts to overview the metabolism of hydrocarbons in microorganisms are confronted with the chemical diversity of such compounds and their reactivities, as well as with various microbial life styles [36]. The study of biodegradation is conventionally treated in separate areas: aliphatic vs. aromatic hydrocarbons, aerobic vs. anaerobic degradation pathways, physiology and overall metabolic pathways vs. enzymatic mechanisms and structures, often with limited knowledge and data exchange. Nonetheless, each of these study areas deals with the same central point that is the "metabolic challenge" to guide an apolar, unreactive compound composed only of carbon and hydrogen into the metabolism [36]. The hydrocarbon must be first functionalized and currently it has been recognized that there is a surprisingly diversity of reactions of activation that had evolved in microorganisms (Table 2).

Mechanisms for hydrocarbon activation		
	Aerobic	**Anaerobic**
Short-Chain non-methane alkanes C2- C10	• Non-heme iron monooxygenase similar to sMMO (C2-C9) • Copper-containing monooxygenase similar to pMMO (C2-C9) • Heme-iron monooxygenases (also refered as soluble Cytochrome P450 (C5-C12)	• Fumarate addition
Long-Chain alkanes >C10	• Heme-Monooxygenase (P450 type) • [Fe2]-Monooxygenase • Non-heme iron monooxygenase (AlkB-related) (C3-C13 or C10-C20) • Flavin-binding monooxygenase (AlmA) (C20- C36) • Thermophilic flavin-dependent monooxygenase (LadA) (C10-C30)	• Fumarate addition • Carboxylation
Aromatic hydrocarbons	• [Fe]-Dioxygenase • [Fe2]-Monooxygenase • [Flavin]-Monooxygenase	• Fumarate addition • Hydroxylation • Carboxylation

Table 2. Overview of aerobic and anaerobic mechanisms for hydrocarbon activation in bacteria.

Mechanisms for hydrocarbon activation are basically different in aerobic and anaerobic microorganisms. Under oxic conditions, hydrocarbon metabolism is always initiated using molecular oxygen as a co-substrate in mono- or dioxygenase reactions that enable the terminal or sub-terminal hydroxylation of aliphatic alkane chains or the mono or dihydroxylation of aromatic rings [37]. In the hydrocarbon activation under anoxic conditions, some proposed reactions comprise: (1) addition to fumarate by glycyl-radical enzymes, (2) methylation of unsubstituted aromatics, (3) hydroxylation with water by molybdenum cofactor containing enzymes of an alkyl substituent via dehydrogenase, and (4) carboxylation catalyzed by yet-uncharacterized enzymes which may actually represent a combination of reaction (2) followed by reaction (1) [38; 37]. Although all these mechanisms of hydrocarbon anaerobic activation have been proposed, the required signature metabolites and enzymes involved have been characterized only for (1) addition to fumarate (demonstrated for toluene, xylene, ethylbenzene, methylnaphthalene, alkanes and alicyclic alkanes); for (3) hydroxylation (demonstrated for ethylbenzene); and for (4) carboxylation (demonstrated for benzene and naphtalene) [39].

5. Biochemical and genetic pathways of microbial hydrocarbon degradation

The enzymatic reactions involved in the aerobic degradation of hydrocarbons by bacteria have been extensively studied for several decades [37]. Genes encoding enzymes for degradation are relatively well understood for aerobic and easily cultivable microorganisms, particularly for a *Pseudomonas* strain, known as *P. putida* GPo1, as well as for the strains *Acinetobacter* sp. ADP1 and *Mycobacterium tuberculosis* H37Rv [39, 40]. On the other hand, the anaerobic hydrocarbon degradation has gained more attention since is supposed to be the predominant mechanism occurring in several polluted environments and oil reservoirs. However, its study is an incipient area because of the peculiarities of the reservoir environment and difficulties that arise from attempts to characterize these communities. Nevertheless, several bacteria from other environments able to use alkanes as carbon source in the absence of oxygen have been described in the last few years [41], but anaerobic bacteria able to degrade hydrocarbons under conditions found in deep petroleum reservoirs have not been isolated so far [2]. Figure 1 represents an overview of the main mechanisms and pathways used by microorganisms to degrade hydrocarbon compounds under aerobic and anaerobic conditions.

5.1. Aerobic degradation

5.1.1. Aliphatic hydrocarbons

In most degradation pathways described, the substrate n-alkane is oxidized to the corresponding alcohol by substrate-specific terminal monooxygenases/hydroxylases. The alcohol is then oxidized to the corresponding aldehyde, and finally converted into a fatty acid. Fatty acids are conjugated to CoA and subsequently processed by β – oxidation to generate acetyl-CoA [42, 40]. Subterminal oxidation has also been described for both short and long-chain alkanes [40]. Both terminal and sub-terminal oxidation can coexist in some microorganisms [41]. Initial terminal hydroxylation of n-alkanes in bacteria can be carried out by enzymes belonging to different classes, named: (1) propane monooxygenase (C3), (2) different classes of butane monooxygenase (C2-C9), (3) CYP153 monooxygenases (C5-C12), (4) AlkB-related non-heme iron monooxigenase (C3-C10 or C10-C20), (5) flavin-binding monooxigenase AlmA (C20-C36), (6) flavin-dependent monooxygenase LadA (C10-C30), (7) copper flavin-dependent dioxygenase (C10-C30) [43].

Among all the alkane activating enzymes, the integral membrane non-heme iron monooxy-genase (AlkB) is the best characterized one. Microorganisms degrading medium (C5-C11) and long (>C12)-length alkanes have been frequently related to the presence of *alk*B genes and that is why the presence of such genes have been widely used as functional biomarker for the characterization of aerobic alkane-degrading bacterial populations in several environmental

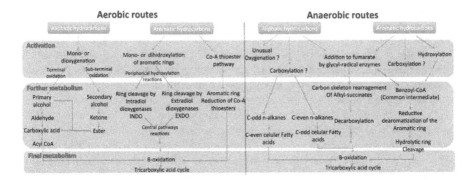

Figure 1. Pathways for aerobic and anaerobic bacterial degradation of hydrocarbon compounds. Two arrows represent more than one reaction.

samples [44, 45] and in bioremediation experiments [46, 47]. The degradation pathway of the *alk* system was first described in *Pseudomonas putida* GPo1 (formerly identified as *P. oleovorans* GPo1), where it is located on the OCT plasmid. In this model system, OCT plasmid contains two operons: *alk*BFGHJKL and *alk*ST [48]. The first operon encodes two components of the *alk* system, a particulate non-heme integral membrane alkane monooxygenase (AlkB) and the soluble protein rubredoxin (AlkG), as well as other enzymes involved in further steps. The second operon encodes for a rubredoxin reductase (AlkT and AlkS), which regulates the expression of the *alk*BFGHJKL operon [48, 49]. Since this system was described, AlkB homologous have been found in many alkane-degrading α- β – and γ –Proteobacteria and high G + C content Gram-positive bacteria (Actinobacteria) [39] and an increasing collection of alkane hydroxylase gene sequences has allowed the diversity analysis of hydrocarbon-degrading microbial populations in different ecosystems. However, comparisons of cloned *alk*B genes or gene fragments have showed that sequence diversity is very high, even among *alk*B genes within the same species [50].

In despite of the relevance of *alk*B genes as a functional biomarker of alkane-degrading bacterial communities, knowledge on the presence and diversity of *alk*B genes in oil reservoirs is scarce. Tourova et al. [51] analysed *alk*B diversity in thermophilic bacterial strains of the genus *Geobacillus* isolated from oil reservoirs or hot springs. They detected, for the first time, sets of *alk*B gene homologous in thermophilic bacteria, and some strains showed different homologous within the same genome. This fact was explained by the occurrence of horizontal gene transfer among these bacteria. Recently, Li et al. [52] aimed to evaluate *alk*B gene diversity and distribution in production water from 3 oilfields in China through a specific PCR-DGGE method. Results showed that sequences found in the water samples were similar to *alk*B genes from other corresponding alkane-degrading strains. But at the same time, they showed the presence of a considerable genetic diversity of alkB genes in the wastewater as evidenced by a total of 13 unique DNA bands detected. Studies on the degradation of alkanes in oil reservoirs are currently in a start point, but in the future they certainly will help to understand the process of degradation in oil reservoir.

In comparison to the few efforts in studying *alk*B system in oil reservoirs, much less is known about the presence of the other enzymatic systems previously listed, which have been described for aerobic degradation of n-alkanes in isolated bacteria or laboratory microcosms. For the most recent elucidated systems for alkane oxidation, named *almA* and *ladA* genes, nothing is known about the environmental distribution of these type of genes in petroleum contaminated sites [53] or oil fields, although the LadA complete degradation pathway has been characterized through genome and proteome analysis of *Geobacillus thermodenitrificans* NG80-2, a thermophilic strain isolated from a deep oil reservoir in Northern China [54]. Currently, it is believed that there are enzyme systems for alkane degradation which have still not been characterized and that may include new proteins unrelated to those already known [41]. Moreover, in many alkane degraders more than one alkane oxidation system have been observed, which have been reported exhibiting overlapping substrate ranges [39, 40]. These observations point out that in order to characterize and explore metabolic diversity and functions involved in alkane degradation one should take into consideration the high diversity of enzymes capable of initiating such metabolism.

5.1.2. Aromatic hydrocarbons

The aerobic bacterial catabolism of aromatic compounds involves a wide variety of peripheral pathways that activate structurally diverse substrates into a limited number of common intermediates that are further cleaved and processed by a few central pathways to the central metabolism of the cell [55]. Metabolic pathways and encoding genes responsible for the degradation of specific members of a highly diverse range of aromatic compounds have been characterized for many isolated bacterial strains, predominantly from the Proteobacteria and Actinobacteria phyla [56]. Degradation by such isolates is typically initiated by members of one of the three superfamilies: the Rieske non-heme iron oxygenases (RNHO), the flavoprotein monooxygenases (FPM) and the soluble diiron multicomponent monooxygenases (SDM). Further metabolism is achieved through di- or trihydroxylated aromatic intermediates. Alternatively, activation is mediated by CoA ligases where the formed CoA derivates are subjected to selective hydroxylation [58, 53]. In the case of hydrophobic pollutants, such as benzene, toluene, naphthalene, biphenyl or polycyclic aromatics, aerobic degradation is usually initiated by activation of the aromatic ring through oxygenation reactions catalyzed by RNHO enzymes or, as intensively described for toluene degradation, through members of SDM enzymes [56].

Further intermediates can be catalyzed by two kinds of enzyme, intradiol and extradiol dioxygenases, which represent two classes of phylogenetically unrelated proteins [58]. These enzymes are key enzymes in the degradation of aromatic compounds, and many of such proteins and their encoding sequences have been described, purified and characterized in the last decades [56]. While all intradiol dioxygenases described so far belong to the same superfamily, the extradiol dioxygenases include at least three members of different families. Type I extradiol dioxygenases (e.g. catechol 2,3-dioxygenases and 1,2-dioxygenases) belong to the vicinal oxygen chelate superfamily enzymes. Type II extradiol dioxygenases are related to LigB superfamily (e.g. protocatechuate 4,5-dioxygenases) and the type III enzymes belongs to the cupin superfamily (e.g. gentisate dioxygenases) [53]. However, members of novel superfamilies performing crucial steps in aromatic metabolic pathways are still being discovered [56, 53].

The knowledge of metabolic properties of isolates has allowed the monitoring of the ability of microorganisms to mineralize aromatic hydrocarbons in soils. Typically, these studies have used primers designed based on conserved gene regions and focused on RNHO or SDM as targets for initiating degradation, or on Extradiol dioxygenases (EXDO) cleaving the aromatic ring [59]. These studies range from those searching for a narrow range of genes similar or identical to those observed in type strains using non-degenerated primers to those searching for subfamilies of homologous genes using degenerated primers [59]. However, due to the immense heterogeneity of such enzymes [57], there will never be a pair of primers that will reliably cover the huge diversity of a catabolic gene family in nature [53].

5.2. Anaerobic degradation

5.2.1. Aromatic hydrocarbons

We have already described the main mechanism for degradation of aromatic compounds in aerobic conditions, where oxygen is not only the final electron acceptor but also co-substrate of two key processes: hydroxylation and cleavage of the aromatic ring by oxygenases. In contrast, in the absence of oxygen, microorganisms use a complete different pathway, based in reductive reactions to attack the aromatic ring [61].

The biochemistry of some anaerobic degradation pathways of aromatic compounds has been studied to some extent; however, the genetic determinants of all these processes and the mechanisms involved in their regulation are much less studied [55]. Recent advances in genome sequencing have led to the complete genetic information for six bacterial strains that are able to anaerobically degrade aromatic compounds using different electron acceptors and that belong to different taxonomic groups of bacteria: denitrifying betaproteobacteria, *Thauera aromatica* and *Azoarcus* sp. EbN1, two alphaproteobacteria, the phototroph *Rhodopseudomonas palustris* strain CGA009 and the denitrifying *Magnetospirillum magneticum* strain AMB-1, and two obligate anaerobic deltaproteobacteria, the iron reducer *Geobacillus metallireducens* GS-15 and the fermenter *Syntrophus aciditrophicus* strain SB [55]. It is worth remembering that, in recent years, important inferences and generalizations have been made about the genetics involved in hydrocarbon metabolism based on these isolated bacteria under conventional laboratory conditions. However, potential novel genes, enzymes and metabolic pathways responsible for degradation processes are probably harbored by yet uncultivated bacteria.

The best understood and apparently the most widespread of these anaerobic mechanisms is the radical-catalyzed addition of fumarate to hydrocarbons, yielding substituted succinate derivatives. This reaction has been recognized for the activation of several alkyl-substituted benzenes as well for n-alkanes [62]. However, understanding of this fumarate-dependent hydrocarbon activation is most advanced in the case of toluene. The key enzyme in this process is the enzyme benzylsuccinate synthase. All enzymes required for β-oxidation of benzylsuccinate are encoded by the *bbs* operon. Subsequent degradation of benzoyl-CoA proceeds via reductive dearomatization, hydrolytic ring cleavage, β-oxidation to acetyl-CoA units and terminal oxidation to Co_2 [63]. In contrast to the anaerobic metabolism of toluene, degradation of ethylbenzene (and probably other alkylbenzenes with carbon chain of at least 2) is entirely different, despite the chemical and structural similarities between the two compounds, and involves a direct oxidation of the methylene carbon via (S)-1-phenylethanol to acetophenone [55]. Ethylbenzene is anaerobically hydroxylated and dehydrogenated to acetophone, which is then carboxyled and converted to benzoylCoA as the first common intermediate of the two pathways [62].

Genetics of the enzymatic system have been only characterized for these two mechanisms for anaerobic hydrocarbon activation. Genes encoding pathways that involve fumarate addition are typically organized in two operons. One operon includes the three structural genes of the protein catalyzing fumarate addition and the other includes genes required for converting succinate derivates to benzoyl-CoA [64]. Gene sequences and organization are relatively conserved among nitrate-reducing bacteria but differ somewhat from those of the iron reducer G. metallireducens [64] and substantially from those of the hexane-degrading nitrate reducer strain HxN1 [65]. Hydrocarbon dehydrogenation pathway is also organized in two operons. One operon contains the structural genes for the first two reactions (ethylbenzene dehydrogenase and 1-phenylethanol dehydrogenase) and the other contains the structural genes for acetophone carboxylase [64].

Kane et al. [66] developed the first real-time polymerase chain reaction (PCR) method to quantify hydrocarbon utilizers based on bssA genes of nitrate-reducing Betaproteobacteria. Since then, there have been several additional studies investigating the presence and/or distribution of anaerobic hydrocarbon utilizers in anaerobic environments via functional gene surveys of bssA, extending the range of detectable hydrocarbon-degrading microbes to iron and sulfate-reducing Deltaproteobacteria and revealing partially novel, site specific degrader populations [67, 68]. Other bssA-based detection studies in impacted environments, as well as studies that combine field metabolomics and molecular tools, are described by other authors [69, 70, 71]. Despite of the role of benzylsuccinate synthase in aromatic hydrocarbon degradation and its use as a biomarker are well documented, there is no study on the presence of this gene in oil reservoirs.

5.2.2. Aliphatic hydrocarbons

Anaerobic degradation of alkanes has not been extensively studied as for some aromatic compounds. The presumable reasons include the greater attention given to BTEX compounds (benzene, toluene, ethylbenzene and xylenes) because of their classification as priority pollutants [71], also the fact that anaerobic growth with n-alkanes is even slower than that with the alkylbenzenes, and finally the fact that long chain alkanes are poorly soluble and often prevents the cultivation of cells homogeneously in the medium [72]. However, anaerobic degradation of alkanes is equally relevant, since alkanes are quantitatively the most important hydrocarbon components of petroleum, and some are acutely toxic and difficult to remediate [71]. Several anaerobic bacteria capable of degrading n-alkanes with 6 or more carbons in length, particularly hexadecane (C16), using sulfate or nitrate as electron acceptors have been isolated [72, 73].

The two main mechanisms of anaerobic degradation of n-alkanes described involve unprecedented biochemical reactions that differ completely from those employed in aerobic hydrocarbon metabolism [73]. The first involves activation at the subterminal carbon of the alkane by the addition of fumarate, analogously to the formation of benzyl succinate during anaerobic degradation of toluene, however further reactions are completely different involving dehydrogenation and hydration [72]. Studies conducted with established axenic cultures have indicated that anaerobic metabolism of oil allkanes predominantly proceeds via addition of fumarate to the double bound [72]. Although alkylsuccinate metabolites have rarely been detected in oil reservoir fluids [74, 75], they have been reported in oil-contaminated environments as well as in oilfield facilities, where their detection is indicative of *in situ* microbial degradation of oil alkanes [71, 75]. Alkylsuccinic acids as intermediates of anaerobic alkane oxidation were first studied by Gieg and Suflita [76] when surveying these metabolites in aquifers contaminated with condensate gas, natural gas liquids, gasoline, diesel, alkanes and BTEX. They found alkylsuccinates originating from C3 to C11 alkanes, as well as putative metabolites originating from compounds with one degree of unsaturation, such as alkenes or alicyclic alkanes. Since this report, other studies have detected alkylsuccinate derivates in petroleum contaminated groundwater systems [76], coal beds [70] and oil fields [74, 77]. The formation of alkylsuccinates is catalyzed by a strictly anaerobic glycyl radical enzyme which has been termed as alkylsuccinate synthase or (1-methyl-alkyl)succinate synthase (Ass or Mas). The genes encoding Ass have recently been identified in the alkane degrading sulfidogenic bacteria *D. alkenivoras* AK-01 [78] and *Desulfoglaeba alkanedexens* ALDC^T [71], as well as in nitrate reducing strains HxN1 [65] and OcN1 [79], all affiliated to the Proteobacteria phylum [80]. Recently, Callaghan et al. [71] detected *ass*A genes in a propane-utilizing mixed culture and in a paraffin-degrading enrichment culture maintained under sulfate-reducing conditions. Despite of no genes for benzyl-and alkylsuccinate synthase were found when environmental metagenome datasets of uncontaminated sites were analyzed in Callaghan et al [71], the authors consider that *ass*A gene could be a useful biomarker for anaerobic alkane metabolism.

The second mechanism for alkane anaerobic degradation is the carboxylation, mainly developed from the growth pattern of the sulfate-reducing strain Hxd3 [81], tentatively named as *Desulfococcus oleovorans*. This strain differs from other alkane degraders for converting C-even alkanes into C-odd cellular fatty acids whereas growth on C-odd alkanes resulted in C-even cellular fatty acids [81, 72]. More recently, Callaghan et al. [82] suggested that a carboxylation-like mechanism analogous to the activation strategy previously proposed by So et al. [81] was the probable route for the anaerobic biodegradation of hexadecane in an alkane-degrading, nitrate-reducing consortium. However, in both cases, the hypothetical fatty acid intermediate (2-ethylalkanoate) that should result from the incorporation of inorganic carbon at C-3 of the alkane has never been detected. There is an on-going debate about this initial activation mechanism. From an energetic point of view, the carboxylation of alkanes is not feasible under physiological conditions, unless the concentration of the fatty acid (2-ethylalkanoate) is in the micromolar order of magnitude or less [80].

Other alternative activation mechanisms are proposed for the anaerobic degradation of alkanes. For instance, the mechanism referred as "unusual oxygenation" is used by the strain *Pseudomonas chloritidismutans* AW-1[T], that is assumed to produce its own oxygen via chlorate respiration used for subsequent metabolism of alkanes [60]. Other alternative mechanism considers that activation in the anaerobic methanogenic system may be initiated by an anaerobic hydroxylation reaction [83].

6. Mechanisms involved in oil biodegradation in petroleum reservoirs

From those microorganisms studied in oilfields, methanogens have received particular attention since they have been isolated and molecularly detected in both low- and high-temperature reservoirs [88, 89]. Their physiological characteristics and potential activity possibly involved in methanogenesis occurring in oil reservoirs have been demonstrated [90]. Furthermore, recently, Jones et al. [20] provided evidence that the patterns of hydrocarbon degradation observed in biodegraded petroleum reservoirs were the result of methanogenic processes. Therefore, microbiological and biogeochemical investigations have indicated that methanogenesis is a widely distributed process in petroleum reservoirs, although still poorly understood [90]. Methanogenesis is the terminal process of biomass degradation. Acetate and hydrogen are the most important immediate precursors for methanogenesis, and are converted into methane by acetoclastic and hydrogenotrophic methanogens, respectively [91]. Acetate can also be a precursor for methanogenesis through syntrophic acetate oxidation coupled to hydrogenotrophic methanogenesis, which is mediated by syntrophic bacteria and methanogenic archaea [92, 93, 94, 95]. Interestingly, acetate is generally abundant in many petroleum reservoirs, at concentrations ranging between 0.3 and 20 mM [96] hence, acetate metabolism is considered an important methane production process in those environments [90].

Cultivation-dependent and -independent approaches have shown the presence of acetoclastic and hydrogenotrophic methanogens and putative syntrophic acetate-oxidizing bacteria in reservoirs [88, 89, 102], indicating that there should be two different pathways of acetate metabolism in the environment, namely acetoclastic methanogenesis and syntrophic acetate oxidation coupled with hydrogenotrophic methanogenesis. Some previous studies suggested that syntrophic acetate oxidation was most likely to occur in petroleum reservoirs, based on molecular biological analysis [89] and thermodynamic calculations [98]. In Jones et al. [20], the composition of oil in microcosms exhibiting methanogenic oil degradation is compared to patterns observed in biodegraded oils from the Gullfaks field in the North Sea. Analysis of the methanogenic communities from oil-degrading microcosms revealed a strong selection for CO_2-reducing methanogens against acetoclastic methanogens, and gas isotope modeling also revealed that to match the d13C of methane and carbon dioxide from biodegraded petroleum reservoirs 75–92% of methanogenesis should be via the CO_2 reduction pathway [20, 11].

The reason why syntrophic acetate oxidation predominates over acetoclastic methanogenesis in oil reservoirs remains unclear. There is evidence from studies of oil contaminated aquifers that crude oil can have a detrimental effect on acetoclastic methanogenesis and, in situations where acetoclastic methanogenesis is inhibited, methanogenic alkane degradation via syntrophic acetate oxidation may be thermodynamically the most favorable alternative pathway [11]. Nonetheless, one recent report suggests that acetoclastic methanogenesis may predominate in some methanogenic oil-degrading systems [19]. Although there is currently great interest in how much each of the two pathways contributes to methane production in petroleum reservoirs, no studies are being conducted to address this question [90].

7. Metagenomics as a tool for a better comprehension of biodegradation

As stated previously, cultivation-based methods have traditionally been utilized for studying the microbiology in oil fields and have yielded valuable information about microbial interactions and their relations with hydrocarbons [42]. However, nowadays, it is known that only a small fraction of the microbial diversity in nature (1-10%) can be grown in the laboratory [84, 85, 86]. Therefore, it is assumed that the ecological functions of the majority of microorganisms in nature and their potential applications in biotechnology remain obscure [87].

In metagenomics, total DNA is extracted from appropriately chosen environmental samples, propagated in the laboratory by cloning techniques, submitted to sequence or function-based screenings and/or subjected to large-scale sequence analysis (Fig. 2). Functional screening of metagenomic libraries offer the advantage that it does not rely on sequence homology to known genes, and for this reason, has allowed the isolation of different enzyme classes from several environments. The probability (hit rate) of identifying a certain gene depends on multiple factors that are intrinsically linked to each other: the host–vector system, size of the target gene, its abundance in the source metagenome, the assay method, and the efficiency of heterologous gene expression in a surrogate host [99].

One of the first studies using metagenomics to study microbial degradation of aromatic compounds was performed by Suenaga and colleagues [100], who constructed a metagenomic library from activated sludge for industrial wastewater. The library was functionally screened for extradiol dioxygenase activities (enzymes for aromatic degradation) and 38 clones were subjected to sequencing analysis [101]. As a result, various types of gene subsets were identified that were not similar to the previously reported pathways performing complete degradation. Moreover, the authors discussed the fact that aromatic compounds in the environment may be degraded through the concerted action of various fragmented pathways. Sierra-Garcia [101] reported the organization of hydrocarbon degradation genes of selected metagenomic fosmid clones derived from a metagenomic library from Brazilian petroleum reservoir and functional screening for hydrocarbon degradation activities. The author found many putative proteins of different aerobic and anaerobic well described catabolic pathways, however the complete catabolic pathways described for hydrocarbon degradation in previous studies were absent in the fosmid clones. Instead, the metagenomic fragments comprised genes belonging to different pathways, showing novel gene arrangements where hydrocarbon compounds were degraded through the concerted actions of these fragmented pathways. These results suggest that there are marked differences between the degradation genes found

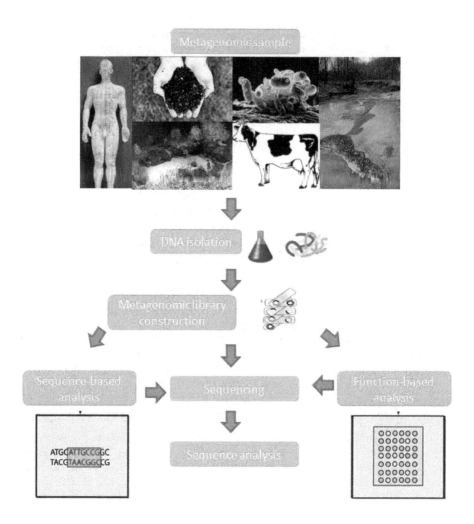

Figure 2. Schematic representation of the different steps for metagenomic analysis.

in microbial communities derived from enrichments of oil reservoir sample and those that have been previously identified in bacteria isolated from contaminated or pristine environments.

However, function-based screening of metagenomic libraries for xenobiotic degradation genes is often considered problematic because of insufficient and biased expression of the heterologous genes in the host *Escherichia coli* [99]. Only a few efforts have been made to solve these problems. In Uchiyama et al. [103], a novel method for function-driven screening is described, which was termed substrate-induced gene expression screening (SIGEX). This high-through-put screening approach employs an operon trap gfp expression vector in combination with fluorescence-activated cell sorting. The screening is based on the fact that catabolic-gene expression is induced mainly by specific substrates and is often controlled by regulatory elements located close to catabolic genes [103]. Using this approach, Uchiyama et al. [103] isolated aromatic-hydrocarbon-induced genes from a metagenomic library derived from groundwater. In Ono et al. [104] another screening strategy was based on functional complementation of a *Pseudomonas putida* host strain containing a naphthalene degrading pathway devoid of the naphthalene dioxygenase (NDO) encoding gene. Two clones were able to restore the ability of the host strain to use naphthalene as a sole carbon source and their genes were similar but no identical to already known operons. The authors refer to the use of other host strains for the construction of metagenomic libraries instead of the well-established *E. coli* as a simpler and economical way to perform function-driven screening in comparison to other reported systems such as SIGEX [103].

In the context of this chapter, several aspects of the hydrocarbon degradation need to be studied to obtain a comprehensive overview of the biodegradation processes that take place in oil reservoirs or petroleum impacted environments. These studies should take into consideration the high diversity of enzymes capable of initiating such metabolism as well as the implementation of integrated studies combining culture and molecular techniques, linking with metabolomics or compound-specific isotope analysis and microcosm studies for a better resolution of in situ microbial activity in petroleum reservoirs.

8. Conclusions and research needs

The understanding about biodegraded petroleum reservoirs have advanced considerably in recent years, but the organisms responsible for the *in situ* activity and a quantitative understanding of the factors which control in-reservoir oil biodegradation remain far from complete. The inaccessibility of petroleum reservoirs and inherent difficulties of microbiological sampling from commercially operating oil wells have required a multidisciplinary approach to delineating the study of subsurface petroleum biodegradation, and to date there are still prevailing paradigms relating to hydrocarbon biodegradation processes. This multidisciplinary approach to study *in situ* petroleum degradation should consider molecular biology, microbiology, and geological and geochemical parameters in order to establish the key organisms, biochemical reactions and mechanisms involved in such complex associations. Indeed, the isolation of anaerobic microorganisms capable of utilizing hydrocarbons is essential for a comprehensive understanding of their role and behavior in anoxic habitats and their complex interactions within methanogenic hydrocarbon-degrading communities. In addition, novel approaches, combining functional metagenomics, transcriptomics, metabolomics and other molecular surveys in microcosms are urgently required to better allow access to a more realistic phylogenetic and metabolic diversity governing oil biodegradation in petroleum reservoirs.

Author details

Isabel Natalia Sierra-Garcia and Valéria Maia de Oliveira*

*Address all correspondence to: vmaia@cpqba.unicamp.br

Microbial Resources Division, Research Center for Chemistry, Biology and Agriculture (CPQBA), University of Campinas, Campinas, Sao Paulo, Brazil

References

[1] Roling, W. (2003). The microbiology of hydrocarbon degradation in subsurface petroleum reservoirs: perspectives and prospects. *Res Microbiol*. 154(5), 321-328.

[2] Head, I. M, Jones, D. M, & Larter, S. R. (2003). Biological activity in the deep subsurface and the origin of heavy oil. *Nature*, 426(6964), 344-52.

[3] Ismail, W, & Gescher, J. (2012). Epoxy coenzyme a thioester pathways for degradation of aromatic compounds. *Appl Environ Microbiol*. , 78(15), 5043-51.

[4] Chandler, D. P, Li, S. M, Spadoni, C. M, Drake, G. R, Balkwill, D. L, Fredrickson, J. K, & Brockman, F. J. (1997). A molecular comparison of culturable aerobic heterotrophic bacteria and 16S rDNA clones derived from a deep subsurface sediment. *FEMS Microbiol Ecol*. , 23, 131-144.

[5] Leigh, M. B, Pellizari, V. H, Uhlik, O, Sutka, R, Rodrigues, J, & Ostrom, N. E. (2007). Biphenyl-utilizing bacteria and their functional genes in a pine root zone contaminated with polychlorinated biphenyls (PCBs). *ISME J 1*, 134-148.

[6] De Lorenzo, V. (2008). Systems biology approaches to bioremediation. *Curr Opin Biotechnol 19*, 579-589.

[7] Jeon, C, Park, W, Padmanabhan, P, Derito, C, Snape, J, & Madsen, E. (2003). Discovery of a bacterium, with distinctive dioxygenase, that is responsible for *in situ* biodegradation in contaminated sediment. *Proc Natl Acad Sci USA 100*, 13591-13596.

[8] Witzig, R, Junca, H, Hecht, H. J, & Pieper, D. H. (2006). Assessment of toluene/biphenyl dioxygenase gene diversity in benzene-polluted soils: links between benzene biodegradation and genes similar to those encoding isopropylbenzene dioxygenases. *Appl Environ Microbiol 72*, 3504-3514.

[9] Foght, J. (2010). Microbial comminities in oil shales, biodegraded and heavy oil reservoirs, and bitumen deposits. In: K. N. Timmis (Ed.) *Handbook of Hydrocarbon and Lipid Microbiology*. Berlin, Heidelberg: Springer Berlin Heidelberg.

[10] Birkeland, N. K. (2004). The microbial diversity of deep subsurface oil reservoirs. *Stud Surface Sci Catal 151*, 385-403.

[11] Head, I. M, Aitken, C. M, Gray, N. D, Sherry, A, Adams, J. J, Jones, D. M, Rowan, A. K, et al. (2010). Hydrocarbon degradation in petroleum reservoirs. In: K. N. Timmis (Ed.) *Handbook of Hydrocarbon and Lipid Microbiology*. Berlin, Heidelberg: Springer Berlin Heidelberg.

[12] Haridon, L, Reysenbach, S, Glenat, A. L, Prieur, P, & Jeanthon, D. C. ((1995). Hot subterranean biosphere in a continental oil reservoir. *Nature*, 377, 223-224.

[13] Grassia, G. S, Mclean, K. M, Glenat, P, Bauld, J, & Sheehy, A. J. (1996). A systematic survey for thermophilic fermentative bacteria and archaea in high temperature petroleum reservoirs. *FEMS Microbiol Ecol*, 21, 47-58.

[14] Magot, M, Ollivier, B, & Patel, B. K. C. (2000). Microbiology of petroleum reservoirs. *Antonie van Leeuwenhoek.*, 77(2), 103-116.

[15] Orphan, V. J, Taylor, L. T, Hafenbradl, D, & Delong, E. F. (2000). Culture-dependent and culture-independent characterization of microbial assemblages associated with high-temperature petroleum reservoirs. *Appl Environ Microbiol.*, 66(2), 700-11.

[16] Aitken, C. M, Jones, D. M, & Larter, S. R. (2004). Anaerobic hydrocarbon biodegradation in deep subsurface oil reservoirs. *Nature,*, 431(7006), 291-4.

[17] Gray, N. D, Sherry, A, Hubert, C, Dolfing, J, & Head, I. M. (2010). Methanogenic degradation of petroleum hydrocarbons in subsurface environments remediation, heavy oil formation, and energy recovery. *Adv Appl Microbiol.*, 72, 137-61.

[18] Widdel, F, & Rabus, R. (2001). Anaerobic biodegradation of saturated and aromatic hydrocarbons. *Curr Opin Biotechnol 12*, 259-276.

[19] Gieg, L. M, Duncan, K. E, & Suflita, J. M. (2008). Bioenergy production via microbial conversion of residual oil to natural gas. *Appl Environ Microbiol*, 74, 3022-3029.

[20] Jones, D, Head, I, Gray, N, Adams, J, Rowan, A, Aitken, C, Bennett, B, et al. (2007). Crude-oil biodegradation via methanogenesis in subsurface petroleum reservoirs. *Nature*, 451(7175), 176-180.

[21] Zengler, K, Richnow, H. H, Rossello-mora, R, Michaelis, W, & Widdel, F. (1999). Methane formation from long chain alkanes by anaerobic microorganisms. *Nature*, 401, 266-269.

[22] Beeder, J, Torsvik, T, & Lien, T. (1995). *Thermodesulforhabdus norvegicus* gen. nov., sp. nov., a novel thermophilic sulfate-reducing bacterium from oil field water. *Arch. Microbiol*, 164, 331-336.

[23] Rees, G. N, Grassia, G. S, Sheehy, A. J, Dwivedi, P. P, & Patel, B. K. C. (1995). *Desulfacinum infernum* gen. nov., sp. nov., a thermophilic sulfate-reducing bacterium from a petroleum reservoir. *Int. J. Syst. Bacteriol*, 45, 85-89.

[24] [24] Grabowski, A, Nercessian, O, Fayolle, F, Blanchet, D, & Jeanthon, C. (2005). Microbial diversity in production waters of a low-temperature biodegraded oil reservoir. *FEMS microbiology ecology*, 54(3), 427-43.

[25] Magot, M, Ravot, G, Campaignolle, X, Ollivier, B, Patel, B. K, Fardeau, M. L, Thomas, P, Crolet, J. L, & Garcia, J. L. (1997). Dethiosulfovibrio peptidovorans gen. nov., sp. nov., a new anaerobic, slightly halophilic, thiosulfate-reducing bacterium from corroding offshore oil wells. *Int. J. Syst. Bacteriol.* , 47, 818-824.

[26] Nilsen, R. K, Torsvik, T, & Lien, T. (1996). Desulfotomaculum thermocisternum sp. nov., a sulfate reducer isolated from a hot North Sea oil reservoir. *Int. J. Syst. Bacteriol.* , 46, 397-402.

[27] Ravot, G, Magot, M, Ollivier, B, Patel, B. K. C, Ageron, E, Grimont, P. A. D, Thomas, P, & Garcia, J. L. (1997). *Haloanaerobium congolense* sp. nov., an anaerobic, moderately halophilic, thiosulfate- and sulfur-reducing bacterium from an African oil field. *FEMS Microbiol. Lett.* , 147, 81-88.

[28] Miranda-tello, E, Fardeau, M. L, Fernandez, L, Ramirez, F, Cayol, J. L, Thomas, P, Garcia, J. L, & Ollivier, B. (2003). *Desulfovibrio capillatus* sp. nov., a novel sulfatereducing bacterium isolated from an oil field separator located in the Gulf of Mexico. *Anaerobe* , 9, 97-103.

[29] Nazina, T. N, Tourova, T. P, Poltaraus, A. B, Novikova, E. V, Grigoryan, A. A, Ivanova, A. E, et al. (2001). Taxonomic study of aerobic thermophilic bacilli: Descriptions of *Geobacillus subterraneus* gen. nov., sp. nov. and *Geobacillus uzenensis* sp. nov. from petroleum reservoirs and transfer of *Bacillus stearothermophilus, Bacillus hermocatenulatus, Bacillus thermoleovorans, Bacillus kaustophilus, Bacillus thermoglucosidasius* and *Bacillus thermodenitrificans* to *Geobacillus* as the new combinations G. *stearothermophilus, G. thermocatenulatus, G. thermoleovorans, G. kaustophilus, G. thermoglucosidasius* and *G. thermodenitrificans. Int. J. Syst. Evol. Microbiol.* , 51, 433-446.

[30] Miroshnichenko, M. L, Hippe, H, Stackebrandt, E, Kostrikina, N. A, Chernyh, N. A, Jeanthon, C, Nazina, T. N, Belyaev, S. S, & Bonch-osmolovskaya, E. A. (2001). Isolation and characterization of *Thermococcus sibiricus* sp. nov. from a Western Siberia high-temperature oil reservoir. *Extremophiles.* , 5, 85-91.

[31] Ravot, G, Magot, M, Fardeau, M. L, Patel, B. K. C, Thomas, P, Garcia, J. L, & Ollivier, B. (1999). *Fusibacter paucivorans* gen. nov., sp. nov., an anaerobic, thiosulfate-reducing bacterium from an oil-producing well. *Int. J. Syst. Bacteriol.* , 49, 1141-1147.

[32] Dahle, H, & Birkeland, N. K. (2006). *Thermovirga lienii* gen. nov. sp. nov., a novel moderately thermophilic, anaerobic, amino-acid-degrading bacterium isolated from a North Sea oil well. *Int. J. Syst. Evol. Microbiol.* , 56, 1539-1545.

[33] Nilsen, R. K, & Torsvik, T. (1996). *Methanococcus thermolithotrophicus* isolated from North sea oil field reservoir water. *Appl. Environ. Microbiol.* , 62, 728-731.

[34]   Magot, M. (2005). Indigenous microbial communities in oil fields. In B. Ollivier and M. Magot, (Eds.) Petroleum microbiology. ASM, Washington, DC., 21-34.

[35]   Söhngen, N. L. (1913). Benzin, Petroleum, Paraffinöl und Paraffin als Kohlenstoff- und Energiequelle für Mikroben. *Zentr Bacteriol Parasitenk Abt II 37*, 595-609.

[36]   Widdel, F, & Musat, F. (2010). Diversity and common principles in enzymatic activa- tion of hydrocarbons. In: K. N. Timmis (Ed.) *Handbook of Hydrocarbon and Lipid Micro- biology*. Berlin, Heidelberg: Springer Berlin Heidelberg.

[37]   Boll, M, & Heider, J. (2010). Anaerobic Degradation of Hydrocarbons: Mechanisms of C-H-Bond activation in the absence of oxygen. In: K. N. Timmis (Ed.) *Handbook of Hy- drocarbon and Lipid Microbiology*. Berlin, Heidelberg: Springer Berlin Heidelberg.

[38]   Foght, J. (2008). Anaerobic biodegradation of aromatic hydrocarbons: pathways and prospects. *J Mol Microbiol Biotechnol.* 15(2-3): 93-120.

[39]   Van Beilen, J. B, & Funhoff, E. G. (2007). Alkane hydroxylases involved in microbial alkane degradation. *Appl Microbiol Biotechnol.*, 74(1), 13-21.

[40]   Wentzel, A, Ellingsen, T. E, Kotlar, H. K, Zotchev, S. B, & Throne-holst, M. (2007). Bacterial metabolism of long-chain n-alkanes. *Appl Microbiol Biotechnol.*, 76(6), 1209-1221.

[41]   Rojo, F. (2009). *Degradation of alkanes by bacteria. Environmental microbiology.*

[42]   Van Hamme, J. D, Singh, A, & Ward, O. P. (2003). Recent advances in petroleum mi- crobiology. *Microbiol Mol Biol Rev.*, 67(4), 503-549.

[43]   Rojo, F. (2010). Enzymes for Aerobic Degradation of Alkanes. In K. N. Timmis (Ed.), *Handbook of Hydrocarbon and Lipid Microbiology* (Berlin, Heidelberg: Springer Berlin Heidelberg., 781.

[44]   Margesin, R, Labbe, D, Schinner, F, Greer, C, & Whyte, L. (2003). Characterization of hydrocarbon-degrading microbial populations in contaminated and pristine alpine soils. *Appl Environ Microbiol.*, 69(6), 3085-3092.

[45]   Kuhn, E, Bellicanta, G. S, & Pellizari, V. H. (2009). New alk genes detected in Antarc- tic marine sediments. *Environ Microbiol.*, 11(3), 669-673.

[46]   Salminen, J. M, Tuomi, P. M, & Jorgensen, K. S. (2008). Functional gene abundances (*nahAc, alkB, xylE*) in the assessment of the efficacy of bioremediation. *Appl Biochem Biotechnol 151*, 638-652.

[47]   Hamamura, N, Fukui, M, Ward, D. M, & Inskeep, W. P. (2008). Assessing soil micro- bial populations responding to crude-oil amendment at different temperatures using phylogenetic, functional gene (*alkB*) and physiological analyses. *Environ Sci Technol 42*, 7580-7586.

[48] Van Beilen, J. B, Wubbolts, M. G, & Witholt, B. (1994). Genetics of alkane oxidation by *Pseudomonas oleovorans*. *Biodegradation 5*, 61-174.

[49] Marchant, R, Sharkey, F. H, Banat, I. M, Rahman, T. J, & Perfumo, A. (2006). The degradation of n-hexadecane in soil by thermophilic geobacilli. *FEMS Microbiol Ecol.*, 56(1), 44-4.

[50] Van Beilen, J. B, Li, Z, Duetz, W. A, Smits, T. H. M, & Witholt, B. (2003). Diversity of Alkane Hydroxylase Systems in the Environment. *Oil Gas Sci Technol.*, 58(4), 427-440.

[51] Tourova, T. P, Nazina, T. N, Mikhailova, E. M, Rodionova, T. A, Ekimov, A. N, Maуhukova, А. V, & Ρoliakus, A. D, (2008). alkB homologs in thermophilic bacteria of the genus *Geobacillus*. *Mol Biol.*, 42(2), 217-226.

[52] Li, W, Wang, L. Y, Duan, R. Y, Liu, J. F, Gu, J. D, & Mu, B. Z. (2012). Microbial community characteristics of petroleum reservoir production water amended with n-alkanes and incubated under nitrate-, sulfate-reducing and methanogenic conditions. *Inter Biodeterior Biodegradation.*, 69, 87-96.

[53] Vilchez-vargas, R, Junca, H, & Pieper, D. H. (2010). Metabolic networks, microbial ecology and "omics" technologies: towards understanding in situ biodegradation processes. *Environ Microbiol.*, 12, 3089-3104.

[54] Feng, L, Wang, W, Cheng, J, Ren, Y, Zhao, G, Gao, C, Tang, Y, et al. (2007). Genome and proteome of long-chain alkane degrading Geobacillus thermodenitrificans NG80-2 isolated from a deep-subsurface oil reservoir. *Proc Natl Acad Sci U S A.*, 104(13), 5602-7.

[55] Carmona, M, Zamarro, M, Blazquez, B, Durante-rodriguez, G, Juarez, J, & Valderrama, J. (2009). Anaerobic catabolism of aromatic compounds: a genetic and genomic view. *Microbiol Mol Biol Rev. 73*, 71-133.

[56] Brennerova, M. V, Josefiova, J, Brenner, V, Pieper, D. H, & Junca, H. (2009). Metagenomics reveals diversity and abundance of meta-cleavage pathways in microbial communities from soil highly contaminated with jet fuel under air-sparging bioremediation. *Environ Microbiol.*, 11(9), 2216-27.

[57] Pérez-pantoja, D, González, B, & Pieper, D. H. (2010). Aerobic degradation of aromatic hydrocarbons. In: K. N. Timmis (Ed.) *Handbook of Hydrocarbon and Lipid Microbiology*. Berlin, Heidelberg: Springer Berlin Heidelberg.

[58] Jouanneau, Y. (2010). Oxidative inactivation of ring cleavage extradiol dioxigenases: mechanism and ferredoxin mediated reactivation. In: K. N. Timmis (Ed.) *Handbook of Hydrocarbon and Lipid Microbiology*. Berlin, Heidelberg: Springer Berlin Heidelberg.

[59] Junca, H, & Pieper, D. H. (2003). Functional gene diversity analysis in BTEX contaminated soils by means of PCR-SSCP DNA fingerprinting: comparative diversity assessment against bacterial isolates and PCR-DNA clone libraries. *Environ Microbiol.*, 6(2), 95-110.

[60] Mehboob, F, Junca, H, Schraa, G, & Stams, A. J. M. (2009). Growth of Pseudomonas chloritidismutans AW-1(T) on n-alkanes with chlorate as electron acceptor. *Appl Microbiol Biotechnol.*, 83(4), 739-47.

[61] Fuchs, G. (2008). Anaerobic metabolism of aromatic compounds. *Ann N Y Acad Sci.*, 1125, 82-99.

[62] Kube, M, Heider, J, Amann, J, Hufnagel, P, Kühner, S, Beck, A, Reinhardt, R, et al. (2004). Genes involved in the anaerobic degradation of toluene in a denitrifying bacterium, strain EbN1. *Arch Microbiol.*, 181(3), 182-94.

[63] Boll, M, Fuchs, G, & Heider, J. (2002). Anaerobic oxidation of aromatic compounds and hydrocarbons. *Curr Opin Chem Biol.*, 6(5), 604-11.

[64] Kaser, f. M, & Coates, J. D. (2010). Nitrate, Perchlorate and Metal respirers. In: K. N. Timmis (Ed.) *Handbook of Hydrocarbon and Lipid Microbiology*. Berlin, Heidelberg: Springer Berlin Heidelberg.

[65] Grundmann, O, Behrends, A, Rabus, R, Amann, J, Halder, T, Heider, J, & Widdel, F. (2008). Genes encoding the candidate enzyme for anaerobic activation of n-alkanes in the denitrifying bacterium, strain HxN1. *Environ Microbiol.*, 10(2), 376-85.

[66] Kane, S. R, Beller, H. R, Legler, T. C, & Anderson, R. T. (2002). Biochemical and genetic evidence of benzylsuccinate synthase in toluene-degrading, ferric iron-reducing *Geobacter metallireducens*. *Biodegradation*, , 13(2), 149-54.

[67] Winderl, C, Schaefer, S, & Lueders, T. (2007). Detection of anaerobic toluene and hydrocarbon degraders in contaminated aquifers using benzylsuccinate synthase (bssA) genes as a functional marker. *Environ Microbiol*, 9, 1035-1046.

[68] Winderl, C, Anneser, B, Griebler, C, Meckenstock, R. U, & Lueders, T. (2008). Depth resolved quantification of anaerobic toluene degraders and aquifer microbial community patterns in distinct redox zones of a tar oil contaminant plume. *Appl Environ Microbiol*, 74, 792-801.

[69] Staats, M, Braster, M, & Roling, W. F. M. (2011). Molecular diversity and distribution of aromatic hydrocarbon-degrading anaerobes across a landfill leachate plume. *Environ Microbiol*, 13, 1216-1227.

[70] Wawrik, B, Mendivelso, M, Parisi, V. A, Suflita, J. M, Davidova, I. A, Marks, C. R, Van Nostrand, J. D, Liang, Y, Zhou, J, Huizinga, B. J, et al. (2012). Field and laboratory studies on the bioconversion of coal to methane in the San Juan Basin. *FEMS Microbiol Ecol.*, 81, 26-42.

[71] Callaghan, A. V, Davidova, I. A, Savage-ashlock, K, Parisi, V. A, Gieg, L. M, Suflita, J. M, Kukor, J. J, et al. (2010). Diversity of benzyl- and alkylsuccinate synthase genes in hydrocarbon-impacted environments and enrichment cultures. *Environ Sci Technol.*, 44(19), 7287-94.

[72] Widdel, F, & Grundmann, O. (2010). Biochemistry of the anaerobic degradation of non-methane alkanes. In: K. N. Timmis (Ed.) *Handbook of Hydrocarbon and Lipid Microbiology*. Berlin, Heidelberg: Springer Berlin Heidelberg.

[73] Grossi, V, Cravolaureau, C, Guyoneaud, R, Ranchoupeyruse, A, & Hirschlerrea, A. (2008). Metabolism of n-alkanes and n-alkenes by anaerobic bacteria: A summary. *Org Geochem.* , 39(8), 1197-1203.

[74] Gieg, L. M, Davidova, I. A, Duncan, K. E, & Suflita, J. M. (2010). Methanogenesis, sulfate reduction and crude oil biodegradation in hot Alaskan oilfields. *Environ Microbiol* , 12(11), 3074-86.

[75] Mbadinga, S. M, Li, K. P, Zhou, L, Wang, L. Y, Yang, S, Liu, Z, Gu, J. F, et al. (2012). Analysis of alkane-dependent methanogenic community derived from production water of a high-temperature petroleum reservoir. *Appl Microbiol Biotechnol.* , 96(2), 531-42.

[76] Gieg, L. M, & Suflita, J. M. (2002). Detection of anaerobic metabolites of saturated and aromatic hydrocarbons in petroleum-contaminated aquifers. *Environ. Sci. Technol.* , 36(17), 3755-3762.

[77] Duncan, K. E, Gieg, L. M, Parisi, V. A, Tanner, R. S, & Suflita, J. M. Green Tringe, S., Bristow, J. ((2009). Biocorrosive thermophilic microbial communities in Alaskan North Slope oil facilities. *Environ Sci Technol 43*, 7977-7984.

[78] Callaghan, A. V, & Wawrik, B. NiChadhain, S.M., Young, L.Y., Zylstra, G.J. ((2008). Anaerobic alkane-degrading strain AK-01 contains two alkylsuccinate synthase genes. *Biochem Biophys Res Commun.* , 366, 142-148.

[79] Zedelius, J, Rabus, R, Grundmann, O, Werner, I, Brodkorb, D, Schreiber, F, Ehrenreich, P, Behrends, A, Wilkes, H, Kube, M, Reinhardt, R, & Widdel, F. (2010). Alkane degradation under anoxic conditions by a nitrate-reducing bacterium with possible involvement of the electron acceptor in substrate activation. *Environ Microbiol Rep.* 3(1), 125-135.

[80] Mbadinga, S. M, Wang, L. Y, Zhou, L, Liu, J. F, Gu, J. D, & Mu, B. Z. (2011). Microbial communities involved in anaerobic degradation of alkanes. *Inter Biodeterior Biodegradation.* , 65(1), 1-13.

[81] So, C, Phelps, C, & Young, L. (2003). Anaerobic transformation of alkanes to fatty acids by a sulfate-reducing bacterium, strain Hxd3. *Appl Environ. 69*(7), 3892-3900.

[82] Callaghan, A. V, Tierney, M, Phelps, C. D, & Young, L. Y. (2009). Anaerobic biodegradation of n-hexadecane by a nitrate-reducing consortium. *Appl Environ Microbiol* , 75, 1339-1344.

[83] (Head, I., Gray, N., Aitken, C., Sherry, A., Jones, M., Larter, S. (2010). Hydrocarbon activation under sulfate-reducing and methanogenic conditions proceeds by different mechanisms. Geophysical Research Abstracts 12 (EGU General Assembly 2010).

[84] Torsvik, V, Goksoyr, J, & Daae, F. L. (1990). High diversity in DNA of soil bacteria. *Appl Environ Microbiol* , 56, 782-787.

[85] Amann, R. I, Ludwig, W, & Schleifer, K. H. (1995). Phylogenetic identification and in situ detection of individual microbial cells without cultivation. *Microbiol Rev* , 59, 143-169.

[86] Torsvik, V, Daae, F. L, Sandaa, R. A, & Øvreås, L. (1998). Novel techniques for analyzing microbial diversity in natural and perturbed environments. *J Biotechnol* , 64, 53-62.

[87] Kellenberger, E. (2001). Exploring the unknown: the silent revolution of microbiology. *EMBO reports*, 2(1), 2-5.

[88] Orphan, V. J, Goffredi, S. K, Delong, E. F, & Boles, J. R. (2003). Geochemical influence on diversity and microbial processes in high temperature oil reservoirs. *Geomicrobiol J* 20, 295-311.

[89] Nazina, T. N, & Shestakova, N. M. Grigor'yan, A.A., Mikhailova, E.M., Tourova, T.P., Poltaraus, A.B., et al. ((2006). Phylogenetic diversity and activity of anaerobic microorganisms of high-temperature horizons of the Dagang oil field (P.R. China). *Microbiology 75*, 55-65.

[90] Mayumi, D, Mochimaru, H, Yoshioka, H, Sakata, S, Maeda, H, Miyagawa, Y, Ikarashi, M, et al. (2011). Evidence for syntrophic acetate oxidation coupled to hydrogenotrophic methanogenesis in the high-temperature petroleum reservoir of Yabase oil field (Japan). *Environ Microbiol.* , 13(8), 1995-2006.

[91] Garcia, J. L, Patel, B. K, & Ollivier, B. (2000). Taxonomic, phylogenetic, and ecological diversity of methanogenic Archaea. *Anaerobe 6*, 205-226.

[92] Zinder, S. H, & Koch, M. (1984). Non-acetoclastic methanogenesis from acetate: acetate oxidation by a thermophilic syntrophic coculture. *Arch Microbiol 138*, 263-272.

[93] Schnurer, A, Houwen, F. P, & Svensson, B. H. (1994). Mesophilic syntrophic acetate oxidation during methane formation by a triculture at high ammonium concentration. *Arch Microbiol 162*, 70-74.

[94] Hattori, S, Kamagata, Y, Hanada, S, & Shoun, H. (2000). *Thermacetogenium phaeum* gen. nov., sp. nov., a strictly anaerobic, thermophilic, syntrophic acetate-oxidizing bacterium. *Int J Syst Evol Microbiol 50*, 1601-1609.

[95] Balk, M, Weijma, J, & Stams, A. J. (2002). *Thermotoga lettingae* sp. nov., a novel thermophilic, methanoldegrading bacterium isolated from a thermophilic anaerobic reactor. *Int J Syst Evol Microbiol 52*, 1361-1368.

[96] Barth, T. (1991). Organic-acids and inorganic-ions in waters from petroleum reservoirs, Norwegian continental-shelf: a multivariate statistical-analysis and comparison with American reservoir formation waters. *Appl Geochem 6*, 1-15.

[97]   Silva, T. R, & Verde, L. C. L. Santos Neto, E.V., Oliveira, V.M. ((2012). Diversity anal-
       yses of microbial communities in petroleum samples from Brazilian oil fields. Inter
       Biodeterior Biodegradation doi:10.1016/j.ibiod.2012.05.005.

[98]   Dolfing, J, Larter, S. R, & Head, I. M. (2008). Thermodynamic constraints on metha-
       nogenic crude oil biodegradation. *ISME J 2*, 442-452.

[99]   Uchiyama, T, & Miyazaki, K. (2009). Functional metagenomics for enzyme discovery:
       challenges to efficient screening. *Curr Opin Biotechnol.* , 20(6), 616-622.

[100] Suenaga, H, Ohnuki, T, & Miyazaki, K. (2007). Functional screening of a metagenom-
       ic library for genes involved in microbial degradation of aromatic compounds. *Envi-
       ron Microbiol.* , 9(9), 2289-2297.

[101] Suenaga, H, Koyama, Y, Miyakoshi, M, Miyazaki, R, Yano, H, Sota, M, Ohtsubo, Y, et
       al. (2009). Novel organization of aromatic degradation pathway genes in a microbial
       community as revealed by metagenomic analysis. *ISME J.* , 3(12), 1335-48.

[102] Sierra-garcia, I. N. Caracterização estrutural e funcional de genes de degradação de
       hidrocarbonetos originados de metagenoma microbiano de reservatório de petróleo.
       M SC. Thesis. Universidade Estadual de Campinas; (2011).

[103] Uchiyama, T, Abe, T, Ikemura, T, & Watanabe, K. (2005). Substrate-induced gene-ex-
       pression screening of environmental metagenome libraries for isolation of catabolic
       genes. *Nat Biotechnol.* , 23(1), 88-93.

[104] Ono, A, Miyazaki, R, Sota, M, Ohtsubo, Y, Nagata, Y, & Tsuda, M. (2007). Isolation
       and characterization of naphthalene-catabolic genes and plasmids from oil-contami-
       nated soil by using two cultivation-independent approaches. *Appl Microbiol Biotech-
       nol.* , 74(2), 501-10.

Chapter 4

Emulsification of Hydrocarbons Using Biosurfactant Producing Strains Isolated from Contaminated Soil in Puebla, Mexico

Beatriz Pérez-Armendáriz,
Amparo Mauricio-Gutiérrez,
Teresita Jiménez-Salgado,
Armando Tapia-Hernández and
Angélica Santiesteban-López

Additional information is available at the end of the chapter

1. Introduction

Among Mexico's main riches are its oil and the great expanses of land used to grow food. A large number of pipelines pass through Mexico's agricultural region carrying diesel, gasoline or crude oil, however, lack of maintenance of the pipeline installations, fuel theft, vehicle transport and even the topographical, terrain and hydrological conditions of the site cause a high incidence of contamination.

Petrolic activities have generated extensive pollution of soils worldwide, mainly in those regions where petroleum is explored, extracted, and refined. The composition of hydrocarbons on polluted soil varies according to environmental conditions and natural degradation processes. In México there are soil impacted by weathered hydrocarbons, which are predominantly saturated and aromatic, become more recalcitrant if polluted soils are not remediated, affecting the underground water, food chains, and diverse human activities.

Hydrocarbon spills on agricultural soil have direct repercussions on soil quality and its function. Some authors [1] indicate that hydrocarbon contamination reduces food crop growth by preventing water and nutrient absorption through the roots, and reducing the transport of metabolites and respiration rate.

The recovery of hydrocarbon-contaminated agricultural soil in Mexico is a complex theme because the producers harvest the crops for sustenance or sale. A remedy is therefore needed that uses sustainable biological technologies which do not pose a risk for the products of the harvests. The production of biosurfactants to recover agricultural soil used for food production is a viable alternative because of their biodegradability. Furthermore, biosurfactants have been used in the oil industry to recover oils from hydrocarbons, in the emulsification of heavy hydrocarbon fractions and in the degradation of polychlorinated biphenyls [2] and polycyclic aromatic hydrocarbons (PAH's) [3].

2. Approach to the problem

In the agricultural fields of Puebla, Mexico two hydrocarbon spills have been reported due to lack of pipeline maintenance. In 2002, a crude oil spill in the town of Acatzingo, Puebla affected a large expanse of agricultural land (approximately 50 hectares) [4]. And in San Martin Texmelucan, Puebla on December 19, 2010, the explosion caused by a crude oil spill took 30 human lives and greatly affected the agricultural land of the population [5]. The inhabitants of the affected regions still perceive damage to the soil and do not consider the land to be fully recovered [4].

In Mexico, the environmental impact of oil industry activities is rigorously controlled by the authorities (Federal Environmental Protection Agency, *Procuraduría Federal de Protección al Medio Ambiente*, PROFEPA) and therefore recuperation should take only a short time. Bioremediation processes have not given the expected results: expanses of contaminated land are heterogeneous as far as climate, water availability and oxygen availability, and the biostimulation of microbial populations is insufficient due to competing autochthonous microorganisms and inadequate nutritional balance [6, 7].

Mexico relies mainly on micro-encapsulation technology for the restoration of hydrocarbon-contaminated land, according to the National Ecology Institute (*Instituto Nacional de Ecologia*, INE) [6] using chemical substances which encapsulate hydrocarbons and prevent biodegradation. Surfactants have also been used to restore marine sediment with a recovery of 45,000 t [8]. Chemical surfactants, however, are not always environmentally biodegradable [9] and so there is a need to use biosurfactants to recover oil hydrocarbons in impacted soils.

3. Area of application

Biosurfactants are molecules with a polar region and a non-polar region, and are hence considered amphipathic, produced by extracellular or intracellular microorganisms, also can reduce surface tension at the air-water interface between two immiscible liquids or between the solid-water interface [10].

Biosurfactants have other advantages over chemical detergents since they are non-toxic and ecologically acceptable [10]. They are also highly effective at breaking down surface tension

[11]. Several authors have reported bacterial strains isolated from hydrocarbon-contaminated soil and water which present emulsifying activity and which are capable of growing in oil using it as sole carbon source. The reported microorganisms are: *Pseudomonas aeruginosa, P. mendocina, P. aureofasciens, Listonella damsela, Bacillus sphaericus, B. brevis, Enterobacter cloacae, Acinetobacter calcoaceticus* var. *anitratus, Hafnia alvei, Citrobacter freundii, C. amalonaticus, Sphingobacterium multivorum, Staphylococcus* sp, *Neisseria* sp, *Micrococcus* sp, *Serratia rubidae, Alcaligenes, Flavobacterium, Nocardia, Achromobacter, Arthrobacter* [12-16]. There has been a recent rise in the study of biosurfactant for their antimicrobial characteristics as fungicide [17, 18] and as, zoospore inhibitors [19].

The use of biosurfactants for the bioremediation of hydrocarbon contaminated soil has been studied intensely since the last decade [2-3, 20]. Biosurfactants have been used by the oil industry to enhanced oil recovery [21, 22], in the emulsification of heavy hydrocarbon fractions [23], and in the treatment of wastewater with insoluble substances. They have also been used in the degradation of polychlorinated biphenyls [2]. Chemical surfactants have the advantage of being non-toxic, environmentally friendly, and biodegradable and can be produced from agricultural substrates [10].

Biosurfactants can be used as additives to stimulate bioremediation; however, the concentration of these can also be increased by the addition of bioemulsifier-producing bacteria. Bioemulsifier-producing bacteria can participate in the biodegradation of hydrocarbons and, alternatively, function as a family of bacteria that supply emulsifiers to another group of bacteria that degrade the contaminants [24].

A mixture of biosurfactants including cellular lipids produced during the degradation of heavy hydrocarbons, and additives increases solubility and facilitates hydrocarbon degradation. Cellular lipids have excellent surfactant properties and can form micelles at low concentrations, but these surfactants do not release the solubilized organic compounds to degrade them [25]. An increase in the apparent solubility of naphthalene has been observed when the concentration of glycolipids excreted by *Pseudomonas areuginosa* 19SJ exceeds the critical micellar concentration (CMC) [26].

Biosurfactants have different chemical compositions depending on the microorganism that produces them and may be lipopeptides, lipoproteins, fatty acids or phospholipids [27]. The production of biosurfactants depends on physicochemical factors (aeration, pH, substrate availability) and their evaluation will depend on kinetic factors (substrate consumption, product formation, and biomass production). Knowing the kinetics of biosurfactant production will allow the proposal of sustainable oil hydrocarbon recovery technologies for aqueous or solid systems.

Mexico has large areas of soil contaminated by oil activities; especially agricultural soils have few alternatives of sustainable technologies, therefore in this work different microorganisms were isolated from hydrocarbons-contaminated soil and the kinetics of biosurfactant production was studied to generate a proposal for the recovery of oil hydrocarbons as Maya crude oil.

4. Materials and methods

4.1. Isolation of biosurfactant-producing strains

Soil sampling was done in an agricultural area of Acatzingo, Puebla, Mexico with the following geographical coordinates 18° 57' 03.0" N 97° 46' 20.5" W. Biosurfactant-producing strains were isolated using 1 g of soil in 10 mL of pre-sterilized distilled water. The culture medium was composed of (g / L): $(NH_4)_2SO_4$ 7.7, KH_2PO_4 5.7, K_2HPO_4 2, $MgSO_47H_2O$ 2, $CaCl_22H_2O$ 0.005, $FeCl_36H_2O$ 0.0025, agar 15; distilled water 1,000 mL and preadapted to a petroleum environment using the Maya petroleum provided by the Mexican State company (PEMEX). Maya petroleum was added on sterilized filter paper (3 cm²; with 2 g petroleum) to every lid in order to develop an atmosphere of volatile hydrocarbons inside the petri dish.

The bacteria were then isolated and grown in a liquid mineral medium (g / L): $(NH_4)_2SO_4$ 7, KH_2PO_4 5.7, K_2HPO_4 2, $MgSO_47H_2O$ 2, $CaCl_22H_2O$ 0.005, $FeCl_36H_2O$ 0.0025, Yeast extract 0.1, glucose 20. Strains presenting biosurfactant production were identified as UPAEP 6, UPAEP 8, UPAEP 9, UPAEP 10, UPAEP 12 and UPAEP 15. The following bacteria were also bought *Arthrobacter* sp ATCC 31012, *Bacillus subtilis* ATCC 21332, *Candida petrophilum* ATCC 20226.

4.2. Strain selection

The selected strains were grown in 50 mL of Lebac medium (g / L): $(NH_4)_2SO_4$ 7, KH_2PO_4 5.7, K_2HPO_4 2, $MgSO_47H_2O$ 2, $CaCl_22H_2O$ 0.005, $FeCl_36H_2O$ 0.0025, Yeast extract 0.1, glucose 20, pH 7.0; in 200 mL Erlenmeyer flasks with a 200 µL aliquot of microorganisms. Twenty-four flasks of each strain were placed in an incubator (FELISA) at 37 °C under constant agitation at 200 rpm. Three flasks were removed at each interval over a 44 and 48 h kinetic.

The parameters evaluated over time were: biomass production, pH, emulsification activity on engine oil and glucose consumption.

Biomass production was determined by taking 2 mL of culture medium and passing it through a pre-dried and pre-weighed cellulose nitrate membrane filter (0.22 µm in diameter). The filter with the biomass was then dried at 100°C for 24 h until constant weight was attained; the biomass was reported in g obtained by weight difference.

4.3. Emulsification index

Emulsification activity was determined by placing 6 mL of engine oil and 4 mL of culture medium with the biosurfactant-producing strains in a vortex [28]. They were agitated for 2 minutes and left to rest for 24 h. The percentage of emulsification was estimated according the following expression:

% Emulsifier = ((Total height of the mixture - Height of emulsified oil) / Total height of the mixture) * 100

4.4. Glucose consumption, pH and Critical Micelle Concentration (CMC)

The glucose was determined by the AOAC 969.39 method taking a 2 mL aliquot of culture medium. If necessary it was diluted with distilled water.

The pH was determined with a potentiometer (Conductronic pH 10). In this investigation, pH was maintained close to neutrality by adding 0.1N NaOH.

The Critical Micelle Concentration (CMC) was determined according to [29].

4.5. Statistical analysis

The results were adjusted to a linear model to obtain the rate of substrate consumption (g glucose h^{-1}), the rate of biomass production (g biomass h^{-1}) and emulsification activity (% emulsifier h^{-1}). The slopes (rates) and correlation coefficients were obtained from regression linear model.

In addition, the average initial and final samples of emulsification activity were analyzed by variant analysis to find significant differences and Duncan-Waller multiple comparison tests. The statistical package used was Minitab version 13 (licensed to UPAEP, Mexico).

4.6. Biodegradation tests of maya crude oil

A preculture of selected strains was grown in Banat broth at 30 °C under constant agitation (200 rpm) for 24 h. An aliquot of the selected strains was taken at an absorbance of 70 UK, inoculated in flasks with 50 mL of medium at a pH of 6.5 with 20,000 ppm of crude oil and incubated at 30 °C for 15 days. Following the incubation process, the samples were put in contact with HPLC grade hexane and agitated for 2 minutes. The mixture was then sonicated (Branson 1210 Ultrasonic Cleaner) for 10 minutes before being transferred to a 250 mL separatory funnel leaving the aqueous phase to decant for later use (Figure 1A). The organic phase, in which the hydrocarbons are found, was recovered by means of an asbestos filter and Na_2SO_4 anhydrous as a desiccant in a 50 mL balloon flask. The organic phase was then distilled using a Büchi Rotavapor R11 with operating temperature of 45 °C (Figure 1B).

4.7. Viability of microorganisms

In addition to the hydrocarbon degradation capacity, the viability of the strains was determined at 8, 16 and 24 days of incubation. The organic phase was therefore eliminated by centrifugation (3000 rpm for 5 minutes) and successive serial dilutions made of 10^{-6} and cultivated on plates of Lebac medium. Isolates strains were grown overnight in Lebac broth at 37 °C under constant agitation at 200 rpm. The biochemical characterization was carried out by the API 20 E, API 20 NE and API 50 CH systems (references No. 20160, 20050 and 50300; bioMérieux) following the manufacturer's recommendations. The identification was assessed by APIweb™ identification software (bioMerieux).

4.8. Biosurfactant recovery

The purification of biosurfactant was performed according to a modified technique described in [30]. With the strains with highest percentage of emulsifier. The strains were previously grown in 500 mL of Lebac medium. The biosurfactant was then extracted from the bacteria with isopropanol-ethanol (3:1) analytical grade (Merck, México) in a separatory flask. It was centrifuged at 1200 rpm for 30 minutes (Solbat), and the supernatant was eliminated. The sample was then filtered using cellulose paper grade 101 (Millipore 2.5µ M). The precipitate obtained was dried for 24 h at 60 °C in an oven (FELISA) and stored in an Eppendorf vial to determine the yield.

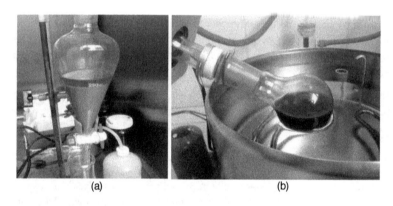

(a)   (b)

Figure 1. a) Emulsification of hydrocarbons. (b) Oil recovery.

5. Results

5.1. Presumptive identification of isolated microorganisms

Six microorganisms were isolated and identified according to their morphology. Table 1 shows the results of the presumptive tests for the identification of bacteria and yeasts by API galleries. The strains UPAEP 8 and UPAEP 15 were related to *Klebsiella pneumoniae* (99 and 97.6 % likelihood respectively). UPAEP 6 strain was closely related to *Klebsiella ornithinolytica* (99 %) and UPAEP 9 strain to *Klebsiella* sp (97 %). Whereas UPAEP 10 strain showed high likelihood (99 %) to *Serratia marcescens* and UPAEP 12 strain to *Candida inconspicua* (75 %).

5.2. Glucose consumption and biomass production

The kinetic characteristics of the bacteria showed similar behavior regarding rapid growth, good adaptation to hydrocarbons and rapid glucose consumption.

All strains consumed glucose in a range of 92 to 100 %. However, the glucose consumption percentage of the commercial strains was lower than the isolates studied; with the exception of *Bacillus subtilis* ATCC 21332 which consumed 93.6 % of glucose (Table 2). Nevertheless, the glucose consumption was inversely proportional to the biomass production during the cell growth (data not shown). The emulsification index was directly proportional at production of biomass, except *Candida petrophilum* ATCC 20226 which showed no relation. Biosurfactant synthesis and biomass production by UPAEP 6, 9, 10, and 15 (Figures 2, 4,5 and 10) strains began during the first few hours (4 to 8) as a response to substrate consumption; UPAEP 8, 12 (Figure 3 and 6) and *Arthrobacter* sp ATCC 31012 (figure 8) strains began at 20, 28 and 50 h. In contrast, *Bacillus subtilis* ATCC 21332 strain biosurfactant production occurred at the end of microbial growth (after of 76 h).

UPAEP 6 strain showed the highest increase in biomass and biosurfactant production at 24 h. Maximum biomass production occurs at 44 h and with a maximum value of 5.3 g L^{-1}. The maximum value of the biosurfactant production (80 %) at 40 h was high considering that crude oil is heavy with a density of 0.92-1.01 g mL^{-1} and an API gravity of 10.1-22.3 and viscosity can reach 10,000 cP [31] (Figure 2).

On the other hand, UPAEP 9, UPAEP 10 and UPAEP 15 strains (Figures 4, 5 and 7) showed maximum biomass production at 28, 24, and 77 h with values of 3.6, 5.3 and 9.5 g L^{-1} respectively. Biosurfactant production started from the first couple of hours and up to 49 h by UPAEP 9 and UPAEP 15 strains reached emulsification of 58 and 69 %; and at 20 h UPAEP 10 strain showed 70 % of biosurfactant production.

UPAEP 8 (Figure 3), UPAEP 12 (Figure 6) and *Arthrobacter* sp ATCC 31012 (Figure 8) strains showed slow biosurfactant production in contrast with the isolated strains. Biosurfactant production started only at 20, 28 and 50 h, and reached a maximum value of 65, 37 and 30 % respectively (at 40, 49 and 72 h). *Arthrobacter* sp ATCC 31012 showed slow growth, the highest biomass production was of 8.5 g L^{-1} at 55 h. Anyhow, UPAEP 12 strain showed a highest increase in biomass production between 28 and 46 h with a final value of 7 g L^{-1} (77 h) and UPAEP 8 strain showed maximum biomass production of 6.6 g L^{-1} at 24 h.

However, *Bacillus subtilis* ATCC 21332 strain (Figure 9) showed maximum biomass production in the first 10 h with 4.6 g L^{-1}. Maximum biosurfactant production (27 %) is observed at the end of the kinetic (70 h).

The *Candida petrophilum* ATCC 20226 strain (Figure 10) showed an important decrease in glucose up to 70 h (data not shown). Biosurfactant production began at 20 h. No relation to substrate consumption or to biomass production was observed. The maximum emulsification percentage obtained was 80 % after 70 h.

The initial pH of the culture medium was 7.0 and lowers during the cellular growth of the studied isolates, therefore was adjusted with NaOH 0.1N to obtain a pH closer to neutrality (data not shown). Thus, the final pH of the culture medium ranged from 6.07 to 7.37 (Table 2). It is interesting to observe, that the drop in pH occurred just before the biosurfactant synthesis, possibly due to a prior synthesis of organic acids as precursors of biosurfactants by UPAEP 6, UPAEP 8, UPAEP 9, UPAEP 10 and UPAEP 15 strains. Yet, the pH was maintained between 6.5 and 6 with few changes during the entire kinetic by UPAEP 12 strain, and *Arthrobacter* sp ATCC 31012 showed only a small drop at 49 h. *Bacillus subtilis* ATCC 21332 and

Candida petrophilum ATCC 20226 strains remained the pH close to neutrality during the entire

kinetic.

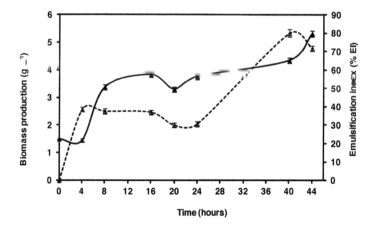

Figure 2. Bacterial growth by bacteria strain UPAEP 6 associated to biomass production (▲), and Emulsification Index EI (%) (Δ). Results are the averages of triplicate experiments ± standard deviation.

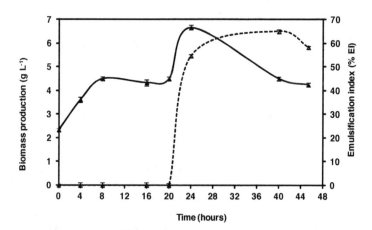

Figure 3. Bacterial growth by bacteria strain UPAEP 8 associated to biomass production (▲), and Emulsification Index EI (%) (Δ). Results are the averages of triplicate experiments ± standard deviation.

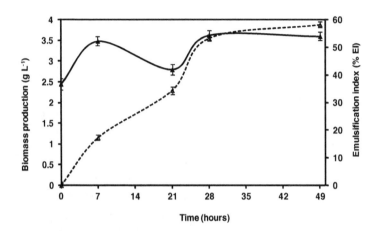

Figure 4. Bacterial growth by bacteria strain UPAEP 9 associated to biomass production (▲), and Emulsification Index
EI (%) (Δ). Results are the averages of triplicate experiments ± standard deviation.

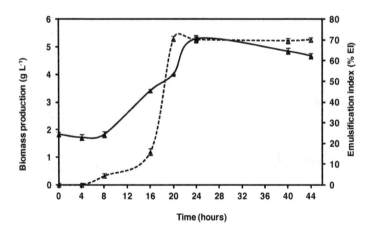

Figure 5. Bacterial growth by bacteria strain UPAEP 10 associated to biomass production (▲), and Emulsification Index
EI (%) (Δ). Results are the averages of triplicate experiments ± standard deviation.

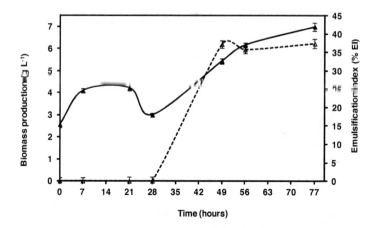

Figure 6. Bacterial growth by bacteria strain UPAEP 12 associated to biomass production (▲), and Emulsification Index EI (%) (Δ). Results are the averages of triplicate experiments ± standard deviation.

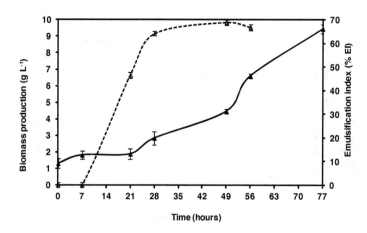

Figure 7. Bacterial growth by bacteria strain UPAEP 15 associated to biomass production (▲), and Emulsification Index EI (%) (Δ). Results are the averages of triplicate experiments ± standard deviation.

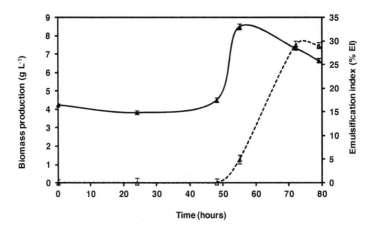

Figure 8. Bacterial growth by bacteria strain commercial *Arthrobacter* sp ATCC 31012 associated to biomass production (▲), and Emulsification Index EI (%) (Δ). Results are the averages of triplicate experiments ± standard deviation.

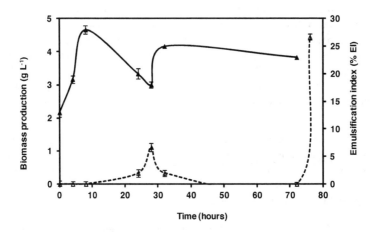

Figure 9. Bacterial growth by bacteria strain commercial *Bacillus subtilis* ATCC 21332 associated to biomass production (▲), and Emulsification Index EI (%) (Δ). Results are the averages of triplicate experiments ± standard deviation.

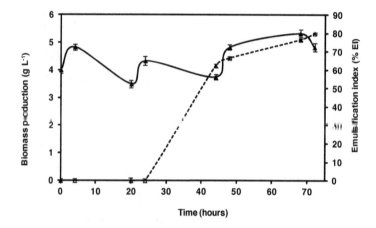

Figure 10. Bacterial growth by bacteria strain commercial *Candida petrophilum* ATCC 20226 associated to biomass production (▲), and Emulsification Index EI (%) (Δ). Results are the averages of triplicate experiments ± standard deviation.

5.3. Production rates

Table 3 shows the results of the estimated rates. The UPAEP 6 strain showed the highest biomass production rate with 0.178 g h⁻¹. The strains with best biosurfactant production rates were UPAEP 10 and UPAEP 8 with 2.5 and 2.39 % h⁻¹, respectively. Significant differences were found in the variance analysis of the emulsification final values with 70% (*Serratia marcescens*) and 80% (*Klebsiella pneumonia*). The highest rates of emulsification were for UPAEP 8 and the yeast *Candida petrophilum* ATCC 20226 (80%). CMC results of the selected strains are similar to that reported for Tergitol (0.0149 mg L⁻¹) and 10 times less than *Serratia marcescens* subsp. *marcescens*.

The capacity of these bacteria to degrade toxic compounds depends on the contact time with the compound, the environmental conditions in which they develop and their physiological versatility.

5.4. Biodegradation tests of maya crude oil

Once the strains had been evaluated, the next step was to evaluate the removal percentage of Maya crude oil (20,000 ppm) using UPAEP 8 (*Klebsiella pneumoniae*) and UPAEP 10 (*Serratia marcescens*). These two bacteria showed a greater than 80 % degradation for Maya crude oil (Figure 11,12 and 13).

Bacterial strain UPAEP	Classification	% likelihood
6	*Klebsiella ornithinolytica*	99
8	*Klebsiella pneumoniae*	99
9	*Klebsiella* sp	75
10	*Serratia marcescens*	99
12	*Candida inconspicua*	75
15	*Klebsiella pneumoniae*	97.6

Table 1. Identification of the bacterial strains was by the API galleries.

Strain UAPEP	Initial pH		Final pH *		% Glucose consumption **	
6	7.0	± 0.2	7.37	± 0.09	99.7	± 0.9
8	7.0	± 0.1	6.65	± 0.11	99.9	± 0.9
9	7.0	± 0.1	6.25	± 0.14	95.8	± 1.0
10	7.0	± 0.1	6.64	± 0.06	99.7	± 0.9
12	7.0	± 0.1	6.07	± 0.14	92.0	± 0.5
15	7.0	± 0.2	6.86	± 0.24	100	± 0.1
Strain ATCC						
31012	7.0	± 0.1	6.22	± 0.15	66.96	± 0.6
20226	7.0	± 0.1	6.45	± 0.12	76.48	± 0.5
21332	7.0	± 0.1	6.64	± 0.13	93.61	± 0.5

* pH values for isolates incubates in Lebac medium for 44 and 48 h at 37°C under constant agitation at 200 rpm (see Methods); each value represents the average of three replicates ± standard deviation.

** Glucose consumption percentage is the difference between initial and final glucose concentration; each value represents the average of three replicates ± standard deviation.

Table 2. Changes of pH and Glucose consumption by Biosurfactants-producing bacterial strains during the bacterial growth.

Figure 11. Maya oil Bioemulsification. Experiment with 20,000 ppm of petroleum and biosurfactan-producing micro-organisms.

Strain UAPEP	Rate Biomass production (g h⁻¹)	R²	Emulsification Activity (% h⁻¹)	R²	Rate substrate consumption (g glucose h⁻¹)	R²	Emulsification Index Final value * (%)	CMC (mg L⁻¹)
6	0.178	0.76	1.72	0.51	0.86	87.1	65^b,c	0.0016
8	0.074	0.68	2.39	0.93	0.277	88.5	80^a	0.0047
9	0.018	0.80	1.13	0.86	0.336	70.0	49^c	0.0014
10	0.074	0.81	2.5	0.82	N.D.**	N.D	70^b	0.0014
12	0.05	0.72	0.01	0.41	0.218	87.0	58^c	0.0010
15	0.100	0.86	1.39	0.64	0.404	78.0	70^b	0.062
Strain ATCC								
31012	0.071	0.83	1.16	0.66	0.428	84.2	40^c	0.005
20226	0	0.21	1.32	0.88	0.390	97.6	80^a	0.005
21332	0.031	0.78	0.19	0.74	0.380	80.0	27^d	0.0015

* Final value Means with different letters are significantly different ($P<0.05$).

* * It was not determined.

Table 3. Biosurfactants-producing bacterial strains isolated from polluted soil with hydrocarbons.

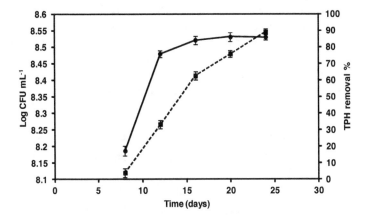

Figure 12. Removal of TPH by bacteria *Klebsiella pneumoniae* (UPAEP 8 strain) isolated from contaminated soil. Strain was grown at 30 ºC, and 20000 ppm of mayan crude oil. Removal of TPH (■). Cell growth of strain with 20000 ppm of mayan crude oil (●). Results are the average of triplicate experiments ± standard deviation.

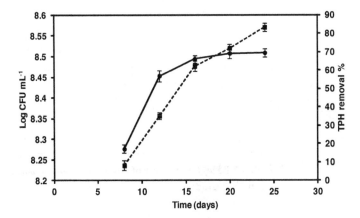

Figure 13. Removal of TPH by bacteria *Serratia marcescens* (UPAEP 10 strain) isolated from contaminated soil. Strain was grown at 30 ºC, and 20000 ppm of mayan crude oil. Removal of TPH (■). Cell growth with 20000 ppm of mayan crude oil (●). Results are the average of triplicate experiments ± standard deviation

6. Discussion

Serratia genus have been reported by other authors as biosurfactants-producing bacterial capable degrader oily compounds [32, 33]. According to [34] bacteria with high capacity to produce biosurfactant promising remain very still, because many companies wish to replace chemical biological to chemical surfactants. The biosurfactant production rate for *Serratia marcescens* and *Klebsiella pneumonia* 2.39 and 2.5 (% h-1) respectively show the significant potential for industrialization of the strains. Biosurfactants-production remains a topic of industrial interest [35] emulsified 20% of 1500 mg / L of octadecane, while the present work with the best strains emulsified 80 and 90% of Mayan crude oil at an initial concentration of 2000 mg/L. According to [34] states that the genus *Pseudomonas* is the most promising from the industrial point of view, among other things because of the chemical nature of the rhapnolipids, in work [35] are employed *Pseudomonas aeruginosa* ATCC 9027, however the strains studied in this work were even better at the emulsification even using oil that is more complex relative to octadecano.

All the selected strains presented emulsifying activity, the majority associated with the growth of microorganisms and a decrease in pH. Some authors [19] reported that for the *Pseudomonas* species, an association has been found in the synthesis of different metabolites (fatty acids, lipopeptides, peptides and amino acids), which can be used for cellular synthesis and biosurfactant production. Although this work is focused on the degradation of recalcitrant hydrocarbons such as Maya crude oil, there is wide interest in biosurfactant production due to its applications in various fields. Other authors [32] performed a chemical and antimicrobial characterization of pseudofactin II, a biosurfactant secreted by *Pseudomonas fluorescens* BD 5 identified as a new cyclic lipopeptide with broad-spectrum bactericidal activity.

The bacteria used the Maya crude oil as sole carbon source, associated with high biomass content and a very high capacity to emulsify hydrocarbon compounds in relatively short operating times (15, 17 and 24 days) compared to those reported by other authors [36-38]. The values of the production kinetics of are very important considering of the scaling the process, *Klebsiella pneumoniae* showed up to 90 % removal and is a promising strain for future biodegradation studies.

The results will allow the use of these cultures as possible inoculants, in real bioremediation experiences where large quantities of inoculants are required. Crude oil biodegradation has been studied extensively because of the high variability of crude oil amount, incubation times and methodologies used to quantify degradation.

7. Future work

In Mexico, particularly on agricultural land, biological techniques which leave no chemical residue and with low-toxicity are required to recover impacted soil. The impact on agricultural soil and its recovery for farmers is a major problem. Sustainable biological techniques may be an alternative and raise the expectations of farmers hoping to plant their crops without risk. Biosurfactants have shown their potential in bioremediation of contaminated soil and water with oil and its derivatives. Because of its low toxicity and biodegradability these are considered as an accepted alternative and environmentally friendly.

However, the *in situ* production of these compounds by microorganisms in natural environments are link to many factors including the type of contaminant, nitrogenous compounds content, interaction with native microorganisms and some others. It is important to perform tests on real soil before the scaling tests since several studies have reported inconsistent results. Therefore the use of microorganisms producing biosurfactants in bioaugmentation processes requires a careful study; new research on the scaling processes to optimize biosurfactants production must be conducted.

The rhamnolipids produced by *Pseudomonas auriginosa* biosurfactants have been extensively studied, but there are other organisms that produce substances with emulsifier, such as those produced by the serrawettin by *Serratia marcescens* this it is a bacteria which has been described as plant growth promoting rhizobacteria (PGPR), which refers to the promotion of growth when plants are inoculated, because it has the ability to produce indole-3-acetic acid (IAA). Due to the activities of the oil industry in Mexico, agricultural soils are contaminated with hydrocarbons, leading to impairment of soil properties and the consequent decline in agricultural production. Technologies should be applied for the recovery of the ground with the least environmental impact. The plant-assisted bioremediation (phytoremediation) is an alternative for the *in situ* treatment of soil contaminated with hydrocarbons. The UPAEP 10 strain of *S. marcescens* is capable of producing biosurfactants and degrades crude oil which is needed for investigating the ability of promoting plant growth in order to develop rhizoremediation technologies.

8. Conclusions

This study showed microorganism isolated of contaminated soils with high capacity of degrading recalcitrant compounds. In México there is a great need to develop clean technologies due to oil spill accidents in agricultural soils. Biosurfactant production by native strains as *Klebsiella pneumoniae* (UPAEP 8 strain) and *Serratia marcescens* (UPAEP 10 strain) showed emulsification rates of up to 80 %, and CMC values were similar than commercial detergents; therefore may be a promising way for recovery of weathered soils with heavy hydrocarbon particles. The implementation of clean technologies will allow farmers to continue producing their products of the harvests harmless and safe for sale and consumption.

Author details

Beatriz Pérez-Armendáriz[1*], Amparo Mauricio-Gutiérrez[1], Teresita Jiménez-Salgado[2], Armando Tapia-Hernández[2] and Angélica Santiesteban-López[1]

*Address all correspondence to: beatriz.perez@upaep.mx

1 Universidad Popular Autónoma del Estado de Puebla, Interdisciplinary Center for Postgraduate Studies, Research and Consulting, Santiago. CP, Puebla, Mexico

2 Benemérita Universidad Autónoma de Puebla. ICUAP, Centro de Investigaciones en Ciencias Microbiológicas, Laboratorio de Microbiología del Suelo. Edificio J 1er. Piso, C.U., Puebla, Mexico

References

[1] Xu JGRJohnson L. 1995. Root growth, microbial activity and phosphatase activity in oil contaminated, remediated and uncontaminated soil planted to barley and field pea. Plant and Soil (1995). , 173(1), 3-10.

[2] Robinson, K, Ghosh, M, & Shi, Z. Mineralization enhancement of non-aqueous phase and soil-bound PCB using biosurfactant. Water Science &Technology (1996).

[3] Deschenes, P, Lafrance, J. P, Villeneuve, R, & Samson, R. Adding sodium dodecyl sulfate and *Pseudomonas aeruginosa* UG2 biosurfactants inhibits polycyclic aromatic hydrocarbon biodegradation in a weathered creosote-contaminated soil. Applied Microbiology & Biotechnology (1996).

[4] Rivera-pineda, F, Ramírez-valverde, B, Juárez-sánchez, J. P, Pérez-armendáriz, B, Estrella-chulim, N, Escobedo-castillo, F, & Ramírez-valverde, G. Implicaciones en la agricultura por el derrame de hidrocarburos en Acatzingo, México. In Recursos naturales y contaminación ambiental. Series Ciencias Ambientales; (2012). , 203-218.

[5] PEMEX Petróleos MexicanosFuga de crudo en el oleoducto nuevo Teapa-Venta de Carpio-Tula. Boletín (2002). http://www.pemex.com/index.cfm?action=news&sectionID=8&catID=40&contentID=269accesed 12 july 2012).(19)

[6] Ortínez-brito, O, Ize-lema, I, & Gavilán-garcía, A. La restauración de suelos contaminados con hidrocarburos en México. Instituto Nacional de Ecología. http://www2.ine.gob.mx/publicaciones/gacetas/422/restauracion.htmlaccesed 14 july (2012).

[7] Pérez-armendáriz, B, Martínez-carrera, D, Calixto-mosqueda, M, Alba, J, & Rodríguez-vázquez, R. Filamentous fungi remove weathered hydrocarbons from polluted

soil of tropical México. Revista Internacional de contaminación ambiental (2010). , 26(3), 193-99.

[8]   Saval, S. La biorremediación como alternativa para la limpieza de sitios contaminados con hidrocarburos. In: seminario internacional sobre restauración de sitios contaminados. Instituto Nacional de Ecología- SERMANAP, agencia de cooperación Internacional del Japón y centro nacional de investigación y Capacitación Ambiental. México (1997).

[9]   Pérez-armendáriz, B, Castañeda-antonio, D, Castellanos, G, Jiménez-salgado, T, Tapia-hernández, A, & Martínez-carrera, D. Efecto del antraceno en la estimulación del maíz y el frijol. Terra Latinoamericana (2011). , 29(1), 95-102.

[10] Yu, H, & Huang, G. H. Isolation and Characterization of Biosurfactant- and Bioemulsifier-Producing Bacteria from Petroleum Contaminated Sites in Western Canada. Soil and Sediment Contamination (2011). , 20(3), 274-88.

[11] Salihu, A, Abdulkadir, I, & Almustapha, M. N. An investigation for potential development on biosurfactants. Biotechnology and Molecular Biology Reviews (2009). , 3(5), 111-17.

[12] Arenas, S. Aislamiento y Caracterización de Bacterias de Ambientes Contaminados por Petróleo en la Refinería "La Pampilla". Collegethesis. PhD Thesis. Universidad Nacional Mayor de San Marcos; (1999).

[13] Rentería, A, & Miranda, H. Aislamiento y Selección Primaria de Microorganismos Capaces de Utilizar Petróleo Como Única Fuente de Carbono. Abstract Book: proceedings of the I Congreso Peruano de Biotecnología y Bioingeniería, ICA, November 1998, Trujillo, Perú. Perú; (1998). , 12-15.

[14] Tantalean, J, & Altamirano, R. Aislamiento y Evaluación del Crecimiento de *Pseudomonas spp.* hidrocarburoclásticas en Petróleo Diesel Abstract Book: proceedings of the I Congreso Peruano de Biotecnología y Bioingeniería, ICA, November 1998, Trujillo, Perú. Perú; (1998). , 12-15.

[15] Merino, F. Estudio de Microorganismos Nativos Productores de Emulsificantes de Petróleo. Masters Degree thesis. Universidad Nacional Mayor de San Marcos; (1998).

[16] Leahy, J. G, & Colwell, R. R. Microbial Degradation of Hydrocarbons in the Environment. Microbiological Reviews (1990). , 54(3), 305-15.

[17] Kiran, T, Nalini, S, Sistla, R, & Sadanandam, M. Surface solid dispersion of glimepiride for enhancement of dissolution rate. International Journal of PharmTech Research (2009). , 1(3), 822-31.

[18] Mukherjee, S, Das, P, Sivapathasekaran, C, & Sen, R. Enhanced production of biosurfactant by a marine bacterium on statistical screening of nutritional parameters. Biochemical Engineering Journal (2008). , 42(3), 254-60.

[19] Tran, H, Kruijt, M, & Raaijmakers, J. M. Diversity and acivity of biosurfactant-producing *Pseudomonas* in the rhizosphere of black pepper in Vietnam. Journal of Applied Microbiology (2008)., 104(3), 839-51.

[20] Banat, I. M, Samarah, N, Murad, M, Horne, R, & Banerjee, S. Biosurfactant Production and Use in Oil Tank Clean- Up. World Journal of Microbiology and Biotechnology (1991)., 7(1), 80-8.

[21] Henry, N. D, Robinson, L, Johnson, E, Cherrier, J, & Abazinge, M. Phenanthrene
Emulsification and Biodegradation Using Rhamnolipid Biosurfactants and *Acinetobacter calcoaceticus* In Vitro. Bioremediation Journal (2011)., 15(2), 109-20.

[22] Zheng, C, Luo, Z, Yu, L, Huang, L, & Bai, X. The Utilization of Lipid Waste for the
surfactant Production and Its Application in Enhancing Oil Recovery. Petroleum Science and Technology (2011)., 29(3), 282-89.

[23] Thavasi, R, Jayalakshmi, S, Balasubramanian, T, & Banat, I. M. Production and characterization of a glycolipid biosurfactant from *Bacillus megaterium* using economically
cheaper sources. World Journal of Microbiology and Biotechnology (2008)., 24(7),
917-25.

[24] Ron, E. Z, & Rosenberg, E. Biosurfactants and oil bioremediation. Current Opinion in
Biotechnology (2002)., 13(3), 249-52.

[25] Falatko, D. M, & Novak, J. T. Effects of biologically produced surfactants on the mo-
bility and biodegradation of petroleum hydrocarbons. Water Environment Research
(1992)., 64(2), 163-9.

[26] Deziel, E, Paquette, G, Villemur, R, Lepine, F, & Bisaillon, J. Biosurfactant Production
by a Soil *Pseudomonas* Strain Growing on Polycyclic Aromatic Hydrocarbons. Applied and Environmental Microbiology (1996)., 62(6), 1908-12.

[27] Das, K, & Mukherjee, A. K. Crude petroleum-oil biodegradation efficiency of *Bacillus
subtilis* and *Pseudomonas aeruginosa* strains isolated from a petroleum-oil contaminated soil from North-East India. Bioresource Technology (2007)., 98(7), 1339-45.

[28] Cooper, D. G, & Goldenberg, D. G. Surface-active agent from two Bacillus species.
Applied and Environmental Microbiology (1987)., 28, 224-229.

[29] Margaritis, A, Kennedy, K, Zajic, J, & Gerson, D. Biosurfactant production *by Nocardia erythropolis*. Dev. Ind. Microbiol. (1979)., 20, 623-630.

[30] Samadi, N, Abadian, N, Akhavan, A, & Fazeli, M. R. Biosurfactant production by the
strain isolated from contaminated soil. Journal of Biological Sciences (2007)., 7(7),
1266-9.

[31] Batzle, M, & Wang, Z. Seismic properties of pore fluids. Geophysics (1992)., 57(11),
1396-1408.

[32] Rajasekar, A, Ganesh-babu, T, Maruthamuthu, S, Karutha-pandian, S T, Mohanan, S,
& Palaniswamy, N. Biodegradation and corrosion behaviour of *Serratia marcescens*

ACE2 isolated from an Indian diesel-transporting pipeline. Journal of Microbiology and Biotechnology. (2007). DOIs11274-006-9332-0, 23(8), 1065-1074.

[33] Janek, T, Lukaszewicz, M, & Krasowska, A. Antiadhesive activity of the biosurfactant pseudofactin II secreted by the Arctic bacterium *Pseudomonas fluorescens* BD5. BMC Microbiology (2012). DOI:10.1186/1471-2180-12-24., 12(24), 1-9.

[34] Marchant, R, & Banat, I. M. Microbial biosurfactants: challenges and opportunities for future exploitation. Trends in Biotechnology (2012). , 30(11), 558-565.

[35] Zhang, Y, & Miller, R M. Enhanced octadecane dispersion and biodegradation by a Pseudomonas rhamnolipid surfactant (biosurfactant). Applied and Environmental Microbiology.(1992). , 58(10), 3276-3282.

[36] Belloso, C, Carrario, J, & Viduzzi, D. Biodegradación de hidrocarburos en suelos contenidos en terrarios: conference proceedings, November 1-5, 1998, XXVI Congreso Interamericano de Ingeniería Sanitaria y Ambiental,Lima, Perú. REPIDISCA: 45167; (1998).

[37] Díaz, I, Favela, E, Gallegos, M, & Gutiérrez, M. Biodegradación de hidrocarburos por un consorcio microbiano de la rizósfera de *Cyperus laxus* lam. Memories proceedings of the VIII Congreso Nacional de Biotecnología y Bioingeniería y IV Congreso Latinoamericano de Biotecnología y Bioingeniería, 12-17 Sep 1999, Huatulco, Oaxaca, México; (1999). http://astonjournals.com/manuscripts/GEBJ-3_Vol2010.pdfaccesed 10july 2012).

[38] Ifeanychukwu Atagana HHaynes R J, Wallis F M. Fungal Bioremediation of Creosote-Contaminated Soil: A Laboratory Scale Bioremediation Study Using Indigenous Soil Fungi. Water, Air and Soil pollution. (2006). 172(1-4):201-219DOI 10.1007/ s11270-005-9074-x

Chapter 5

Biodegradation of PCDDs/PCDFs and PCBs

Magdalena Urbaniak

Additional information is available at the end of the chapter

1. Introduction

As a consequence of the rapid development of modern society during the 20th century, a significant amount of organic chemicals has been dispersed into the environment. Many of them have been used as pesticides, insecticides, defoliants and industrial chemicals or produced as undesirable industrial by-products. A large amount of them show several metabolic and toxic activities including mutagenic, immunotoxic and carcinogenic effects. From this group of substances, the organochlorine compounds include polychlorinated dibenzo-p-dioxins (PCDDs), polychlorinated dibenzofurans (PCDFs) and polychlorinated biphenyls (PCBs), which have received the most attention according to their persistence in the environment, bioaccumulation and hazard for biota [1].

PCDDs/PCDFs

PCDDs and PCDFs are a group of organic chemicals that contain 75 structurally related individual congeners widely distributed in the environment. They were present on Earth for a long time before humans, as they are formed as a result of forest fires and volcanic explosions. They are also manufactured as unwanted by-products in a range of processes, such as municipal waste incineration, metal smelting, chlorine bleaching in the pulp and paper industry, and vehicular emissions. Such a variety of PCDD/PCDF sources causes their widespread occurrence in the environment. They have been detected in soil, surface water, sediments, plants and animal tissue in all regions of the Earth [2,3].

Chlorinated dioxin's precursor is dibenzo-p-dioxin, which consists of two benzene rings bridged by oxygen [4-8] (Fig. 1).

Polychlorinated dibezofurans are similar to polychlorinated dibenzo-p-dioxins, in terms of chemical structure and biological activity (Fig. 2).

Figure 1. The structural formula of 2,3,7,8-tetrachlorodibenzo-*p*-dioxin [9, changed].

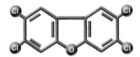

Figure 2. The structural formula 2,3,7,8-tetrachlorodibenzofuran (9, changed).

The physical and chemical properties of toxic congeners of PCDD and PCDF are depicted in Table 1 and 2, respectively.

Compound	Melting point (25°C)	Solubility in water in mg/l (25°C)	Vapour pressure (Pa) in 25°C	Log K_{ow}
2,3,7,8-TCDD	305-306	1.93×10^{-3}	2.0×10^{-7}	6.8
1,2,3,7,8-PeCDD	240-241	1.93×10^{-3}	5.8×10^{-8}	6.64
1,2,3,4,7,8-HxCDD	273-275	4.42×10^{-6}	5.1×10^{-9}	7.8
1,2,3,6,7,8-HxCDD	283-286	4.42×10^{-6}	4.8×10^{-9}	7.8
1,2,3,7,8,9-HxCDD	243-244	4.42×10^{-6}	6.5×10^{-9}	7.8
1,2,3,4,6,7,8-HpCDD	264-265	2.4×10^{-6}	7.5×10^{-10}	8.0
OCDD	325-326	0.75×10^{-7}	$1.1 \times 10^{6,8}$	8.2

Table 1. Physical and chemical properties of PCDDs [10, changed].

PCBs, in turn, due to their stable properties such as low dielectric constant, chemical inertness, non-flammability, high heat capacity, high electrical resistivity and low acute toxicity, were found to be ideal for industrial applications and thus were produced and used in many countries including the United States, Russia, Japan, France and Czechoslovakia. Global PCBs use is estimated to be 1.2 to 1.5 million tonnes. Although the production and use of PCBs was banned almost all over the world more than 30 years ago due to their toxic effects on humans and biota, they are still detected in many ecosystem compartments [11-14]. The PCB molecule consists of two phenyl rings, in which the chlorine atoms are substituted in place of hydrogen atoms. Theoretically, there could be 209 individual PCB congeners (Fig. 3).

Compound	Melting point (25°C)	Solubility in water in mg/l (22.7°C)	Vapour pressure (Pa) in 25°C	Log K$_{ow}$
2,3,7,8-TCDF	227-228	4.19×10^{-4}	2.0×10^{-6}	6.53
1,2,3,7,8-PeCDF	225-227	4.19×10^{-4}	2.3×10^{-7}	6.79
2,3,4,7,8-PeCDF	196-196.5	2.36×10^{-4}	3.5×10^{-7}	6.92
1,2,3,4,7,8-HxCDF	225.5-226.5	8.25×10^{-6}	3.2×10^{-8}	6.92
1,2,3,6,7,8-HxCDF	232-234	1.77×10^{-6}	2.9×10^{-8}	6.92
1,2,3,7,8,9-HxCDF	246-249	1.77×10^{-6}	2.4×10^{-8}	6.92
2,3,4,6,7,8-HxCDF	239-240	1.77×10^{-6}	2.6×10^{-8}	6.92
1,2,3,4,6,7,8-HpCDF	236-237	1.35×10^{-6}	4.7×10^{-9}	7.92
1,2,3,4,7,8,9-HpCDD	221-223	1.35×10^{-6}	6.2×10^{-9}	7.92
OCDF	258-260	1.16×10^{-6} (in 25°C)	5×10^{-9}	8.78

Table 2. Physical and chemical properties of PCDFs [10, changed].

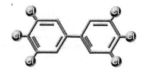

Figure 3. The structural formula of 2,2 ', 3,3', 4,4 '-hexachlorobiphenyl [9, changed].

PCBs have been produced under several trade names, e.g., Clophen (Bayer, Germany), Aroclor (Monsanto, USA), Kanechlor (Kanegafuchi, Japan), Santothrem (Mitsubishi, Japan), Phenoclor and Pyralene (Prodolec, France) (Table 3).

Apirolio	Diaclor	No-Flamol
Areclor	Duconol	Pydraul
Aroclor	Dykanol	Pyralene
Arubren	Elemex	Pyranol
Asbestol	Euracel	Pyroclor
Askarel	Fenchlor	Phenoclor
Bakola	Hivar	Saf-T-Khul
Biclor	Hydol	Santotherm
Chlorextol	Inclor	Santovac
Chlorinol	Iterteen	Siclonyl
Chlorphen	Kennechlor	Solvol
Clophen	Montar	Sovol
Delor	Nepolin	Therminol

Table 3. Major trade names of PCBs [15].

Commercial PCBs are complex mixtures of 30–60 congeners, which are the major PCB components of most environmental extracts. Each individual compound shows a unique combination of physico-chemical and biological properties dependent on the degree of chlorination (Table 4).

Aroclor compound	Water solubility (mg/l) 25°C	Vapour pressure 25°C	Density 25°C [g/cm3]	Appearance	Boiling point [°C]
Aroclor 1016	0.4200	4.0×10^{-4}	1.33	Clear oil	325–356
Aroclor 1221	0.5900	6.7×10^{-3}	1.15	Clear oil	275–320
Aroclor 1232	0.4500	4.1×10^{-3}	1.24	Clear oil	290–325
Aroclor 1242	0.2400	4.1×10^{-3}	1.35	Clear oil	325–366
Aroclor 1248	0.0540	4.9×10^{-4}	1.41	Clear oil	340–375
Aroclor 1254	0.0210	7.7×10^{-5}	1.50	Light, yellow, viscous oil	365–390
Aroclor 1260	0.0027	4.0×10^{-5}	1.58	Light, yellow, viscous oil	385–420

Table 4. Physical and chemical properties of selected Aroclors [15, after 16].

Currently, many countries impose strict controls on the use and release of PCDDs/PCDFs and PCBs. As a result their input into the environment has decreased significantly. Nevertheless, their release from contaminated sites and their redistribution on a global scale is still observed [17-18]. Their slow decomposition in the environment and the hazards they pose for living organisms makes PCDDs/PCDFs and PCBs large-scale environmental degraders, especially because their toxicity can be further enhanced by their ability to accumulate in the soil and sediments and their bioaccumulation and biomagnification within aquatic and land food chains (Fig. 4).

It should also be underlined that PCDDs/PCDFs and PCBs also pose a risk to human health. They have been shown to produce toxic responses similar to those caused by 2,3,7,8-TCDD, the most potent congener within this group. Studies on animals demonstrate that PCDDs/PCDFs and PCBs are implicated in mutagenic and carcinogenic effects such as liver damage, malignant melanoma and preneoplastic and neoplastic changes [1, 19]. Other manifestations related to PCDDs/PCDFs and PCBs are gastrointestinal (gastric hyperplasia, ulceration, necrosis), respiratory (chronic bronchitis and coughs), dermal (chloracne, oedema, alopecia, hyperkeratosis of epithelium), neurotoxic (impaired behavioural responses, depressed motor activity, developmental deficits, numbness) and immunotoxic (lymphoid tissue atrophy, leukocyte and lymphocyte reduction, suppressed antibody responses), hepatotoxic (hepatomegaly, hyperplasia of the bile duct, necrosis, fatty degeneration, porphyria) and reproductive problems (decreased sperm motility and number, increased miscarriages, decreased survival and mating success) [1, 19].

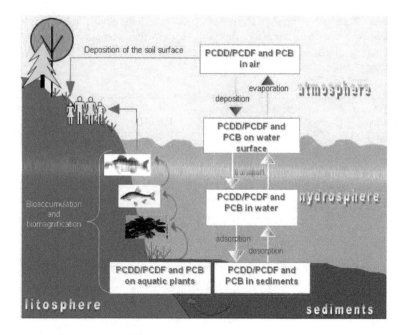

Figure 4. Transport and circulation of PCDDs/PCDFs and PCBs in the environment.

2. Microbiological transformation of PCDDs/PCDFs and PCBs

The degradation of PCDDs/PCDFs and PCBs is classified into two sections: biological transformation by microorganism activity and physico-chemical transformation.

The first group includes anaerobic, aerobic and sequential anaerobic-aerobic transformation. The latter can be classified into photochemical and thermal degradation.

Microbiological transformation depends on enzymes produced by microorganisms which enable modification of toxic compounds into less toxic forms. Biological degradation can carry on as mineralization when microorganisms use the organic compound as a source of carbon and energy, or as co-metabolism where microorganisms need other sources of carbon and energy and the transformation of pollutants occurs as a concurrent process. Products of this process can be further mineralized, otherwise incomplete degradation occurs, leading to the formation and accumulation of more toxic metabolites than parent substrates.

The effectiveness of degradation rates varies depending on the conditions present in the environment and comprises: 1) input of pollutants, 2) physical parameters (oxygen content, temperature, light intensity, pH, conductivity) and 3) biological parameters (presence of microorganisms able to degrade a given pollutant and the availability of carbon and/or other sources of energy). All of the above variables determine the rate of biological and physical transformation of analysed compounds.

2.1. Aerobic conditions

Bacterial cometabolism

Aerobic transformation occurs in environments that are rich in oxygen and involves the use of microbial molecules, such as mono- and dichlorinated PCDDs/PCDFs and PCBs, as a source of carbon and energy. It should be noted that in about 90% of cases, the process takes place as co-metabolism, which means that the microorganisms need an additional source of carbon apart from PCDDs/PCDFs or PCBs.

Data from the literature confirms the aerobic biodegradation of PCDD/PCDF and PCB compounds and the rate of this process increases with the reduction of PCDD/PCDF and PCB chlorination [20-23]. Thus, for example, molecules containing five or more chlorine atoms are not susceptible to the effects of aerobic microorganisms.

PCDDs/PCDFs

In the case of PCDDs and PCDFs the research conducted over the last 30 years has widely described their aerobic biodegradation [19-22, 24]. Worldwide studies have demonstrated that many isolated strains of bacteria, such as *Rhodococus opacus* SAO101, *Beijerinckia* sp. B8/36, *Psudomonas veronii* PH-03, *Psudomonas* sp. HH69, CA10, EE41, *Bacillus megaterium* AL4V, *Sphingomonas* sp. RWI and HL7, are capable of the biodegradation of slightly chlorinated PCDDs/PCDFs under aerobic conditions [21, 24-29]. To increase the rate of aerobic biodegradation of PCDDs/PCDFs and PCBs an additional source of carbon, for example a small amount of un-substituted PCDD or biphenyls [20], carbazole [30], o-dichlorobenzene [25] or benzoic acid or 3-methoxybenzoic [30] can be used.

PCBs

The first data on the aerobic degradation of PCBs was reported by Ahmed and Focht [31] in 1973 and the respective study was devoted to the degradation of biphenyl and monochloro-biphenyl to chlorobenzoic acid by two species of *Achromobacter*. Furukawa et al. [32] demonstrated that a species of *Acinetobacter* and *Alcaligenes* can rapidly adsorb 2,5,2' trichlorobiphenyl onto the cell surface, then metabolize and release metabolic compounds from the cell. Since then numerous investigations have focused on the occurrence and distribution of PCB-degrading microorganisms and their capability to biodegrade PCBs. For example, Clark et al. [33] reported that *Alcalegenes denitrificants* and *A. odorans* can degradate Aroclor 1242 (a mixture of PCB containing 42% chlorine) by co-metabolism. A study by Novakova et al. [34] showed the results of the degradation of Delor 103 by *Psudomonas* sp. P2 and *Alcaligenes eutropha*. Optimal PCB degradation was obtained by the addition of biphenyl, saccharose, agar or an amino acid mixture as the source of carbon. A reduction of degradation efficiency was observed by the addition of glycerol or pyruvate. To completely degrade PCBs by aerobic bacteria, various microbial strains with specific congener preferences are required.

Bacterial mineralization

According to data described by Field and Sierra-Alvarez [35] there are few well documented examples of chlorinated PCDDs/PCDFs and PCBs serving as the sole source of carbon and energy for pure bacterial strains. This is shown by the research of Hong et al. [28] wherein the *Pseudomonas veronii* PH-03 has been used to utilize 1-CDD and 2-CDD growing on aliphatic acids generated from ring cleavage. The mentioned strain of *Pseudomonas veronii* accumulated the dead products 3-chlorocatchol and 4-chlorocatchol from the chlorinated rings. Similar results were also obtained by Arfmann et al. [36] by using a *Sphingomonas* sp. strain RW1 growing on 4CDF. The substrate of carbon and energy was a 5-carbon aliphatic acid and a 2-hydroxypenta-2,4dienoate released from the ring cleavage and the dead-end products were 3-chlorosalicylic acid.

The complete mineralization of PCDDs/PCDFs was also achieved by using co-cultures including a PCDD/PCDF-degrader and a 3-chlorosalicylic acid degrader. For example, a study by Wittich et al. [37] showed that use of *Sphingomonas* sp RW16 and *Pseudomonas* sp. RW10 enabled the complete degradation of 2-CDF and 3CDF. The co-culture mixture combined with *Sphingomonas* sp. RW1 and *Burkholderia* sp. JWS was shown to completely degrade 4-CDF [36]. The above research demonstrates that *Sphingomonas* sp. RW16 and *Sphingomonas* sp. RW1 were capable of degrading the CDF and the *Pseudomonas* sp. RW10. *Burkholderia* sp. JWS utilized the 3-chlorosalicylic acid as the released as dead-end product.

Fungal cometabolism

It should also be mentioned that fungi, similarly to bacteria, are capable of PCDD/PCDF degradation in aerobic conditions, in both mineralization and the co-metabolism process.

Fungi use enzymes (lignin peroxidase or manganese peroxidase) to oxidise the molecule of the compound. The first described case of use of the fungal biodegradation is the work of Bumpus et al. [38], in which the authors documented the mineralization of [^{14}C] 2,3,7,8-TCDD to $^{14}CO_2$ within 30 days by the fungi of *Phanerochaete chrysosporium*. *P. chrysosporium* has also been successfully used to degrade 2,7-DCDD [39].

The biodegradation activity of fungi is not limited to less chlorinated congeners. There is evidence that *P. chrysosporium* is able to remove 34% and 48% of a mixture of congeners containing from 5 to 8 chlorine atoms in the molecule during 7 and 14 days [40].

2.2. Anaerobic conditions

Anaerobic microorganisms are well adapted to pollutants with a high carbon concentration due to the diffusional limitation of oxygen. Anaerobic transformations of PCDDs/PCDFs and PCBs include reductive dehalogenation using PCDDs/PCDFs and PCBs as electron acceptors. During this process a substituent chlorine atom is replaced with a hydrogen atom.

Reductive dehalogenation occurs in soils and sediments, where different microorganisms possessing dehalogenation enzymes responsible for dechlorination and dehalogenation processes exist. The rate, extent and route of dechlorination are dependent on environmental factors, such as carbon availability, electron donors, presence of electron acceptors other than PCDDs/PCDFs and PCBs, temperature and pH. All of these factors influence the composition of a microorganism's community and their activity.

PCDDs/PCDFs

The first evidence of degradation of PCDDs/PCDFs under anaerobic conditions was obtained by spiking sediment microcosms with highly chlorinated congeners of HpCDD, HxCDD and PeCDD [40].The rate of removal of those compounds in biologically active sediments was from 19% to 56% higher in comparison to heat-killed sediments. The products of such biodegradation processes were TCDD and TCDF congeners [40, 41]. The main microorganisms capable of efficient degradation of these compounds were mainly bacteria of the genus *Dehalococcoides* [43-45]. Experiments with the use of OCDD (8 chlorine atoms) at a concentration of 5.3 ml/L applied into sediment microcosms, showed that after 7 months the congener was distributed into forms that contain only 1 to 3 chlorine atoms [46-47].

PCBs

The first evidence of anaerobic degradation of PCBs was reported based on the observed modification of Hudson River and Silver Lakes sediments contaminated by commercially produced PCBs. The increase of low-chlorinated PCBs in comparison to the high-chlorinated congeners was consistent with reductive dechlorination [48]. Furukawa et al. [49] demonstrated that species of *Acinetobacter* and *Alcaligenes* may rapidly adsorb 2,5,2'-trichlorobiphenyl onto the cell surface and then metabolise and release metabolic compounds from the cell. From that time many of investigations were devoted to the occurrence and distribution of PCB-degrading microorganisms and their capability to biodegrade PCBs.

Master et al. [48] showed that many commercial PCB mixtures can be reductively dechlorinated under anaerobic conditions, for example, Aroclor was dechlorinated at rates of 3 μg Cl/g of sediment per week. The dechlorination occurs at temperatures of 12°C and PCB concentrations of 100–1000ppm [49]. Fava et al. [50] described the degradation of Aroclor 1242 by three strains: *Comamonas testosteroni*, *Rhodococcus rhodochrus* and *Psudomonas putida* with total losses of 13.8%, 19.1% and 24.6%, respectively. In both experiments, the favoured positions for dechlorination were (in order) meta>para>ortho and preference was shown for "open" sites 2 and 3, indicative of the action of 2,3-dioxygenase enzymes [50]. Fava et al. [50] reported that the dechlorination of Fenclor 54 primarily occurred from the meta- and para positions, while ortho-substituted congeners accumulated in the medium. Other studies showed an inability of anaerobic microorganisms to degrade the low chlorinated biphenyls. The occurrence of diortho- and monoorthochlorobiphenyls, as well as the biphenyl rings, was identified even after a one year incubation [31].

2.3. Sequential anaerobic-aerobic conditions

Laboratory experiments showed that microbial degradation of lower chlorinated PCDDs/PCDFs and PCBs occurs at a faster rate than in higher chlorinated ones. Lower chlorinated congeners produced by dechlorination can be readily degraded by indigenous bacteria, which in consequence, reduces the potential bioconcentration risk and the exposure to PCDDs/PCDFs and PCBs by conversion to congeners with a low bioaccumulation potential in the food chain [35, 51]. The lightly chlorinated PCDDs/PCDFs and PCBs congeners produced during the anaerobic dechlorination may then be substrates for oxidative destruction by aerobic microorganisms, which leads to the production of chlorobenzoic acid, which is easily degraded by bacteria.

The findings described above indicate that a complete degradation of PCDDs/PCDFs and PCBs can be achieved by sequential exposure to anaerobic and aerobic biodegradation. Highly chlorinated congeners can be transformed to compounds of lower chlorination during reductive dechlorination under anaerobic conditions. Lightly chlorinated congeners, produced during anaerobic dechlorination, might then become substrates for oxidative destruction by aerobic microorganisms, which can lead to the production of chlorobenzoic acid, which is further easily degraded by bacteria [34, 51].

3. Physical transformation of PCDDs/PCDFs and PCBs

There is also a division of degradation processes that takes into account the physicochemical degradation of PCDD/PCDF and PCB compounds.

3.1. Photochemical degradation

Photochemical degradation called photolysis also depends on the degree of chlorination, the position of chlorine atoms in the biphenyl ring and the solvent used for PCDD/PCDF and PCB dissolution. The primary process in photoreaction is reductive dechlorination, but examples of photo-induced isomerization and condensation of individual chlorobiphenyls have been also reported.

The first laboratory experiments on photolysis were conducted with mercury lamps as the UV source, with a wavelength of about 254nm, which results in the dechlorination of PCBs. Later, sunlight simulating lamps were used, which also confirmed the degradation of the chlorinated compounds [52-54].

It should also be mentioned that the higher chlorinated biphenyls undergo photolysis faster than less chlorinated ones. For example, the exposure of PCB to a 310nm wavelength causes of reduction of about 70% tetra-, 96% of hexa- and 99% of octachlorobiphenyl. Experiments with tetrachlorobiphenyls showed that the major products after irradiation at 300nm are di- and trichlorinated biphenyls [52]. Bunce et al. [53] reported intensified photodegradation with increased irradiation time.

Photolysis is regarded as one of the major processes reducing PCDDs/PCDFs and PCBs in the environment. Bunce at al. [53] estimated the loss of PCBs in natural waters at the magnitude of 10 to 1000g/Km^{-2}/year. In shallow water bodies at least one chlorine atom from mostly chlorinated PCB molecules is photodegradated per year. Zepp et al. [54] reported that humic acids and suspended materials may induce and accelerate PCB photodegradation.

Several researchers described accelerated in-situ photolysis by the addition of various organics, such as isooctane, hexane and cyclohexane, on the surface of contaminated soil [56-58]. Doughtery et al. [59] found that solar-induced photolysis reactions can be a principal mechanism for the transformation of PCDD/PCDF to less toxic forms.

3.2. Thermal degradation

The last group of PCDD/PCDF and PCB transformations is thermal degradation, leading to the complete destruction of toxic substances at temperatures above 700°C or producing more toxic congeners such as TCDD at temperatures below 700°C. This kind of PCDD/PCDF and PCB destruction is well adapted on an industrial scale for the safe disposal of waste products containing PCDDs/PCDFs and PCBs.

4. Environmental biodegradation of PCDD/PCDF and PCB

PCDDs/PCDFs and PCBs are substances that are created during different types of natural and industrial processes. Their appearance in the environment and in consequence in food products creates a serious threat to human health and ecosystem functioning as far as their genotoxic and toxic effects on living organisms are concerned [59]. Therefore, natural trans-formation of PCDDs/PCDFs and PCBs is a critical event in determining their fate in the environment.

4.1. Phytoremediation

Phytoremediation is defined, according to Macek et al. [61], after Cunningham and Betri [62] and Cunningham et al. [63], as the use of green plants to remove, contain, or render harmless environmental contaminants. According to other authors, phytotechnology is a set of tech-nologies that use plants to remediate contaminated sites [64-68].

Phytoremediation uses living plants for the remediation of contaminated mediums, such as soil, sediment, sludge and water (in situ as well as ex situ) by the removal, degradation or stabilization of a given contaminant [64].

According to Macek et al. [61], after Salt et al. [69], phytoremediation is currently divided into several subtypes:

- phytoextraction

- phytodegradation

- rhizofiltration

- phytostabilization

- phytovolatilization

These techniques are an alternative to the widely used methods of physical, physico-chemical and thermal remediation. Their advantages include the possibility of application ex-situ and in-situ, low investment and operating costs with high effectiveness and non-invasiveness in the environment [70-72].

The main problem with the use of phytoremediation techniques is their long operational time and the fact that many of the bioremediation techniques are still in the experimental stage [70-72].

The genesis of the phytoremediation process was observed by the rate of degradation of organic chemicals in the soil with and without vegetation cover. On the basis of the obtained results it was stated that vegetation cover promotes the reduction of organic compounds in soil. Currently, a variety of research indicates the positive effects of using higher plants to degrade organic compounds [73-81].

Siciliano et al. [73] demonstrated the reduction of organochlorine compounds by about 30% during 2 years of plant cultivation, whereas on the soil without plants, the reduction was 2 times lower. Nedunuri et al. [74] reported the reduction of aromatic compounds by about 42% and 50% by using fibre flax (*Lolium annual*) and St. Augustine grass (*Stenotaphrum secundatum*), respectively, over a period of 21 months. Other examples showed remediation of soil contaminated by crude oil using a combination of grass and fertilizers [74-77]. Vervaeke et al. [78] reported a 57% reduction of aromatic compounds and mineral oils during 1.5 years of willow (*Salix viminalis*) cultivation. Pradham et al. [79] demonstrated the usage of phytoremediation as a primary remediation technology and as a final step for treatment of soil contaminated with PAHs. The authors recorded a 57% reduction in PAHs after 6 months of alfalfa (*Medicago sativa*), switch grass (*Panicum virgatum*) and little bluestem grass (*Schizachyrian scoparium*) growth.

A study by Gregor and Fletcher [80] demonstrated the ability of plant cells to metabolize PCBs. While, research by Jou et al. [81] showed the uptake of PCDDs/PCDFs by *Boussonetia papyrifera* growing on highly contaminated soil. The authors reported similar concentrations and distributions of PCDD/PCDF and PCB congeners in plant tissues and soils. Other research demonstrated that several plants of the genus *Cucurbita* (e.g., courgette, pumpkin and squash) can readily take up PCDD and PCDF from soil and translocate them to leaves and fruits [82-84]. It was also found that *Cucurbita* plants can phytoextract PCBs from soil and translocate some quantities to aerial tissues [85, 86]. This confirms that the PCDD/PCDF and PCB contents in plants may closely relate to the surrounding environments where plants grow [81]. Nevertheless, Uegaki et al. [87] reported no concentration differences in brown rice grown in three different soils: dioxin-contaminated soil, paddy soil and upland soil. The authors assumed that growing rice in soil contaminated with high concentrations of dioxins has no influence of the PCDD/PCDF levels in rice tissue [87].

4.2. Rhizoremediation

Rhizoremediation of organic micropollutants is one of the most effective remediation processes due to existing interactions in the rhizosphere between plant roots, plant exudates, soil and microorganisms. Mackova et al. [64] reported that plants support bioremediation by the release of exudates and enzymes that stimulate both microbial and biochemical activity in the surrounding soil and mineralization in the rhizosphere. Plants can also accelerate bioremediation in surface soils by stimulating the growth and metabolism of soil microorganisms through the release of nutrients and the transport of oxygen to their roots [61-62, 67]. Moreover, the fact that up to 40% of carbohydrates, amino acids and other photosynthesis products are stored in the plant rhizosphere, plays an important role in the availability of carbon used by microorganisms in the co-metabolism process.

A study by Whipps [88] demonstrated that 1g of rhizosphere soil contains a 10^{12} higher amount of microorganisms in comparison to non-planted soil. Microorganisms settling in the rhizosphere also play a role in the protection of plants against pathogens and stress induced by too high a concentration of contaminants and facilitate nutrient uptake by a given plant [89-93].

Bacteria present in the rhizosphere soil serve remediation functions by secreting the appropriate enzymes (e.g., peroxidase, phosphatase, dioxygenase, P450 monooxygenase, dehalogenaza, nitrylases and nitroreductase) involved in the degradation of organic pollutants. Such enzymes are also found in plants and fungi that colonize plant roots. This led to a thesis on the interaction of plants and microorganisms in order to completely destroy a given pollutant [93-99]. This process is called rhizodegradation and is defined as the degradation of pollutants in the root zones of plants (rhizosphere).

The effectiveness of rhizosphere biodegradation depends on the ability of microorganisms to adapt to a given pollution concentration and the effectiveness of root colonization [97]. The interactions between plants, soil and rhizosphere microorganisms are multifaceted and according to Macek et al. [61] can give mutual benefit to both organisms. This mutualistic relationship is responsible for the accelerated degradation of soil contaminants in the presence of plants [101]. Research on this issue is ongoing. Already existing publications confirm the validity of the use of rhizoremediation to reduce PCDDs/PCDFs and dl-PCBs. For example, an article by Kuiper et al. [98] demonstrated that naturally occurring rhizosphere biodegradation can be enhanced by the addition of microorganisms to the rhizosphere.

The important group of substances present in the rhizosphere are complexes of aromatic compounds such as flavonoids and coumarins. These compounds are used by bacterial microflora as a source of carbon and nitrogen [73, 98-99, 102-103]. They are structurally similar to organic compounds such as PCBs and PAHs. This indicates the potential of using such evolutionary established metabolic pathways of rhizosphere microorganisms for the remediation of organic pollutants [104]. Thus, many researchers are interested in the ability of microorganisms inhabiting the rhizosphere to degrade organochlorine pollutants and the role of flavonoids and coumarins produced by plants [99, 103, 105-108].

Worldwide studies describe many kinds of pollutants including PCBs, PAH, petroleum hydrocarbons, chlorinated pesticides like Pentachlorophenol and 2,4-Dichlorophenoxyacetic acid, which were more rapidly degraded in the rhizosphere compared to the bulk soil [64, 109-111]. Research by Betts [112] conducted on soil contaminated by petroleum hydrocarbons showed its considerable improvement by using several plants species such as Bermuda grass, rye grass, white clover and tall fescue. A study by Burken and Schnoor [113] described the positive role of root exudates on atrazine uptake by plants (poplar trees). The research also showed that phenolics, flavonoids and terpenes present in root exudates can induce the bacterial degradation of PCBs [61, 114—115]. A study by Mackova et al. [116] showed the effect of tobacco, nightshade, alfalfa and horseradish on PCB removal from contaminated soil. The obtained results showed 6% to 33.7% removal of PCBs during 6 months of experimentation. The authors also underline the role of the studied plants as a source of bacterial consortia capable of PCB degradation.

5. Perspectives in environmental biodegradation of PCDDs/PCDFs and PCBs

PCDDs/PCDFs and PCBs are compounds that occur in all types and structures of ecosystems. Their transfer takes place through biogeochemical cycles but it is their long half-life in the environment, their accumulation and biomagnification in aquatic and terrestrial food chains and their toxicity that determine their long-term and large-scale threat to the environment and humans. As a result, one of the priority tasks of recent research on PCDDs/PCDFs and PCBs is to characterize the processes that determine their transport and deposition in ecosystems, in order to regulate their allocation and diminish their concentration. Reversing ecosystem degradation and reducing PCDD/PCDF and PCB concentrations in the environment requires solutions based on integrative problem-solving science, such as ecological engineering and ecohydrology [117].

A key element of the ecohydrology theory is the assumption that an excess amount of pollutants including PCDDs/PCDFs and dl-PCBs and their negative effects on the environment can be limited by so-called "dual regulation". Until now, the above methodology was used to reduce the occurrence of toxic cyanobacterial blooms resulting from excessive inflow of phosphorus into water. This concept involves the use of biological and hydrological processes to control the amount and allocation of phosphorus in the ecosystem through increasing biofiltration and by the formation of ecosystem biota [118-120].

Similarly, in order to diminish the concentration of PCDDs/PCDFs and PCBs in the environment there is a need to not only reduce the pollutant load from point and non-point sources but also to develop and apply in-situ bioremediation strategies [72, 117-123]. The application of bioremediation technologies should focus on the possibilities of exploiting and strengthening the functioning of the given ecosystem to reduce the recorded concentrations of PCDDs/PCDFs and PCBs.

The phyto-and rhizoremediation techniques described above are examples of the use of the natural properties of the ecosystem to reduce the environmental PCDD/PCDF and PCB contamination.

Currently, in order to improve the rate and efficiency of such remediation processes a number of advantages have been developed and applied. Some of them are focused on the stimulation of growth and activity in microbial communities in order to accelerate remediation efficiency and diminish the concentration of PCDDs/PCDFs and PCBs in environment.

It should be underlined that there are two main types of microorganism: indigenous and exogenous. Indigenous ones are those that are found already living at a given site. To stimulate the growth of these indigenous microorganisms, the proper soil temperature, oxygen and nutrient content may need to be provided. If the biological activity needed to degrade a particular contaminant is not present in the soil at the site, microorganisms from other locations, whose effectiveness has been tested, can be added to the contaminated soil. These are called exogenous microorganisms [56]. Research has shown that the stimulation of an indigenous microbial population, by injecting methanol and acetate as an electron donor, enhances the removal of tetrachloroethane (PCE) to ethane [124]. Nevertheless until now, scientists have been faced with the problem of the application of isolated microorganisms in situ, as they are often unable to adapt and compete with microorganisms naturally occurring at contaminated sites. This is mainly due to the inability to grow a culture of microorganisms below a certain depth, the lack of sufficient amounts of nitrogen, phosphorus and carbon in the environment, the low bioavailability of pollutants and the preferential use of carbon from non-toxic substrates rather than toxic. An important role is played by the presence of contaminants that inhibit the growth of microorganisms. Currently, in order to avoid such a situation the analogues of the natural soil contaminant are added to the remediated soil. This stimulates the micropollutants' degradation pathways in the microorganisms' cells [99,105,125].

Another problem with bioremediation is the availability of the contaminant to the degrading organisms. To solve this problem research has been conducted on the use of surfactants as potential agents for enhancing solubility and removing contaminants from soil and sediments [126-128]. As reported by Nakajima et al. [129], the addition of sodium dodecyl sulphate, Triton-100 and sodium taurocholate increases the bioavailability of PCBs and PAHs.

Bioaugmentation is another method used in order to improve the microbial degradation of pollutants. This process is based on the introduction of appropriate species for the degradation of specific contaminants. The efficacy of bioaugmentation is contradictory, as far as both positive and negative results have been obtained. A successful bioaugmentation was observed for the remediation of PAHs in sediments [124]. Nevertheless, other studies have achieved no positive results [130].

On the basis of the above data, contemporary bioremediation strategies should be implemented in combination, for example phytoremediation and biostimulation or rhizoremediation and bioaugmentation. This would accelerate the usage of plants and enhance the activity of degrading microorganisms in order to minimize the risk played by PCDDs/PCDFs and PCBs.

It is also possible to remediate soil by using transgenic organisms. Currently, most of the research into the use of transgenic organisms is carried out on a laboratory scale. These experiments are mainly concerned with the introduction of genes encoding biosynthetic pathways of biosurfactants (in order to increase the bioavailability of contaminants), the introduction of genes that enable increased resistance to given contaminants in microbial communities or genes encoding the enzymes' degradative pathways (e.g., cytochrome P450) [131-136].

The latest research by Lan Chun et al. [136] demonstrated the positive role of the electrical stimulation of microbial PCB degradation. The authors found a 40-60% reduction in total PCB concentration in weathered sediments exposed to electric currents, while no significant decrease in PCB concentration was observed in control sediments.

The techniques described above and their advantages, such as biostimulation and bioaugmentation, can be adopted and used in large-scale remediation processes. Examples of such an approach include the utilization of wetlands and biofilters.

Wetlands are often described as "the kidneys of the landscape" owing to their the intrinsic function to transform and store organic matter and nutrients [138] and associated micropollutants such as PCDDs/PCDFs and PCBs. This ability has been exploited for water quality improvement [138]. Constructed wetlands were first used for wastewater treatment in the 1950s. In recent years constructed wetlands have been widely used for urban and agricultural runoff treatment. They utilize natural processes to purify water in a sustainable, cost and energy effective way with minimal operation and maintenance cost [140]. Furthermore, the usage of constructed wetlands as tools in the treatment of polluted waters, has been gaining popularity as an ecological engineering alternative over conventional, chemical based methods [141-142]. Several scholars have shown successful utilizations of constructed wetlands for the treatment of a wide variety of wastewaters including industrial effluents [142-144], urban storm water, agricultural runoff [146-147], domestic wastewater [148] and animal wastewaters [149]. Schulz and Peall [150] determined the effectiveness of constructed wetlands in retaining agricultural pesticide pollution as 89% during runoff. Several researchers have proven the ability of constructed wetlands to mitigate pesticide pollution derived from various agricultural nonpoint sources [151-155].Considering the above, it appears that the use of constructed wetlands to purify water from organochlorine compounds is a promising challenge.

Furthermore, the use of land-water ecotones constructed in a river valley with different kinds of plants and microorganisms may partially purify the inflowing surface- and groundwater contamination by PCDDs/PCDFs and PCBs [156-157]. Such structures may capture, immobilize and/or degrade PCDDs/PCDFs and dl-PCBs [61, 96, 103].

The other promising solution involves the use of biofilters for the purification of inflowing water, wastewater, leachate etc. Such biofilters combined with areas of intensive sedimentation, which enable the deposition of matter, nutrients and micropollutants and their further biodegradation by existing microbial consortia and areas of macrophyte growth, wherein intensive phytodegradation processes occur, are considered to be one of the most effective solutions for pollutant removal. Results obtained by Urbaniak et al. [158] in the Asella Demonstration Project demonstrated changes in the Toxic Equivalent (TEQ) of PCDDs/PCDFs in the sediments of the Asella river and lake taken before and after biofilter construction. Authors showed a 70% reduction in sediment toxicity after one year of biofilter implementation. This indicates the positive role of biofiltration in the quality of lake ecosystems and in consequence on human health. The implementation of such biofiltration system enabled a reduction in the input of PCDDs/PCDFs into the lake through sedimentation and due to acceleration of photo- and biodegradation processes the quality of the whole river-lake system was improved.

6. Conclusions

PCDDs/PCDFs and PCBs pose one of the most challenging problems in environmental science and technology. Their fate, transport and biodegradation in the environment occur via complex networks, involving complicated interactions with other contaminants and with various physiological, chemical and biological processes. Those processes can be used and modified in order to diminish their environmental concentration. The promising results of such activities performed by researchers worldwide were described in this chapter. Nevertheless, the still existing challenge is to develop a bioremediation strategy that involves and integrates different types of solutions, on the scale of the whole ecosystem, in order to optimize the effectiveness of pollutant removal.

Acknowledgements

This chapter has been carried out as a part of the following projects:

- "Innovative resources and effective methods of safety improvement and durability of buildings and transport infrastructure in the sustainable development" financed by the European Union, from the European Fund of Regional Development based on the Operational Programme of the Innovative Economy, POIG.01.01.02-10-106/09

- The Polish Ministry of Science and Higher Education, Project: N N305 365738 "Analysis of point source pollution of nutrients, dioxins and dioxin-like compounds in the Pilica River catchment and draw up of reclamation methods";

- Ministry of Foreign Affairs of the Republic of Poland within the Polish Aid Programme 2012, project no. 62/2012: "Implementation of Ecohydrology – a transdisciplinary science for integrated water management and sustainable development in Ethiopia".

Author details

Magdalena Urbaniak[1,2*]

1 European Regional Centre for Ecohydrology under the auspices of UNESCO, Łódź, Poland

2 University of Łódź, Department of Applied Ecology, Łódź, Poland

References

[1] Schecter A, Birnbaum L,. Ryan JJ, Constable JD, Dioxins. An overview. Environmental Research 2006, 101 419–428.

[2] Im SH, Kannan K, Matsuda M, Giesy JP, Wakimoto T. Sources and distribution of polychlorinated dibenzo-p-dioxins and dibenzofurans in sediment from Masay Bay, Korea. Environmental Toxicology and Chemistry 2002; 21 245-252.

[3]   Hilscherova K, Kannan K, Nakata H, Yamashita N, Bradley P, Maccabe J.M, Taylor
      A. B, Giesy J.P, Polychlorinated dibenzo-p-dioxin and dibenzofuran concentration
      profiles in sediments and flood-plain soils of the Tittabawssee River, Michigan. Envi-
      ronmental Science Technology 2003; 37 468-474.

[4]   Grochowalski A. Sources of dioxins and ways of their entering into the environment.
      Problems of waste combustion. I Symposium "Dioxin-man-environment"
      22-23.09.1994. Cracow University of Technology 1994.

[5]   Makles Z, Świątkowski A, Grybowska S. Hazardous dioxins. Arkady Publisher War-
      saw 2001.

[6]   Martinez D, Muller RK. Gifte in unsere. Hand, II AUFL. Urania Verlag Leipzig, Jen
      Berlin 1988.

[7]   Rappe C. Dioxin chemistry – on overview. Herbicides in war. The long-term ecologi-
      cal and human consequences. Westing A.H, SIPRI, Stockholm, Taylor and Francis.
      London, Philadelphia 1984.

[8]   Sokołowski M, Śliwakowski M. Sources of dioxin formation outside the combustion
      process. I National Symposium - Dioxin-Man-Environment. Cracow University of
      Technology. Krakow 22-23.09 1994.

[9]   Kołodziejak-Nieckuła E. Poison targeted Wiedza i życie. 2001; 6 (in Polish).

[10] Wasiela T, Tam I, Krajewski J, Tarkowski S. Environmental health risks, Dioxins.
      IMP, Łódź 1999.

[11] Novell LH, Capel PD, Dilenis PD. Pesticides in stream sediment and aquatic biota.
      Distribution, trends, and governing factors. Pesticides in the hydrologic system ser-
      ies. CRC Press, Boca Raton, FL. 1999.

[12] Brasner AMD, Wolff RH. Relationships between land use and organochlorine pesti-
      cides, PCBs, and semivolatile organic compounds in streambed sediment and fish on
      the Island of Oahu, Hawaii. Archives of Environmental Contamination Toxicology
      2004; 46, 385-398.

[13] Smith JA, Witkowski PJ, Fusillo TV. Manmade organic compounds in the surface wa-
      ters of the United States – a review of current understanding. U.S. Geological Survey
      Circular 1988; 1007 92.

[14] Lulek J. Polychlorinated biphenyls in Poland: history, fate, and occurrence. In: R.L.
      Lipnick, J.L.M. Hermens, K.C. Jones and D.C. Muir, (Eds), ACS Symposium Series
      772 2001; 85.

[15] Urbaniak M. Polychlorinated biphenyls: sources, distribution and transportation in
      the environment – a literature review, Acta Toxicologica 2007; 15(2) 83-93.

[16] WHO/EURO, PCBs, PCDDs, PCDFs: Prevention and control of accidental and environmental exposures. Environmental Health Series 23. Copenhagen: World Health Organization, Regional Office for Europe 1987.

[17] Bletchly JD. Polychlorinated biphenyls. Production, current use and possible rates of further disposal in OECD member countries. Barres MC, Koeman H, Visser R, [Eds.] Proceedings of PCB seminar. Amsterdam: Ministry of Housing, Physical Planning, and Environment 1984.

[18] Hansen LG. Environmental toxicology of polychlorinated biphenyls. Safe S, Hutzinger O, (Eds.) Environmental Toxin Series. New York: Springer-Verlag 1987; 15–48.

[19] De Vito M, Birnbaum LS 1994.Toxicology of dioxins and related chemicals. In: Schecter A, Dioxins and Health (Ed), New York: Plenum Press 1994; 139-162

[20] Parsons JR, Storms MCM. Biodegradation of chlorinated dibenzo-para-dioxins in batch and continuous cultures of strain JB1. Chemosphere 1989; 19 1297–1308.

[21] Wilkes H, Wittich RM, Timmis KN, Fortnagel P, Francke W. Degradation of chlorinated dibenzofurans and dibenzo-p-dioxins by Sphingomonas sp. strain RW1. Applied Environmental Microbiology 1996; 62 367–371.

[22] Schreiner G, Wiedmann T, Schimmel H, Ballschmiter K. Influence of the substitution pattern on the microbial degradation of mono- to tetrachlorinated dibenzo-p-dioxins and dibenzofurans. Chemosphere 1997; 34 1315–1331.

[23] Keim T, Francke W, Schmidt S, Fortnagel P. Catabolism of 2,7-dichloro- and 2,4,8-trichlorodibenzofuran by Sphingomonas sp. strain RW1. Journal of Industrial Microbiology and Biotechnology 1999; 23 359–363.

[24] Klecka GM, Gibson DT. Metabolism of dibenzo-para-dioxin and chlorinated dibenzo-para-dioxins by a *Beijerinckia* species. Applied Environmental Microbiology 1980; 39 288–296.

[25] Du XY, Zhu NK, Xia XJ, Bao ZC, Xu XB. Enhancement of biodegradability of polychlorinated dibenzo-p-dioxins. Journal Environmental Science Health Part A-Toxic/Hazard. Subst. Environmental Engineering 2001; 36 1589–1595.

[26] Kimura N, Urushigawa Y. Metabolism of dibenzo-p-dioxin and chlorinated dibenzo-p-dioxin by a gram-positive bacterium, *Rhodococcus opacus* SAO 101. Journal of Bioscience and Bioengineering 2001; 92 138–143.

[27] Habe H, Chung JS, Lee JH, Kasuga K, Yoshida T, Nojiri H, Omori T. Degradation of chlorinated bibenzofurans and dibenzo-p-dioxins by two types of bacteria having angular dioxygenases with different features. Applied Environment microbiology 2001; 67 3610-3617.

[28] Hong HB, Nam IH, Murugesan K, Kim YM, Chang YS. Biodegradation of dibenzo-p-dioxin, dibenzofuran, and chlorodibenzo-p-dioxins by Pseudomonas veronii PH-03. Biodegradation 2004; 15 303–313.

[29] Sulistyaningdyah WT, Ogawa J, Li QS, Shinkyo R, Sakaki T, Inouye K, Schmid RD, Shimizu S. Metabolism of polychlorinated dibenzo-p-dioxins by cytochrome P450BM-3 and its mutant. Journal of Biotechnological Letters 2004; 26 1857–1860.

[30] Habe H, Ashikawa Y, Saiki Y, Yoshida T, Nojiri H, Omori T. *Sphingomonas* sp. strain KA1, carrying a carbazole dioxygenase gene homologue, degrades chlorinated dibenzo-p-dioxins in soil. FEMS Microbiological Letters 2002; 211 43 49.

[31] Ahmed M, Focht DD. Degradation of polychlorinated biphenyls by two species of *Achromobacter*. Canadian Journal of Microbiology 1973; 19 47–52.

[32] Furukawa K, Tonomura K, Kamibayashi A. Effect of chlorine substitution on the biodegradability of polychlorinated biphenyls. Applied and Environmental Microbiology 1978; 35 223–7.

[33] Clark RR, Chian ESK, Griffin RA. Degradation of polychlorinated biphenyls by mixed microbial cultures. Applied and Environmental Microbiology 1979; 37 680–688.

[34] Novakova H, Vosahlikova M, Pazlarova J, Mackova M, Burkhard J, Demnerova K. PCB metabolism by *Peudomonas* sp. P2. *Intern Biodeterior Biodegrad.* 2002, 50, 47–54.

[35] Field JA, Sierra-Alvarez R Microbial degradation of chlorinated dioxins. Chemosphere 2008; 71 1005-1018.

[36] Arfmann HA, Timmis KN, Wittich RM. Mineralization of 4-chlorodibenzofuran by a consortium consisting of *Sphingomonas* sp. strain RW1 and *Burkholderia* sp. strain JWS. Applied Environmental Microbiology 1997; 63 3458–3462.

[37] Wittich RM, Strompl C, Moore ERB, Blasco R, Timmis KN. Interactions of sphingomonas and Pseudomonas strains in the degradation of chlorinated dibenzofurans. Journal of Industrial Microbiology and Biotechnology 1999; 23 353-358.

[38] Bumpus M, Tien D, Wright SD. Oxidation of persistent environmental-pollutants by a white root fungi. Science 1985; 228 1434–1436.

[39] Valli K, Wariishi H, Gold MH. Degradation of 2,7-dichlorodibenzo-para-dioxin by the lignin-degrading basidiomycete *Phanerochaete chrysosporium*. Journal of Bacteriology 1992; 174 2131–2137.

[40] Takada S, Nakamura M, Matsueda T, Kondo R, Sakai K. Degradation of polychlorinated dibenzo-p-dioxins and polychlorinated dibenzofurans by the white root fungus *Phanerochaete sordida* YK-624. Applied Environmental Microbiology 1996; 62 4323–4328.

[41] Adriaens P, Grabic-Galic D. Reductive dechlorination of PCDD/F by anaerobic cultures and sediments. Chemosphere 1994; 29 2253–2259.

[42] Adriaens P, Fu QZ. Grabic-Galic D. Bioavailability and transformation of highly chlorinated dibenzo-p-dioxins and dibenzofurans in anaerobic soils and sediments. Environmental Science and Technology 1995; 29 2252-2260.

[43] Bungie M, Ballerstedt H, Lechner U. Regiospecific dechlorination of spiked tetra- and trichlorodibenzo-p-dioxins by anaerobic bacteria from PCDD/F contaminated Spittelwasser sediments. Chemosphere 2001; 43 675-681.

[44] Bungie M, Adrian L, Kraus A, Opel M, Lorenz WG, Anderssen JR, Gorish H, Lechner U. Reductive dehalogenation of chlorinated dioxins by an anaerobic bacterium. Nature 2003; 421 357-360.

[45] Fennell DE, Nijenhuis I, Wilson SF, Zinder SH, Haggblom MM. *Dehalococcoids ethenogenes* strain 195 reductively dechlorinated diverse chlorinated aromatic pollutants. Environmental Science and Technology 2004; 38 2075-2081.

[46] Barkovskii AL, Adriaens P. Microbial dechlorination of historically present and freshly spiked chlorinated dioxins and diversity of dioxin-dechlorinating populations. Applied Environmental Microbiology 1996; 62 4556-4562.

[47] Barkovskii AL, Adriaens P. Impact of humic constituents on microbial dechlorination of polychlorinated dioxins. Environmental Toxicology and Chemistry 1998; 17 1013-1020.

[48] Master ER, Lai VW, Kuipers B, Cullen WR, Mohn WW. Sequential anaerobic-aerobic treatment of soil contaminated by weathered Aroclor 1260. Environmental Science and Technology 2002; 36 100–3.

[49] Boyle AW, Silvin CJ, Hassett JP, Nakas JP, Tanenbaum SW. Bacterial PCB biodegradation. Biodegradation 1992; 3 285–98.

[50] Fava F, di Gioia D, Cinti S, Marchetti 40. L, Quattroni G. Degradation and dechlorination of low-chlorinated biphenyls by a three-membered bacterial co-culture. Appl Microbiol Biotechnol 1994; 41 117–23.

[51] Bunge M, Lechner U. Anaerobic reductive dehalogenation of polychlorinated dioxins. Appl. Microbial Biotechnol. 2009; 84 429-444.

[52] Ruzo LO, Zabik MJ, Schuetz RD. Photochemistry of bioactive compounds: photoproducts and kinetics of polychlorinated biphenyls. Journal of Agricultural and Food Chemistry 1974; 22 199–202.

[53] Bunce NJ, Kumar Y, Brownlee BG. An assessment of the impact of solar degradation of polychlorinated biphenyls in the aquatic environment. Chemosphere 1978; 7 155–64.

[54] Zepp RG, Baughman GL, Scholtzhauer PF. Comparison of photochemical behavior of various humic substances in water: sunlight induced reactions of aquatic pollutants photosensitized by humic substances. Chemosphere 1981; 10 109–17.

[55] Buekens A, Huang H. Comparative evaluation of techniques for controlling the formation and emission of chlorinated dioxins/furans in municipal waste incineration. Journal of Hazardous Material 1999; 62 1-33.

[56] Kulkarni PS, Crespo JG, Afonso CAM. Dioxins sources and current remediation technologies – a review. Environmental International Journal 2008; 34 139-153.

[57] Balmer ME, Guss KU, Schwarzenbach RP. Photolysis transformation of organic pollutants on soil surfaces – an experimental approach. Environmental Science and Technology 2000; 34 1240-1245.

[58] Goncalves C, Dimou A, Sakkas V, Alpendurda MF, Albanis TA. Photolitic degradation of quinalphos in natural waters and on soil matrices under simulated solar irradiation. Chemosphere 2006; 64 1375-1382

[59] Doughtery EJ, McPeters AL, Overcash MR, Carbonell RG. Theoretical analysis of a method for in situ decontamination of soil containing 2,3,7,8-tetrachlorodibezno-p-dioxin. Environmental Science Technology 1993; 27 505-515.

[60] Van den Berg M, Birnbaum L, Denison M, Farland W. The 2005 World Health Organization reevaluation of human and mammalian toxic equivalency factors for dioxins and dioxin-like compounds. Toxicology Science 2006; 93 223–241.

[61] Macek T, Mackova M, Kas J. Exploitation of plants for the removal of organics in environmental remediation. Biotechnology Advances 2000; 18 23–34.

[62] Cunningham SD, Berti WR. Remediation of contaminated soils with green plants: an overview. In vitro Cell Development Biology 1993; 29 207-212. Cunningham SD, Berti WR, Huang JW. Phytoremediation of contaminated soils. Tibtech Journal 1995; 13 393-397.

[63] Cunningham SD, Anderson TA, Schwab AP, Hsu FC. Phytoremediation of soils contaminated with organic pollutants. Sparks DL (ed.) Advances in Agronomy, Academic Press San Diego Ca. 1996, 56 55-114

[64] Mackova M, Vrchotova B, Francova K, Sylvestre M, Tomaniova M, Lovecka P, Demnerova K, Macek M. Biotransformation of PCBs by plants and bacteria -consequences of plant-microbe interactions .European Journal of Soil Biology 2007; 43 233-241.

[65] Macek T, Mackova M, Kucerova P, Chroma L, Burkhard J, Demnerova K. Phytoremediation,. S.N. Agathos, W. Reineke (Eds.), Biotechnology for the Environment: Soil Remediation, Kluwer Academic Publishers, Brussels 115-137, 2002.

[66] Macek T, Francova K, Kochankova L, Lovecka P, Ryslava E, Rezek J, Sura M, Triska J, Demnerova K, Mackova M. Phytoremediation: biological cleaning of a polluted environment. Reviews on Environmental Health 2004; 19 63-82.

[67] Schnoor JL, Licht L.A, McCutcheon SC, Wolfe NL, Carreira LH, Phytoremediation of organic contaminants, Environmental Science and Technology 1995; 29 318- 323.

[68] Schnoor JL, Phytoremediation of Soil and Ground-water, GWRT Series, E-Series: TE-02-01 2002; 1-45.

[69] Salt DE, Smith RD, Raskin I. Phytoremediation. Ann Rev Plant Physiol Plant Mol Biol 1998; 49 643–68.

[70] Buczkowski R, Kondzielski I, Szymański T. Metody remediacji gleb zanieczyszczonych metalami ciężkimi. Uniwersytet Mikołaja Kopernika w Toruniu; 2002.

[71] Newman LA, Reynolds ChM. Phytodegradation of organic compounds. Current. Opinion in Microbiology 2004; 15 225-230.

[72] Gerhard KE, Huang X-D, Glick BR, Greenberg BM. Phytoremediation and rhizoremediation of organic soil contaminants: Potential and challenges. Plant Science 2009; 176 20-30.

[73] Siciliano SD, Germida JJ, Banks K, Greer CW. Changes in microbial community composition and function during a polyaromatic hydrocarbon phytoremediation field trial. Applied Environmental. Microbiology 2003; 69 483-489.

[74] Nedunuri KV, Govindaraju RS, Banks MK, Schwab AP, Chen Z. Evaluation of phytoteremediation for field-scale degradation of total petroleum hydrocarbons. Journal of Environmental Engineering 2000; 126 483-490.

[75] Robinson SL, Novak JT, Widdowsen MA, Crosswell SB, Fetterolf GJ. Field and laboratory evaluation of the impact of tall fescue on polyaromatic hydrocarbon degradation in aged creosote-contaminated surface oil. Journal of Environmental Engineering 2002; 129 232-240.

[76] White PM Jr, Wolf DC, Thoma GJ, Reynolds CM. Phytoremediation of alkylated polycyclic aromatic hydrocarbons in a crude oil-contaminated soil. Water Air Soil Pollution 2006; 169 207–220.

[77] Banks MK, Kulakow P, Schwab AP, Chen Z, Rathbone K. Degradation of crude oil in the rizosphere of sorghum bicolor. International Journal of Phytoremediation 2003; 5 225-234.

[78] Vervaeke P, Luyssaert S, Mertens J, Meers E, Tack FM, Lust N.. Phytoremediation prospects of willow stands on contaminated sediments: a field trial. Environmental Pollution 2003; 126 27-282.

[79] Pradhan SP, Conrad JR, Paterek JR, Srivastava VJ. Potential of phytoremediation for treatment of PAHs, In: Rainey PB Adaptation of Pseudomonas fluorescens to the plant rhizosphere. Environmental Microbiology 1999; 1 243-257.

[80] Gregor AW, Fletcher JS. The influence of increasing chlorine content on the accumulation and metabolism of polychlorinated biphenyls by Pau's Scarlet Rose cells. Plant Cell Response 1988; 7 329-332.

[81] Jou JJ, Chung JC, Weng YM, Liawc SL, Wang MK: Identification of dioxin and dioxin-like polychlorbiphenyls in plant tissues and contaminated soils. Journal of Hazardous Material 2007; 149174–179.

[82] Hülster A, Marschner H. Transfer of PCDD/PCDF from contaminated soils to food and fodder crop plants. Chemosphere 1993; 27 439–446.

[83] Hülster A, Mueller JF, Marschner H.. Soil–plant transfer of polychlorinated dibenzo-p-dioxins and dibenzofurans to vegetables of the cucumber family (Cucurbitaceae). Environmental Science and Technology 1994; 28 1110–1115.

[84] Engwall M, Hjelm K. Uptake of dioxin-like compounds from sewage sludge into various plant species – assessment of levels using a sensitive bioassay. Chemosphere 2000; 40 1189–1195.

[85] White JC, Parrish ZD, Isleyen M, Gent MP, Iannucci-Berger W, Eitzer BD, Kelsey JW, Mattina MI. Influence of citric acid amendments on the availability of weathered PCBs to plant and earthworm species. International Journal of Phytoremediation 2005; 8 63–79.

[86] Inui H, Wakai T, Gion K, Kim YS, Eun H. Differential uptake for dioxin-like compounds by zucchini subspecies. Chemosphere 2008; 73 1602–1607.

[87] Uegaki R, Seike N, Otani T. Polychlorinated dibenzo-p-dioxins, dibenzofurans and dioxin-like polychlorinated biphenyls in rice plants: possible contaminated pathways. Chemosphere 2006; 65 1537–1543.

[88] Whipps JM. Carbon economy. Lynch JM. (ed.) The rhizosphere. Wiley, New York, 1990; p59–97.

[89] Rainey PB. Adaptation of Pseudomonas fluorescens to the plant rhizosphere. Environmental Microbiology 1999; 1 243-257.

[90] Lugtenberg BJJ, Dekkers L, Bloemberg GV. Molecular determinants of rhizosphere colonization by Pseudomonas. Annual Review Phytopathology 2001; 39 461–490

[91] Gianfreda L, Rao MA, Potential of extra cellular enzymes in remediation of polluted soils: a review. Enzyme Microbiology Technology Journal 2004; 35 339-354.

[92] Liu L, Jiang C-Y, Liu X-Y, Wu J-F, Han J-G, Liu S-J. Plant–microbe association for rhizoremediation of chloronitroaromatic pollutants with Comamonas sp. strain CNB-1. Environmental Microbiology 2007; 9 465–473.

[93] Dams RI, Paton GI, Killham K. Rhizoremediation of pentachlorophenol by Sphino-gobium chlorophenolicum ATCC 39723. Chemosphere 2007; 68 864-870.

[94] Macek T, Mackova M, Brkhar J, Demnerova K. Introduction of green plants for the control of metals andorganics I environmental remediation. Holm FW, (ed) Effluents from alternative demilitaryzation technologies. NATO PS Series 1998; 71-85.

[95] Lamoureux GL, Flear DS. Pesticide metabolism in higher plants: In vitro enzyme studies. Paulson GD, Frear DS, Marks EP (eds.). Xenobiotic metabolism. In vitro methods. American chemical Society Symposium Series, 97, Washington DC, ASC 1979; 263-266.

[96] Susarla S, Medina VF, McCutcheon SC. Phytoremediation: an ecological solution to organic chemical contamination. Ecological Engineering 2002; 18 647–658.

[97] Singer AC. The chemical ecology of pollutant biodegradation. Bioremediation and phytoremediation from mechanistic and ecological perspectives Mackova M, Dowling D, Macek T. (eds). Phytoremediation and rhizoremediation. Theoretical background. focus on biotechnology Springer, Dordrecht, 2004; 5-21.

[98] Kuiper I, Lagendijk EL, Bloemberg GV, Lugtenberg BJJ. Rhizoremediation: a beneficial plant–microbe interaction. Molecular Plant Microbe Interactions 2004; 17 6–15.

[99] Chaudhry Q, Blom-Zandstra M, Gupta S, Joner EJ. Utilizing the synergy between plants and rhizosphere microorganisms to enhance breakdown of organic pollutants in the environment. Environmental Science Pollution Researches 2005; 12 34–48.

[100] Yateem A., Al-Sharrah T., Bin-Haji A. Investigation of microbes in the rhizosphere of selected grasses for rhizoremediation of hydrocarbon-contaminated soils. Soil Sed. Contam, 2007; 16 269–280.

[101] Shimp JF, Tracy JC, Davis LC, Lee E, Huang W, Erickson LE, Schnoor JL. Beneficial effects of plants in the remediation of soil and groundwater contaminated with organic materials. Critical Reviews Environmental Science and Technology 1993; 23 41–77.

[102] Leigh MB, Fletcher JS, Fu X, Schmitz FJ. Root turnover: an important source of microbial substrates in rhizosphere remediation of recalcitrant contaminants. Environmental Science and Technology 2002; 36 1579–1583.

[103] Yateem A, Al-Sharrah T, Bin-Haji A. Investigation of microbes in the rhizosphere of selected grasses for rhizoremediation of hydrocarbon-contaminated soils. Soil and Sedimentation Contamination 2007; 16 269–280.

[104] Holden PA, Firestone MK. Soil microorganisms in soil cleanup: how can we improve our understanding? Journal Environmental Quality 1997; 26 32-40.

[105] Ferro AM, Rock SA, Kennedy J, Herrick JJ, Turner DL. Phytoremediation of soils con-
       taminated with wood preservatives: greenhouse and field evaluations. International
       Journal Phytoremediation 1999; 1 289–306.

[106] Thoma GJ, Lam TB, Wolf DC. A mathematical model of phytoremediation for petro-
       leum contaminated soil: sensitivity analysis. International Journal of Phytoremeda-
       tion 2003; 5 125–136.

[107] Pillai BVS, Swarup S. Elucidation of the flavonoid catabolism pathway in *Pseudomo-
       nas putida* PML2 by comparative metabolic profiling. Appl Environ Microbiol. 2002;
       68143–151.

[108] Leigh MB, Prouzova P, Mackova M, Macek T, Nagle DP, Fletcher JS. Polychlorinated
       biphenyl (PCB)-degrading bacteria associated with trees in a PCB contaminated site.
       Applied Environmental Microbiology 2006; 72 2331–2342.

[109] Mackova M, Macek T, Ocenaskova J, Burkhard J, Demnerova K, Pazlarova J. Selec-
       tion of the potential plant degraders of PCB. Chemické Listy 1996; 90 712–3.

[110] Mackova M, Macek T, Kucerova P, Burkhard J, Tiska J, Demnerova K. Plant tissue
       cultures in model studies of transformation of polychlorinated biphenyls. Chemical
       Papers 1998; 52 599–600.

[111] Nichols TD, Wolf DC, Rogers HB, Beyrouty CA, Reynolds CM. Rhizosphere microbi-
       al populations in contaminated soils. Water, air, Soil Pollution 1997; 95:165-178.

[112] Betts KS. TPH soil cleanup aided by ground cover. Environmental Science Technolo-
       gy 1997; 31 214A.

[113] Burken JG, Schnoor JL. Phytoremediation: plant uptake of atrazine and role of root
       exudates. Journal Environmental Engineering 1996; 122 968-963.

[114] Donelly PK, Fletcher JS. Potential use of fungi as bioremediation agents. In: Ander-
       son TA. (ed.) Bioremediation through rhizosphere technology. ACS Symposium Ser-
       ies no. 563. American Chemical Society 1994; 93-99.

[115] Fletcher JS, Donnelly PK, Hegde RS. Biostimulation of PCB-degrading bacteria by
       compounds released from plant toots. Hinche RE, Anderson DB, Hoeppel RE. (eds.)
       Bioremediation of recalcitrant organics. Battelle Press. Columbus 1995; 131-136.

[116] Mackova M, Pruzova P, Stursa P, Ryslava E, Uhlik O, Beranova K, Rezek J, Kurzawo-
       va V, Demnerova K, Macek T. Phyto/rhizoremediation studies using long-term PCB-
       contaminated soil. Environ. Sci. Pollut. Res. 2009; 16 817-829.

[117] Zalewski M. Ecohydrology for implementation of the Water Framework Directive.
       Water Management 2011; 164 WM1 1-12.

[118] Zalewski M, Janauer, GA, Jolankaj G. Ecohydrology: a new paradigm for the sustain-
       able use of aquatic resources. Conceptual Background, Working Hypothesis, Ration-

ale and Scientific Guidelines for the Implementation of the IHP-V Projects 2.3:2.4. UNESCO, Paris Technical Documents in Hydrology 1997; 7.

[119] Zalewski M., editor. Guidelines for the Integrated Management of the Watershed-Phytotechnotogy and Ecohydrology. UNEP/UNESCO. UNEP IETC Freshwater Management. Series No 5 2002

[120] Zalewski M, Wagner-Lotkowska I, Robarts RD. Integrated Watershed Management - Ecohydrology and Phytotechnology - Manual. Venice Osaka, Shiga, Warsaw, Lodz 2004.

[121] Zalewski M. Ecohydrology for compensation of global change. Brazilian Journal of Biology 2010; 70(3) 689-695.

[122] Zalewski M, Bis B, Łapinska M, Frankiewicz P, Puchalski W. The importance of the riparian ecotone and river hydraulics for sustainable basin-scale restoration scenarios. Aquatic Conservation: Marine and Freshwater Ecosystems 1998; 8 287-307.

[123] Zalewski M. Ecohydrology - The scientific background to use ecosystem properties as management tools toward sustainability of water resources. Guest Editorial, Ecological Engineering 2000; 16 1-8.

[124] Major DW, McMaster ML, Cox EE, Edwards EA, Dworatzek SM, Hendrokson ER, Starr MG, Payne JA, Buonamici lW. Field demonstration of successful bioaugmentation to achieve dechloriantion of tetrachloroethane to ethane. Environmental Science and Technology 2002; 36 5106-5116.

[125] Brunner W, Sutherland FH, Focht DD. Enhanced biodegradation of polychlorinated biphenyls in soil by analogue enrichment and bacterial inoculation. Journal Environmental Quality 1985; 14 324–328

[126] Yeong SW. Evaluation of the use of capillary numbers for quantifying he removal of DNAPL trapped in a porous medium by surfactant and surfactant foam floods. Journal of Colloid Interface Science 2005; 282 182-187

[127] Johnson DN, Pedit JA, Miller CT. Efficient near-complete removal of DNAPL from three-dimensional, heterogeneous porous media using a novel combination of treatment technologies. Environmental Science and Technology 2004; 38 5149-5156.

[128] West CC, Harwell JH. Surfactants and subsurface remediation. Environmental Science and Technology 1992; 26 2324-2330.

[129] Nakajima F, Baun A, Ledin A, Mikkelsen PS. A novel method for evaluating bioavailability of polycyclic aromatic hydrocarbons in sediments of an urban stream. Water Science and Technology 2005; 51275-281.

[130] Tam NFY, Wong YS. Efectivness of bacterial inoculums and mangrove plants on remediation of sediment contaminated with polycyclic aromatic hydrocarbons. Marine Pollution Bulletin 2008; 57 716-726.

[131] Doty SL, James CA, Moore AL, Vajzovic A, Singleton GL. Ma C, Khan Z, Xin Shang TQ, Wilson AM, Tangen J, Westergeen AD, Newman LA, Strand SE, Gordon MP. Enhanced metabolism of halogenated hydrocarbons in transgenic plants containing mammalian cytochrome P450 2E1. PNAS 97, 2000; 6287-6291.

[132] Dua M, Singh A, Sethunathan N, Johri A. Biotechnology and bioremediation: successes and limitations. Applied Microbiology Biotechnology 2002; 59143 -152.

[133] Lovely DR, Cleaning up with genomics: applying molecular biology to bioremediation. Nat. Rev 2003; 1 35-44.

[134] Kawahigashi H. Hirose S. Ohkawa H. Ohkawa Y.Transgenic rice plants expressing human CYP1A1 exude herbicide metabolites from their roots. Plant Science 2003, 165 373–381.

[135] Cherian S, Oliveira MM. Transgenic plants in phytoremediation: recent advances and new possibilities. Environmental Science and Technology 2005; 39 9377-9390.

[136] Kawahigashi H, Hirose S, Ohkawa H, Ohkawa Y. Phytoremediation of the herbicides atrazine and metolachlor by transgenic rice plants expressing human CYP1A1, CYP2B6 and CYP2C19. Journal of Agricultural and Food Chemistry 2006; 54 2985–2991.

[137] Lan Chun Ch, Payne RB, Sowers KR, May HD. Electrical stimulation of microbial PCB degradation in sediment. Water Research Journal 2013; 24 141-151.

[138] Mitsch WJ, Gosselink JG. Wetlands, 4th edn. John Wiley & Sons, New York 2007.

[139] Brix H. Use of Constructed Wetlands In Water Pollution Control: Historical Development, Present Status and Future Perspectives. Water Science and Technology Journal 1994; 30(8) 209-223.

[140] Zhang D, Richard MG and Tan SK. Constructed wetlands in China. Ecological Engineering 2009; 35 1367–1378.

[141] Mitsch WJ, Jorgensen SE. Ecological Engineering and Ecosystem Restoration. John Wiley & Sons, Inc, New York 2004

[142] Kadlec RH, Knight RL.. Treatment Wetlands. Boca Raton (USA). Lewis. 1996

[143] Chen H. Surface-Flow Constructed Treatment Wetlands for Pollutant Removal: Applications and Perspectives. Wetlands 2011; 31 805–814.

[144] Cheng S, Grosse W, Karrenbrock F, Thoennessen M. Efficiency of constructed wetlands in decontamination of water polluted by heavy metals. Ecological Engineering 2002; 18 317–325.

[145] Cronk JK. Constructed wetlands to treat wastewater from dairy and swine operations: a review. Agriculture, Ecosystems and Environment 1996; 58 97-114.

[146] Fenta BG. Constructed Wetland System for Domestic Wastewater Treatment: A Case Study in Addis Ababa, Ethiopia A thesis submitted to the School of Graduate Studies

of the Addis Ababa University in Partial Fulfilment of the Requirements for the Degree of Master of Science in Environmental Science 2007

[147] Koskiaho J, Ekholm P, Raty M. , Riihimaki J. Puustinen M. Retaining agricultural nutrients in constructed wetlands experiences under boreal conditions. Ecological Engineering 2003; 20 89- 103.

[148] Koukia S, M'hirib F., Saidia N, Belaïd S, Hassen A. Performances of a constructed wetland treating domestic wastewaters during a macrophytes life cycle. Ecological Engineering 2000; 15 77–90.

[149] Cronk JK. Constructed wetlands to treat wastewater from dairy and swine operations: a review. Agriculture, Ecosystems and Environment 1996; 58 97-114.

[150] Schulz R, Peall SKC. Effectiveness of a Constructed Wetland for Retention of Nonpoint-Source Pesticide Pollution in the Lourens River Catchment, South Africa. Environmental Science and Technology 2001; 35 422-426.

[151] Scholz M, Lee B-H. 2008. Constructed wetlands: a review. International Jour Environ Stud 2005; 62(4) 421–47.

[152] Schulz R. Field studies on exposure, effects and risk mitigation of aquatic nonpoint-source insecticide pollution: a review. Journal Environmental Quality 2004; 33 419–48.

[153] Destandau F, Martin E. and Rozan A. Potential of artificial wetlands for removing pesticides from water in a cost-effective framework, Working Paper No 5. 2011.

[154] Budd R, O'Geen A, Goh KS, Gan J, Efficacy Of Constructed Wetlands In Pesticide Removal From Tailwaters In The Central Valley, Kalifornia. Environmental Science and Technology 2009; 43 2925–2930.

[155] Tournebize J.E, Passeport C, Chaumont C. Fesneau A, Guenne, B. Pesticide decontamination of surface waters as a wetland ecosystem service in agricultural landscapes. Ecological Engineering 2012.

[156] Naiman, RJ & Decamps, H. (eds.) The Ecology and Management of Aquatic–Terrestrial Ecotones. UNESCO, MAB, Parthenon, Paris 1990.

[157] Schiemer, F, Zalewski, M. & Thorpe, JE (Eds) The Importance of Aquatic–Terrestrial Ecotones for Freshwater Fish. Developments in Hydrobiology, 105.Kluwer Academic Publisher, Dordrecht, Boston, London 1995.

[158] Urbaniak M, Zerihun Negussie Y, Zalewski, M, The ecohydrological biotechnology (SBFS) for reduction of dioxin-induced toxicity in Asella lake, Ethiopia. Geophysical Research Abstracts 2012; 14 EGU2012-14431-1.

Biodegradation and Anaerobic Digestion

Chapter 6

Challenges for Cost-Effective Microalgae Anaerobic Digestion

Álvaro Torres, Fernando G. Fermoso,
Bárbara Rincón, Jan Bartacek, Rafael Borja and
David Jeison

Additional information is available at the end of the chapter

1. Introduction

Microalgae, the common denomination for a broad group of photosynthetic prokaryotes and eukaryotes, are characterized by an efficient conversion of the solar energy into biomass. They are a promising feedstock for the production of third generation biofuels for several reasons:

1. Microalgae photosynthesis allows biological CO_2 fixation, which is expected to mitigate atmospheric CO_2 increase (Amin 2009; Brennan & Owende 2010; Mutanda et al. 2011).

2. Microalgae are 10 – 50 times more efficient than plants in terms of CO_2 fixation (Wang et al. 2008). Thus, microalgae can fix 1.83 tonnes of CO_2 per 1 tonne of produced microalgae (Chisti 2007).

3. Microalgae can be produced on non-arable areas such as lakes, oceans or deserts, thus reducing competition with food production (Mussgnug et al. 2010; Stephens et al. 2010). This advantage is a key factor when energy supply is considered in desert zones near oceans.

4. Some microalgae can grow under saline conditions, which strengthen the use of microalgae as feedstock for biofuel production in desert zones near the ocean when freshwater supply is not feasible.

Most of current efforts to take advantage of microalgae as a source of bioenergy are directed to biodiesel production, considering the ability of certain types of microalgae to accumulate lipids under controlled culture conditions. Microalgae biodiesel produced from microalgae lipids also presents technical advantages compared to lignocellulosic biomass based biodiesel.

Biodiesel from microalgae has a higher calorific value (30 and 29 MJ/kg for *C. protothecoides* and *Microcystis aeruginose*, respectively) and lower viscosity and density than plants-based biodiesel (Costa & de Morais 2011). However, the biodiesel yield from algae is rather low compared to biodiesel from lignocellulose energy (Chisti 2007; Sialve *et al.* 2009; Scott *et al.* 2010; Stephens *et al.* 2010). Indeed, with current technology, a negative energy balance was calculated by Lardon *et al.* (2009) when evaluating biodiesel production from *C. vulgaris*, considering biomass drying and further lipid extraction by solvents. During biodiesel production from microalgae, energy consumption associated with culture mixing and pumping, lipid extraction, nutrients addition, drying is of particular importance (Scott *et al.* 2010). Indeed, Lardon *et al.* (2009) estimated that the necessary energy consumption for drying was near 85% of the total energy consumption in a biodiesel production process from microalgae. Another drawback of biodiesel process is associated with the microalgae cultivation step, as nutrient requirements are 55-111 times higher than for e.g. rapeseed cultivation (Halleux *et al.* 2008). Under these conditions, biodiesel production from microalgae may not be energetically and environmentally sustainable (Sialve *et al.* 2009; Ras *et al.* 2011).

2. Microalgae as a source of biogas

Biogas production through anaerobic digestion is an established technology where a wide variety of residues can be used as substrate. In 2011, 8,760 anaerobic digesters were reported in Europe (IEA, 2011). The contribution of this technology to the reduction of carbon emissions, green energy and green gas policies has generated intense interest, especially over the past decade.

When considering biogas production from microalgae two alternatives can be conceived: Microalgae biodiesel production and further anaerobic digestion of microalgae residues for biogas production (Process 1, Figure 1A) and anaerobic digestion of whole microalgae with biogas as sole biofuel (Process 2,Figure 1B).

Process 1: Biodiesel production and subsequent biogas production from spent microalgae. Two principal drawbacks are identified when biodiesel production from microalgae is considered: high nutrients requirements for microalgae growth and low energy efficiency of biodiesel production process. Anaerobic digestion may contribute to overcome such limitations, by enabling nutrients recovery and biogas production when spent microalgae after lipid extraction is used as substrate. This is based on the fact that biogas can be used as source of renewable energy and that during anaerobic digestion process, nitrogen and phosphorus may be recovered, creating opportunities for their reuse as nutrients. Theoretical energy contribution of anaerobic digestion is presented in Figure 1A, assuming microalgae content of lipids, proteins and carbohydrates to be 30, 45 and 25%, respectively.

Figure 1A shows that an energy yield of 11MJ per kilogram of gross microalgae is reached when biodiesel production is considered. If oil extracted microalgae is used as substrate in anaerobic digestion process, methane produced would have a maximum theoretical contribution of 17MJ per kilogram of gross microalgae (thermal). Such value has been computed

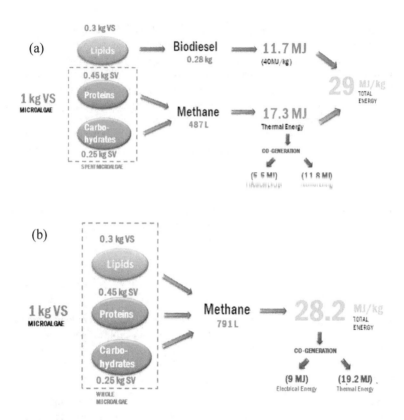

Figure 1. Energy potential of microalgae considering: a) Biodiesel production and further anaerobic digestion of microalgae residues for biogas production or b) Anaerobic digestion of whole microalgae only for biogas production.

assuming carbohydrate and protein methanogenic potentials of 0.415 and 0.851 L CH_4/kg VS, respectively (Angelidaki & Sanders 2004). If the latter thermal energy is transformed into electricity, a maximum energy yield of 5.5 MJ per kilogram of gross microalgae would be achieved (assuming a conversion efficiency of 32%). Thus, a substantial increase in energy yield could be theoretically achieved, representing a considerable contribution to biodiesel sustainability and economic feasibility. Energy contained in biogas can be used for both anaerobic digestion and trans-esterification reactor heating. Electricity obtained via co-generation can be used for different purposes such as photobioreactor mixing, microalgae harvesting and drying (Harun *et al.* 2010; Razon & Tan 2011). Neumann et al. (2011) evaluated energy contribution of biogas production in Process 1 for *Botryococcus braunii* with 30% lipid content. The latter study considered a nutrient recovery step through membrane liquid/solid separation from anaerobic digestion reactor and heptane evaporating step in order to recovery this solvent. Biogas production could theoretically contribute with close to 50% of the overall energy yield of Process 1.

Process 2: Biogas production from whole microalgae. Another alternative to recover energy from microalgae consists of methane production from whole microalgae. In such process, all organic matter (proteins, carbohydrates and lipids) present in microalgae biomass would be converted into methane and carbon dioxide, without considering biodiesel production (De Schamphelaire & Verstraete 2009; Douskova *et al.* 2010; Zamalloa *et al.* 2011). Several advantages are recognized when energy production from whole microalgae through biogas generation is considered: Biogas productions involves high energy yields, biogas production would not require microalgae biomass drying (it involves wet fermentation), biogas can be used to produce heat and electricity through co-generation, microalgae cultures can be used for biogas upgrading (i.e. CO_2 biosequestration), microalgae species not capable of accumulating lipids may be also used as feedstock. Moreover, co-digestion with other types of biomass such as solid or liquid wastes is feasible. Anaerobic digestion of algal and microalgae biomass has been previously studied by some researches (Vergara-Fernández *et al.* 2008; De Schamphelaire & Verstraete 2009; Mussgnug *et al.* 2010; Zamalloa *et al.* 2011). Figure 1B shows the energy potential of Process 2, in which whole microalgae is used as substrate in order to produce biogas. In this estimation, all energy is produced as methane, which allows theoretical maximum energy recovery of 27 MJ per kg of volatile solids of microalgae (8.6MJ of electricity and 18.4 MJ of heat, if co-generation is considered). The lower operational energy demands for biogas production, compared with biodiesel together with biogas, makes Process 2 very promising for energy recovery.

3. Anaerobic digestion of microalgae

Reports of the anaerobic digestion of microalgae go back to the fifties when Golueke *et al.* (1957) was one of the first authors studying the feasibility of sunlight energy conversion to methane by algae sunlight fixation followed by biomass anaerobic fermentation. In this early study, 0.5 m³ of biogas was obtained per volatile kg of algal biomass, with methane content 63%. More than two decades later, Nair *et al.* (1983) reported a lower yield, close to 0.22 m³/kg VSS, at loading rate 1.7 kg/(m³ d). Despite those early reports, biogas production from algae and microalgae has not yet widely researched (Foree & McCarty 1970; Samson & Leduy 1983; Tarwadi & Chauhan 1987; Vergara-Fernández *et al.* 2008; De Schamphelaire & Verstraete 2009; Mussgnug *et al.* 2010; Zamalloa *et al.* 2011).

3.1. Choosing microalgal culture for direct biogas production

The ideal microalgae specie for a maximum biogas production would that presenting:

1.   thin or no cell wall

2.   large cells

3.   high growth rate in non-sterile media

4.   high resistivity against natural contaminants

5.   carbohydrate-based cell wall.

Of the above mentioned factors, the quality of cell wall is crucial for anaerobic digestion of algae. This is because cell walls are hard to degrade biologically and their presence avoids contact of anaerobic bacteria with the readily degradable content of algal cells. Therefore, the influence of cell wall presence is described in detail in the following text.

3.1.1. Composition of algal cell wall

Cell wall in microalgae represents 12-36% of total cell mass (cell wall weight/cell weight) in different microalgae (Table 1). Microalgae cell wall is composed mainly of carbohydrates and proteins which represent 30-76% and 1-37% of cell wall, respectively.

Microalgae	Cell Wall	Cell Wall composition (%)			References
	(% w/w)	Carbohydrates	Protein	n.d.*	
Chlorella vulgaris (F)	20.0	30.00	2.46	67.54	(Abo-Shady et al. 1993)
Chlorella vulgaris (S)	26.0	35.00	1.73	63.27	(Abo-Shady et al. 1993)
Kirchneriella lunaris	23.0	75.00	3.96	21.04	(Abo-Shady et al. 1993)
Klebsormidium flaccidum	36.7	38.00	22.60	39.40	(Domozych et al. 1980)
Ulothrix belkae	25.0	39.00	24.00	37.00	(Domozych et al. 1980)
Pleurastrum terrestre	41.0	31.50	37.30	31.20	(Domozych et al. 1980)
Pseudendoclonium basiliense	12.8	30.00	20.00	50.00	(Domozych et al. 1980)
Chlorella Saccharophila	-	54.00	1.70	44,30	(Blumreisinger et al. 1983)
Chlorella fusca	-	68.00	11.00	20.00	(Blumreisinger et al. 1983)
Chlorella fusca	-	80.00	7.00	13.00	(Loos & Meindl 1982)
Monoraphidium braunii	-	47.00	16.00	37.00	(Blumreisinger et al. 1983)
Ankistrodesmus densus	-	32.00	14.00	54.00	(Blumreisinger et al. 1983)
Scenedesmus obliquos	-	39.00	15.00	46.00	(Blumreisinger et al. 1983)

* not determined.

Table 1. Cell wall composition of microalgae.

Other compounds found in microalgal cell wall are uronic acid, glucosamine, hidroxyproline, proline, sporopollenin, carotenoids and another resistant biopolymers (Punnett & Derrenbacker 1966; Domozych et al. 1980; Blumreisinger et al. 1983; Brown 1991, 1992; Abo-Shady et al. 1993).

In relation to carbohydrates in microalgae cell wall, neutral sugars, cellulose and hemicelluloses are the main components. Blumreisinger et al. (1983) studied five different microalgae in relation to carbohydrate composition in cell wall, obtaining a prominent neutral sugar component. Composition of cellulose and hemicelluloses has ranged between 6-17% and 18-32% for microalgae studied in other researches carried out by Abo-Shady *et al.* (1993) and Domozych *et al.* (1980), respectively. On the other hand, Northcote *et al.* (1958) reported contents of cellulose near to 45% in cell wall of *Chlorella pirenoidosa*. Unlike these researches, Loos and Meindl (1982) found no presence of cellulose in cell wall of *Clhorella fusca*. In relation to proteins, peptides, proline and hidroxyproline are the main components. According to Punnett and Derrenbacker (1966), the cell wall of six different microalgae consisted of peptides (simple amino acid composition) but it contained no protein. In addition, this research revealed the existence of proline in the cell wall of *Chlorella vulgaris* and hidroxyproline in the cell wall of *Chlorella pyrenoidosa* and *Scenedesmus obliquos*.

3.1.2. Degradability of algal cell wall

Although methane yield is dependent on microalgae composition (Sialve *et al.* 2009), the resistance of cell wall is considered to be the limiting factor for the anaerobic digestion of microalgae (Afi *et al.* 1996; Chen & Oswald 1998). The kinetics of anaerobic digestion is highly dependent on the degradability of the given microalgae species (Sialve *et al.* 2009). Mussgnug *et al.* (2010) studied the methane production from six different microalgae, obtaining from 287 to 587 mL CH_4/ g VS. The low levels of methane yield were related to low cell degradation and high amount of indigestible residues. According to these results, easily degradable microalgae had no cell wall or a protein-based cell wall not containing cellulose/hemicellulose. Batch tests with low methane yields, intact cell walls of microalgae were found with light microscopy in this study. Thus, the intracellular content was not available for efficient digestion. The presence of biopolymers resistant to anaerobic degradation has been reported in the outer cell wall of microalgae species such as *Botryococcus braunii* (Templier *et al.* 1992; Banerjee *et al.* 2002). Moreover, microalgae degradability is related to cell wall structures containing these resistant biopolymers. Some microalgae have a protective tri laminar outer wall called tri laminar sheath (*TLS*), which hinders efficient microalgae degradation (Derenne *et al.* 1992). Thus, higher *TLS* resistance to degradation reported by Derenne *et al.* (1992) for microalgae *B. braunii* has been associated to the presence of sporopollenin-like biopolymers (Kadouri *et al.* 1988; Derenne *et al.* 1992). Other indigestible compound found in microalgae cell wall is algaenan, which has been reported as non-hydrolysable resistant biopolymer composed of polyether linked long-chain (up to C36) n-alkyl units (Gelin *et al.* 1997; Blokker *et al.* 1998; Gelin *et al.* 1999; Simpson *et al.* 2003).

3.1.3. Source of methane in algae

Many authors have related methane yield from microalgae to their composition (Sialve *et al.* 2009; Mairet *et al.* 2011; González-Fernández *et al.* 2012; Mairet *et al.* 2012), especially with the content of lipids, carbohydrates and proteins. However, the experimental data collected from literature do not show strong correlation between lipids, carbohydrates and proteins found in various algal species and the methane yield obtained by various authors (Fig. 2).

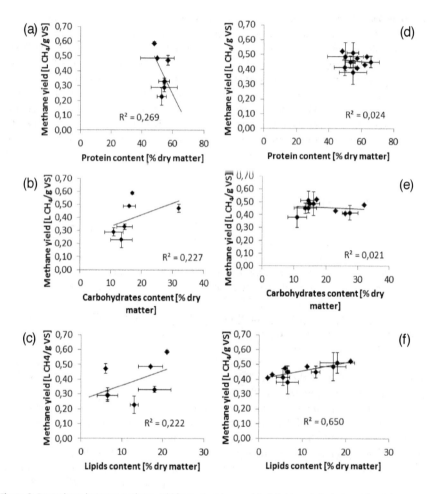

Figure 2. Dependence between methane yield from microalgae and their lipids, carbohydrates and proteins content. Each data point represents one algae species while the error bars show the range found in the literature. Figures (a), (b) and (c) show experimentally obtained methane yields, figures (d), (e) and (f) represent theoretical methane yield for the given algae composition calculated according to Angelidaki and Sanders (2004). Data were extracted from multiple authors (Becker 2007; Griffiths & Harrison 2009; Sialve et al. 2009; Mairet et al. 2011; González-Fernández et al. 2012; Mairet et al. 2012).

Angelidaki and Sanders (2004) presented theoretical methane yields from proteins, carbohydrates and lipids of 0.50, 0.42 and 1.01 L/g VS, respectively (Fig. 3). Even when these values are used for calculation of the potential methane yield from various algal species, no strong correlation can be found (Fig. 2d, e and f). Theoretically, lipids content has the biggest influence on methane yield, but as lipids are usually not the mayor source of methane (Fig. 2), the correlation between lipids content and methane yield is still rather vague (Fig. 2).

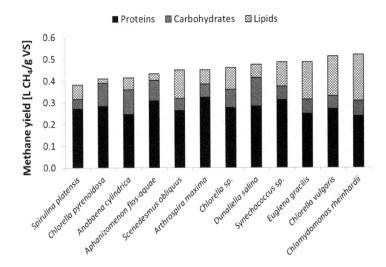

Figure 3. Potential methane yield from proteins, carbohydrates and lipids present in various algae species calculated according to Angelidaki and Sanders (2004). The data on proteins, carbohydrates and lipids content in algae were extracted from Becke (2007), Sialve (2009), Griffiths and Harrison (2009) and González-Fernández et al. (2012).

These facts clearly show that the ration between various macromolecules is not the most important parameter determining the actual methane yield from algae. As it was mentioned before, content of inert organic matter (e.g. cell wall) would play more important role (González-Fernández *et al.* 2012).

These findings show that plain composition of algal biomass indeed cannot be the main factor while choosing the best algal strain for methane production. Biomass production rate and the content of cell-walls will be of higher importance. Moreover, environmental conditions such as the salinity of available water source must be taken into account.

3.2. Pretreatment

In order to overcome limitation caused by cell wall degradability, which is necessary to access the intracellular content, cell disruption (pretreatment) has been pointed out as an important contributor in order to enhance anaerobic digestion efficiency. As mentioned above, cell wall degradability affects both Processes 1 and 2. However, in Process 1, cell wall degradability should not be as critical as in Process 2 since lipid extraction itself may be considered a pretreatment step.

There are different pretreatment techniques applied to microalgae, which can be classified as enzymatic, chemical and mechanical treatments. Mechanical pretreatment include autoclaving, homogenizers, microwaves and sonication, which increases the availability of organic matter (Angelidaki & Ahring 2000). Chemical pretreatment will increase availability of compounds resistant to anaerobic hydrolysis due to the enhanced disintegration (Bonmatí et al. 2001).

Chemical pretreatment can be clasified as acid or alkaline treatment. An increase in soluble hemicellulose present in cell wall is expected when alkaline pre-treatment is used (Abo-Shady et al. 1993). Thus, chemical pre-treatment is suitable when microalgae cell wall is rich on hemicelluloses. Also, enzymatic pretreatment has been used in order to attack cell wall and improve compounds extraction from microalgae. Enzymatic pretreatment with α-amilase, amylo-glucosidase and cellulase have shown a positive effect on cell wall hydrolysis (Choi et al. 2010; Fu et al. 2010). Fu et al. (2010) reported a 62% increase in cell wall hydrolysis, when *Chlorella sp.* was pretreated by immobilized cellulase.

Few studies report the effect of cell disruption pretreatment in anaerobic digestion (Samson & Leduy 1983; Chen & Oswald 1998). Samson and Leduy (1983) reported an increase of 78% in soluble COD when algae *Spirulina maxima* was mechanically pretreated (sonication and mechanical disintegration). However, no increase in methane yield was observed.

Finally, two considerations should be taken into account when cell disruption pretreatment is evaluated in the context of anaerobic digestion: On one hand, energy consumption associated with pretreatment should be low in order to avoid a negative contribution to the energy balance of anaerobic digestion process. On the other hand, contribution to the biodegradability of the given substrate should be a response variable when the effect of pretreatment on anaerobic digestion is evaluated. In other words, some pretreatment techniques increase solubility of organic matter but do not increase its biodegradability.

3.3. Inhibiting factors related to anaerobic digestion

Figure 1B shows the energy potential when microalgae are used as substrate in order to produce biogas. In this estimation, total energy is produced as methane, which allows a theoretical maximum energy recovery of 27MJ per kg of volatile solids of microalgae. As in Process 1, part of energy produced will be spent for supplying the energy necessary for microalgae harvesting and up-concentration, photobioreactor mixing, photobioreactor and anaerobic reactor heating, etc. The theoretical estimations of energy production from anaerobic digestion presented in this review have been so far computed considering 100% of microalgae biodegradability and high performance of anaerobic digestion. However, an energy production lower than ideal can be expected when limiting factors in anaerobic digestion process are considered. For this reason, this book chapter examines different limiting factors of anaerobic digestion, which are necessary to overcome in order to improve performance of this process.

3.3.1. Ammonium inhibition

Ammonium is presented as protonated form (NH_4^+) and deprotonated form (NH_3, ammonia). The latter is considered to be the specie responsible for the inhibition of anaerobic digestion, due to its permeability through cell membrane (De Baere et al. 1984). There are several mechanisms by which ammonia will act as inhibitor of anaerobic bacteria among which are intracellular pH changes, increase in energy requirements for maintenance and inhibition of specific enzymes (Wittmann *et al.* 1995).

Several factors determining ammonia concentration in anaerobic reactor has been reported, but substrate concentration is a major one (Sialve *et al.* 2009). Distribution of total ammonia between protonated and deprotonated forms strongly depends on factors such as pH and temperature. At high pH values ammonium gets deprotonated forming toxic ammonia (NH_3) (Borja *et al.* 1996). Its inhibitory effect can result in volatile fatty acids accumulation due to a decrease in methanogenic activity, which generates a decrease in pH and ammonia concentration (Chen *et al.* 2008). This interaction may generate an inhibited steady-state, in which the process remains stable despite inhibition (Angelidaki & Ahring 1993; Angelidaki *et al.* 1993). Temperature is another variable that determine NH_4^+/NH_3 ratio, which is directly related to the increase of ammonia fraction and thus, inhibition level (Braun *et al.* 1981; Angelidaki & Ahring 1994).

Microalgal biomass can be expected to have low C/N ratio due to the high protein content in microalgae (Becker 2007). Then, anaerobic degradation of these residues is expected to generate a high ammonium concentration that may cause inhibition of anaerobic microbial consortia, especially methanogenic bacteria (Angelidaki & Ahring 1993; Chen *et al.* 2008). In addition, high ammonium concentration may affect biogas quality since ammonia can be stripped into gas phase (Sialve *et al.* 2009).

During anaerobic digestion of oil extracted microalgae (Process 2 on Figure 1), ammonia inhibition is expected to be especially of concern, since oil extraction will decrease C/N ratio. Figure 4 shows an estimation of the effect of substrate concentration and free ammonia levels in a hypothetical anaerobic digestion reactor. Estimation was calculated considering protein content reported by Becker (2007), operation pH value 8, temperature 35º C, ammonia conversion 90% and total lipid extraction efficiency. Figure 4 shows that inhibitory ammonia concentrations will develop whenever solids concentration exceeding 2% are applied during the anaerobic digestion step. This result was evaluated considering free ammonia inhibition at 100 mg/L NH_3 (dotted line in Figure 4).

Results shown in Figure 4 indicate that that either anaerobic digestion has to be performed at very low levels of solids concentration, or mechanisms for ammonia removal must be implemented. It has to be remained that Figure 4 assumes 90% of conversion of proteins. Lower protein conversions will reduce the chances of ammonia inhibition. However it is clear that this phenomena needs to be addressed if high rate digestion of microalgae is of interest.

One way to overcome this drawback is the possibility of co-digestion in order to provide an optimal C/N ratio for anaerobic digestion process (Yen & Brune 2007; Ehimen *et al.* 2011). Thus, a higher C/N ratio co-substrate should be mixed with microalgae in order to increase anaerobic digestion yield. This strategy is more attractive considering the fact that some co-substrate can stimulate enzymatic synthesis and, hence, increase hydrolysis and degradability (Yen & Brune 2007). Also, co-digestion can dilute toxic compounds decreasing their concentration below toxic/inhibition levels (Sialve *et al.* 2009).

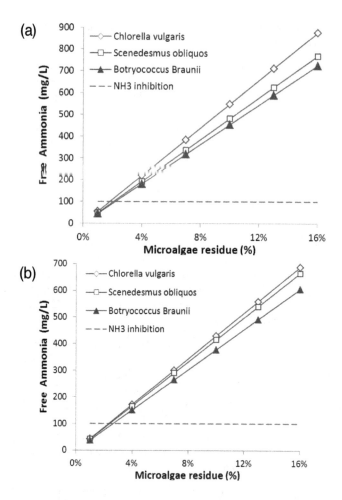

Figure 4. Estimation of free ammonia concentration on anaerobic digestion reactor from substrate level of feedstock, considering (a) processes 1, Biodiesel production and subsequent biogas production from spent microalgae and (b) process 2, Biogas production from whole microalgae.

3.3.2. Salt inhibition

Salt inhibition is expected to be relevant when saline microalgae are used as substrate for biogas production. In those locations where freshwater is not abundant or available, saline microalgae may be of interest, if cultivation takes place close to the sea. In those situations, salinity may even be higher than sea water when open pounds are used, as a result of water evaporation. If biomass is not diluted with fresh water after harvesting, downstream processes such as anaerobic digestion may need to deal with the salinity present in the biomass.

At low concentrations, sodium is essential for methanogenic bacteria. Probably, it is due to its role in ATP formation or NADH oxidation (Dimroth & Thomer 1989). Sodium concentration ranges 100-350mg/L have been reported as beneficial for mesophilic methanogenic growth (McCarty 1965; Patel & Roth 1977). Although moderate concentrations can stimulate bacteria growth, excessive amounts of salt reduce growth rate, and can cause severe inhibition or toxicity (Soto et al. 1991). Moreover, high salt levels can cause dehydration in bacteria due to osmotic pressure (De Baere et al. 1984; Yerkes et al. 1997).

Different levels of saline tolerance in anaerobic bacteria have been reported (Lefebvre & Moletta 2006). Easily degradable substrates seem to increase salt tolerance, most likely as a result of higher energy availability to cope with the energetic requirements of salt tolerance mechanisms (Xiao & Roberts 2010). Rinzema et al. (1988) found non acetoclastic methanogenic activity at 16 g/L of sodium concentration. The concentration that generated 50% of activity reduction (IC50) was 10 g/L and no bacteria adaptation after 12 weeks was observed. Similar saline tolerance was observed by Liu and Boone (1991). Feijoo et al. (1995) analyzed sodium inhibition for anaerobic bacteria from different reactors. A high tolerance in anaerobic bacteria from reactor treating wastewater under salinity conditions was observed, which was interpreted as consequence of bacteria adaptation. IC50 value for these bacteria was 16.3 g Na^+/L and entire inhibition was observed at 21 g Na^+/L.

Several reports indicate that biomass acclimation may significantly increase the activity under saline conditions (Soto et al. 1991; Omil et al. 1995; Chen et al. 2008; Kimata-Kino et al. 2011). However, reports are also available where no or little acclimating was observed (Aspe et al. 1997). Then, selection rather than adaptation is likely to be the mechanisms to provide high activity when big changes in salinity are imposed, requiring the presence of salinity-tolerant microorganisms in the inoculum (Gebauer 2004). It is indeed a common practice to use inoculums containing sources of saline resistant microorganisms, such as marine sediments (Xiao & Roberts 2010).

3.4. Biogas upgrading

Many biogas applications such as vehicle use, household distribution and electricity production, require some level of biogas upgrading to remove impurities or to increase methane content.

CO_2 removal is a key factor in order to obtain a higher calorific value of biogas. Processes such as solvent absorption, activated carbon adsorption and membrane filtration have been used for CO_2 removal (Kapdi et al. 2005; Makaruk et al. 2010; Ryckebosch et al. 2011).

Photosynthetic microorganisms such microalgae can also be used to remove CO_2 from biogas. Microalgae cultures are regarded as an interesting tool for carbon dioxide capture from gases such as flue gases from boilers, combustion engines or thermal power plants. This would not only alleviate impact of CO_2 emissions on the environment, but it would also reduce the cost of microalgae production (Doucha et al. 2005; Ryu et al. 2009). Stabilization ponds have been already recognized as potential CO_2 scrubbers due to their (micro-) algae growth (Shilton et al. 2008). Several authors have reported the successful growth of microalgae using flue gases. Negoro *et al.* (1993) reported productivities similar to those using pure CO_2, and showed that growth was barely influenced by the content of SO_x and NO_x contained in flue gases. Similar results were obtained by Hauck *et al.* (1996) who found no inhibition of *Chlorella sp.* by the levels of NO_x typically contained in flue gases. Doucha *et al.* (2005) reported 50% of flue gas decarbonization when working with a photobioreactor. In this study, 4.4 kg of CO_2 was needed for the production of 1 kg of dried algal biomass.

Conde *et al.* (1993) achieved biogas purification in laboratory experiments up to methane content of 97% with algae grown on synthetic nutrient medium. Mandeno *et al.* (2005) achieved CO₂ reduction from 40 to less than 5% using synthetic biogas, observing little transfer of oxygen to the biogas, so explosive methane/oxygen mixtures would not be formed. Similar results in terms of CO_2 reduction were obtained by Travieso *et al.* (1993) working with real biogas. Several microalgae species such as *Chlorococcum littorale*, *Chlorella sp.*, *Chlorella sp.* UK001, *Chlorella vulgaris*, *Chlorella kessleri*, *Scenedesmus obliquus*, *Spirulina sp.*, *Haematococcus pluvialis* or *Botryococcus braunii* have shown high levels of tolerance to high partial pressures of CO_2 (Wang *et al.* 2008; Brennan & Owende 2010). Mass transfer of carbon dioxide from gas to liquid phase is dependent on several factors highlighting chemical balance in microalgae media, pH and flow pattern of reactor in which culture is growing (Kumar *et al.* 2010). However, no full scale installations are under operation with this concept.

Available publications do not report negative effects of high methane partial pressures over microalgae cultures. Moreover, Meier *et al.* (2011) reported no inhibition effect when exposing a culture of *N. gaditana* to atmospheres containing methane up to 100%.

Hydrogen sulfide is present in biogas at low concentrations although its treatment should be considered. Some studies have reported a hydrogen sulphide decrease after biogas is upgraded in microalgae culture (Conde *et al.* 1993; Heubeck *et al.* 2007; Sialve *et al.* 2009). Most likely, hydrogen sulphide removal should be attributed to relative high solubility in growth medium (Conde *et al.* 1993; Sialve *et al.* 2009). Solubilised hydrogen sulphide can be easily oxidized into sulphate due to oxygen presence in growth medium.

4. Conclusions

Microalgal biomass is a promising substrate for renewable energy production. In this book chapter, direct anaerobic digestion without previous biodiesel extraction was shown to be the most promising method of energy production from microalgae. Lipids used for biodiesel production can also serve as a rich source of biogas with energetic efficiency higher than when microalgae are used for subsequent biodiesel and biogas production. The higher energy efficiency is given mostly by the simple technology with low energy demand used for methane production. These benefits combined with the possibility of CO_2 and nutrients recycling from the anaerobic effluents make anaerobic digestion the best technology for removable energy production from microalgae.

Acknowledgements

The authors would like to thank to Marie Curie's International Research Staff Exchange Scheme (IRSES) "Renewable energy production through microalgae cultivation: closing material cycles" (PIRSES-GA-2011-295165) for providing financial support. Dr. Rincón wishes to express her gratitude to the Ramón y Cajal Program from the Spanish Science and Innovation Ministry for providing financial support. Dr. Jeison wishes to thank CONICYT Chile and FONDECYT Project 1120488.

Author details

Álvaro Torres[1], Fernando G. Fermoso[2], Bárbara Rincón[2], Jan Bartacek[3], Rafael Borja[2] and David Jeison[1]

1 Chemical Engineering Department, Universidad de La Frontera, Temuco, Chile

2 Instituto de la Grasa (CSIC). Avenida Padre García Tejero, Sevilla, Spain

3 Department of Water Technology and Environmental Engineering, Institute of Chemical Technology Prague, Prague, Czech Republic

References

[1] Abo-Shady A. M., Mohamed Y. A. and Lasheen T. (1993). Chemical composition of the cell wall in some green algae species. Biologia Plantarum 35(4), 629-32, ISSN: 1573-8264

[2] Afi L., Metzger P., Largeau C., Connan J., Berkaloff C. and Rousseau B. (1996). Bacterial degradation of green microalgae: Incubation of Chlorella emersonii and Chlorella vulgaris with Pseudomonas oleovorans and Flavobacterium aquatile. Organic Geochemistry 25(1-2), 117-30, ISSN: 0146-6380

[3] Amin S. (2009). Review on biofuel oil and gas production processes from microalgae. Energy Conversion and Management 50(7), 1834-40, ISSN: 0196-8904

[4] Angelidaki I. and Ahring B. K. (1993). Thermophilic anaerobic digestion of livestock waste: The effect of ammonia. Applied Microbiology and Biotechnology 38(4), 560-4, ISSN: 1432-0614

[5] Angelidaki I. and Ahring B. K. (1994). Anaerobic thermophilic digestion of manure at different ammonia loads: Effect of temperature. Water Research 28(3), 727-31, ISSN: 0043-1354

[6] Angelidaki I. and Ahring B. K. (2000). Methods for increasing the biogas potential
     from the recalcitrant organic matter contained in manure. Water Science and Tech-
     nology 41 (3), 189-94, ISSN: 0273-1223

[7] Angelidaki I., Ellegaard L. and Ahring B. K. (1993). A mathemathical model for dy-
     namic simulation of anaerobic digestion of complex substrates, focusing on ammonia
     inhibition. Biotechnology and Bioengineering 42, 159-66, ISSN: 1097-0290

[8] Angelidaki I. and Sanders W. (2004). Assessment of the anaerobic biodegradability of
     macropollutants. Reviews in Environmental Science and Biotechnology 3(2), 117-29,
     ISSN: 1569-1705

[9] Aspe E., Marti M. C. and Roeckel M. (1997). Anaerobic treatment of fishery wastewa-
     ter using a marine sediment inoculum. Water Research 31(9), 2147-60 ISSN:
     0043-1354

[10] Banerjee A., Sharma R., Chisti Y. and Banerjee U. C. (2002). Botryococcus braunii: A
     renewable source of hydrocarbons and other chemicals. Critical Reviews in Biotech-
     nology 22(3), 245-79, ISSN: 1549-7801

[11] Becker E. W. (2007). Micro-algae as a source of protein. Biotechnology Advances
     25(2), 207-10, ISSN: 0734-9750

[12] Blokker P., Schouten S., Van den Ende H., De Leeuw J. W., Hatcher P. G. and Sin-
     ninghe Damsté J. S. (1998). Chemical structure of algaenans from the fresh water al-
     gae Tetraedron minimum, Scenedesmus communis and Pediastrum boryanum.
     Organic Geochemistry 29(5-7 -7 pt 2), 1453-68, ISSN: 0146-6380

[13] Blumreisinger M., Meindl D. and Loos E. (1983). Cell wall composition of chlorococ-
     cal algae. Phytochemistry 22(7), 1603-4, ISSN: 0031-9422

[14] Bonmatí A., Flotats X., Mateu L. and Campos E. (2001). Study of thermal hydrolysis
     as a pretreatment to mesophilic anaerobic digestion of pig slurry. Water Science and
     Technology 44 (4), 109-16, ISSN: 0273-1223

[15] Borja R., Sánchez E. and Durán M. M. (1996). Effect of the clay mineral zeolite on am-
     monia inhibition of anaerobic thermophilic reactors treating cattle manure. Journal of
     Environmental Science and Health - Part A Toxic/Hazardous Substances and Envi-
     ronmental Engineering 31(2), 479-500, ISSN: 1532-4117

[16] Braun R., Huber P. and Meyrath J. (1981). Ammonia toxicity in liquid piggery man-
     ure digestion. Biotechnology Letters 3(4), 159-64, ISSN: 1573-6776

[17] Brennan L. and Owende P. (2010). Biofuels from microalgae-A review of technolo-
     gies for production, processing, and extractions of biofuels and co-products. Renewa-
     ble and Sustainable Energy Reviews 14(2), 557-77, ISSN: 1364-0321

[18] Brown M. R. (1991). The amino-acid and sugar composition of 16 species of microalgae used in mariculture. Journal of Experimental Marine Biology and Ecology 145(1), 79-99, ISSN: 0022-0981

[19] Brown M. R. (1992). Biochemical composition of microalgae from the green algal classes Chlorophyceae and Prasinophyceae. 1. Amino acids Sugars and pigments. Journal of Experimental Marine Biology and Ecology 161(1), 91-113, ISSN: 0022-0981

[20] Conde J. L., Moro L. E., Travieso L., Sanchez E. P., Leiva A., Dupeiron R. and Escobedo R. (1993). Biogas purification process using intensive microalgae cultures. Biotechnology Letters 15(3), 317-20, ISSN: 1573-6776

[21] Costa J. A. V. and de Morais M. G. (2011). The role of biochemical engineering in the production of biofuels from microalgae. Bioresource Technology 102(1), 2-9, ISSN: 0960-8524

[22] Chen P. H. and Oswald W. J. (1998). Thermochemical treatment for algal fermentation. Environment International 24(8), 889-97, ISSN: 0160-4120

[23] Chen Y., Cheng J. J. and Creamer K. S. (2008). Inhibition of anaerobic digestion process: A review. Bioresource Technology 99(10), 4044-64, ISSN: 0960-8524

[24] Chisti Y. (2007). Biodiesel from microalgae. Biotechnology Advances 25(3), 294-306, ISSN: 0734-9750

[25] Choi S. P., Nguyen M. T. and Sim S. J. (2010). Enzymatic pretreatment of Chlamydomonas reinhardtii biomass for ethanol production. Bioresource Technology 101(14), 5330-6, ISSN: 0960-8524

[26] De Baere L. A., Devocht M., Van Assche P. and Verstraete W. (1984). Influence of high NaCl and NH4Cl salt levels on methanogenic associations. Water Research 18(5), 543-8, ISSN: 0043-1354

[27] De Schamphelaire L. and Verstraete W. (2009). Revival of the biological sunlight-to-biogas energy conversion system. Biotechnology and Bioengineering 103(2), 296-304, ISSN: 1097-0290

[28] Derenne S., Largeau C., Berkaloff C., Rousseau B., Wilhelm C. and Hatcher P. G. (1992). Non-hydrolysable macromolecular constituents from outer walls of Chlorella fusca and Nanochlorum eucaryotum. Phytochemistry 31(6), 1923-9, ISSN: 0031-9422

[29] Dimroth P. and Thomer A. (1989). A primary respiratory Na+ pump of an anaerobic bacterium: the Na+-dependent NADH:quinone oxidoreductase of Klebsiella pneumoniae. Archives of Microbiology 151(5), 439-44, ISSN: 1432-072X

[30] Domozych D. S., Stewart K. D. and Mattox K. R. (1980). The comparative aspects of cell wall chemistry in the green algae (Chlorophyta). Journal of Molecular Evolution 15(1), 1-12, ISSN: 1432-1432

[31] Doucha J., Straka F. and Lívanský K. (2005). Utilization of flue gas for cultivation of microalgae (Chlorella sp.) in an outdoor open thin-layer photobioreactor. Journal of Applied Phycology 17(5), 403-12, ISSN: 1573-5176

[32] Douskova I., Kastanek F., Maleterova Y., Kastanek P., Doucha J. and Zachleder V. (2010). Utilization of distillery stillage for energy generation and concurrent production of valuable microalgal biomass in the sequence: Biogas-cogeneration-microalgae-products. Energy Conversion and Management 51(3), 606-11, ISSN: 0196-8904

[33] Ehimen E. A., Sun Z. F., Carrington C. G., Birch E. J. and Eaton-Rye J. J. (2011). Anaerobic digestion of microalgae residues resulting from the biodiesel production process. Applied Energy 00(10), 3454-03, ISSN: 0306-2619

[34] Feijoo G., Soto M., Méndez R. and Lema J. M. (1995). Sodium inhibition in the anaerobic digestion process: Antagonism and adaptation phenomena. Enzyme and Microbial Technology 17(2), 180-8, ISSN: 0141-0229

[35] Foree E. G. and McCarty P. L. (1970). Anaerobic decomposition of algae. Environmental Science and Technology 4(10), 842-9, ISSN: 1520-5851

[36] Fu C. C., Hung T. C., Chen J. Y., Su C. H. and Wu W. T. (2010). Hydrolysis of microalgae cell walls for production of reducing sugar and lipid extraction. Bioresource Technology 101(22), 8750-4, ISSN: 0960-8524

[37] Gebauer R. (2004). Mesophilic anaerobic treatment of sludge from saline fish farm effluents with biogas production. Bioresource Technology 93(2), 155-67, ISSN: 0960-8524

[38] Gelin F., Boogers I., Noordeloos A. A. M., Sinninghe Damsté J. S., Riegman R. and De Leeuw J. W. (1997). Resistant biomacromolecules in marine microalgae of the classes eustigmatophyceae and chlorophyceae: Geochemical implications. Organic Geochemistry 26(11-12), 659-75, ISSN: 0146-6380

[39] Gelin F., Volkman J. K., Largeau C., Derenne S., Sinninghe Damsté J. S. and De Leeuw J. W. (1999). Distribution of aliphatic, nonhydrolyzable biopolymers in marine microalgae. Organic Geochemistry 30(2-3), 147-59, ISSN: 0146-6380

[40] Golueke C. G., Oswald W. J. and Gotaas H. B. (1957). Anaerobic digestion of Algae. Applied microbiology 5(1), 47-55, ISSN: 1365-2672

[41] González-Fernández C., Sialve B., Bernet N. and Steyer J. P. (2012). Impact of microalgae characteristics on their conversion to biofuel. Part II: Focus on biomethane production. Biofuels, Bioproducts and Biorefining 6(2), 205-18, ISSN: 1932-1031

[42] Griffiths M. and Harrison S. (2009). Lipid productivity as a key characteristic for choosing algal species for biodiesel production. Journal of Applied Phycology 21(5), 493-507, ISSN: 0921-8971

[43] Halleux H., Lassaux S., Renzoni R. and Germain A. (2008). Comparative life cycle assessment of two biofuels: Ethanol from sugar beet and rapeseed methyl ester. International Journal of Life Cycle Assessment 13(3), 184-90, ISSN: 1614-7502

[44] Harun R., Davidson M., Doyle M., Gopiraj R., Danquah M. and Forde G. (2010). Technoeconomic analysis of an integrated microalgae photobioreactor, biodiesel and biogas production facility. Biomass and Bioenergy 35(1), 741-7, ISSN: 0961-9534

[45] Hauck J. T., Olson G. J., Scierka S. J., Perry M. B. and Ataai M. M. (1996). Effects of simulated flue gas on growth of microalgae. ACS Division of Fuel Chemistry, Preprints 41(4), 1391-4

[46] Heubeck S., Craggs R. J. and Shilton A. (2007). Influence of CO_2 scrubbing from biogas on the treatment performance of a high rate algal pond. Water Science and Technology 55 (11), 193-200, ISSN: 0273-1223

[47] Kadouri A., Derenne S., Largeau C., Casadevall E. and Berkaloff C. (1988). Resistant biopolymer in the outer walls of Botryococcus braunii, B race. Phytochemistry 27(2), 551-7, ISSN: 0031-9422

[48] Kapdi S. S., Vijay V. K., Rajesh S. K. and Prasad R. (2005). Biogas scrubbing, compression and storage: Perspective and prospectus in Indian context. Renewable Energy 30(8), 1195-202, ISSN: 0960-1481

[49] Kimata-Kino N., Ikeda S., Kurosawa N. and Toda T. (2011). Saline adaptation of granules in mesophilic UASB reactors. International Biodeterioration & Biodegradation 65(1), 65-72, ISSN: 0964-8305

[50] Kumar A., Ergas S., Yuan X., Sahu A., Zhang Q., Dewulf J., Malcata F. X. and van Langenhove H. (2010). Enhanced CO_2 fixation and biofuel production via microalgae: Recent developments and future directions. Trends in Biotechnology 28(7), 371-80, ISSN: 0167-7799

[51] Lardon L., Hélias A., Sialve B., Steyer J. P. and Bernard O. (2009). Life-cycle assessment of biodiesel production from microalgae. Environmental Science and Technology 43(17), 6475 81, ISSN: 1520-5851

[52] Lefebvre O. and Moletta R. (2006). Treatment of organic pollution in industrial saline wastewater: A literature review. Water Research 40(20), 3671-82, ISSN: 0043-1354

[53] Liu Y. and Boone D. R. (1991). Effects of salinity on methanogenic decomposition. Bioresource Technology 35(3), 271-3, ISSN: 0960-8524

[54] Loos E. and Meindl D. (1982). Composition of the cell wall of Chlorella fusca. Planta 156(3), 270-3, ISSN: 1432-2048

[55] Mairet F., Bernard O., Cameron E., Ras M., Lardon L., Steyer J. P. and Chachuat B. (2012). Three-reaction model for the anaerobic digestion of microalgae. Biotechnology and Bioengineering 109(2), 415-25, ISSN: 00063592

[56] Mairet F., Bernard O., Ras M., Lardon L. and Steyer J.-P. (2011). Modeling anaerobic digestion of microalgae using ADM1. Bioresource Technology 102(13), 6823-9, ISSN: 0960-8524

[57] Makaruk A., Miltner M. and Harasek M. (2010). Membrane biogas upgrading processes for the production of natural gas substitute. Separation and Purification Technology 74(1), 83-92, ISSN: 1383-5866

[58] Mandeno G., Craggs R., Tanner C., Sukias J. and Webster-Brown J. (2005). Potential biogas scrubbing using a high rate pond. Water Science and Technology 51(12), 253-6, ISSN: 0273-1223

[59] McCarty P. L. (1965). Thermodynamics of biological synthesis and growth. Air and water pollution 9(10), 621-39, ISSN: 0568-3408

[60] Meier L., Torres A., Azocar L., Neumann P., Vergara C., Rivas M. and Jeison D. (2011). Biogas upgrading through microalgae culture: Effect of methane concentration on microalgae activity. In: X Latinamerican and Symposium on Anaerobic Digestion, Ouro Preto, Brazil.

[61] Mussgnug J. H., Klassen V., Schlüter A. and Kruse O. (2010). Microalgae as substrates for fermentative biogas production in a combined biorefinery concept. Journal of Biotechnology 150(1), 51-6, ISSN: 0168-1656

[62] Mutanda T., Ramesh D., Karthikeyan S., Kumari S., Anandraj A. and Bux F. (2011). Bioprospecting for hyper-lipid producing microalgal strains for sustainable biofuel production. Bioresource Technology 102(1), 57-70, ISSN: 0960-8524

[63] Nair K. V. K., Kannan V. and Sebastian S. (1983). Bio-gas generation using microalgae and macrophytes. Indian Journal of Environmental Health 24(4), 277-84, ISSN: 0367-827X

[64] Negoro M., Hamasaki A., Ikuta Y., Makita T., Hirayama K. and Suzuki S. (1993). Carbon dioxide fixation by microalgae photosynthesis using actual flue gas discharged from a boiler. Applied biochemistry and biotechnology 39-40(1), 643-53, ISSN: 1559-0291

[65] Neumann P., Torres A., Azocar L., Meier L., Vergara C. and Jeison D. (2011). Biogas production as a tool for increasing sustainability of biodiesel production from microalgae Botryococcus braunii. In: X Latinamerican and Symposium on Anaerobic Digestion, Ouro Preto, Brazil.

[66] Northcote D. H., Goulding K. J. and Horne R. W. (1958). The chemical composition and structure of the cell wall of Chlorella pyrenoidosa. The Biochemical journal 70(3), 391-7, ISSN: 1470-8728

[67] Omil F., Mendez R. and Lema J. M. (1995). Anaerobic treatment of saline wastewaters under high sulphide and ammonia content. Bioresource Technology 54(3), 269-78, ISSN: 0960-8524

[68] Patel G. B. and Roth L. A. (1977). Effect of sodium chloride on growth and methane production of methanogens. Canadian Journal of Microbiology 23(7), 893-7, ISSN: 1480-3275

[69] Punnett T. and Derrenbacker E. C. (1966). The amino acid composition of algal cell walls. Journal of General Microbiology 44(1), 105-14, ISSN: 0022-1287

[70] Ras M., Lardon L., Bruno S., Bernet N. and Steyer J.-P. (2011). Experimental study on a coupled process of production and anaerobic digestion of Chlorella vulgaris. Bioresource Technology 102(1), 200-6, ISSN: 0960-8524

[71] Razon L. F. and Tan R. R. (2011). Net energy analysis of the production of biodiesel and biogas from the microalgae: Haematococcus pluvialis and Nannochloropsis. Applied Energy 88(10), 3507-14, ISSN: 0306-2619

[72] Rinzema A., Van Lier J. and Lettinga G. (1988). Sodium inhibition of acetoclastic methanogens in granular sludge from a UASB reactor. Enzyme and Microbial Technology 10(1), 24-32, ISSN: 0141-0229

[73] Ryckebosch E., Drouillon M. and Vervaeren H. (2011). Techniques for transformation of biogas to biomethane. Biomass and Bioenergy 35(5), 1633-45, ISSN: 0961-9534

[74] Ryu H. J., Oh K. K. and Kim Y. S. (2009). Optimization of the influential factors for the improvement of CO2 utilization efficiency and CO2 mass transfer rate. Journal of Industrial and Engineering Chemistry 15(4), 471-5, ISSN: 1226-086X

[75] Samson R. and Leduy A. (1983). Influence of mechanical and thermochemical pretreatments on anaerobic digestion of Spirulinamaxima algal biomass. Biotechnology Letters 5(10), 671-6, ISSN: 1573-6776

[76] Scott S. A., Davey M. P., Dennis J. S., Horst I., Howe C. J., Lea-Smith D. J. and Smith A. G. (2010). Biodiesel from algae: Challenges and prospects. Current Opinion in Biotechnology 21(3), 277-86, ISSN: 0958-1669

[77] Shilton A. N., Mara D. D., Craggs R. and Powell N. (2008). Solar-powered aeration and disinfection, anaerobic co-digestion, biological CO2 scrubbing and biofuel production: The energy and carbon management opportunities of waste stabilisation ponds. Water Science and Technology 58 (1), 253-8, ISSN: 0273-1223

[78] Sialve B., Bernet N. and Bernard O. (2009). Anaerobic digestion of microalgae as a necessary step to make microalgal biodiesel sustainable. Biotechnology Advances 27(4), 409-16, ISSN: 0734-9750

[79] Simpson A. J., Zang X., Kramer R. and Hatcher P. G. (2003). New insights on the structure of algaenan from Botryoccocus braunii race A and its hexane insoluble botryals based on multidimensional NMR spectroscopy and electrospray-mass spectrometry techniques. Phytochemistry 62(5), 783-96, ISSN: 0031-9422

[80] Soto M., Mendez R. and Lema J. M. (1991). Biodegradability and toxicity in the anaerobic treatment of fish canning wastewaters. Environmental Technology 12(8), 669-77, ISSN: 0959-3330

[81] Stephens E., Ross I. L., King Z., Mussgnug J. H., Kruse O., Posten C., Borowitzka M. A. and Hankamer B. (2010). An economic and technical evaluation of microalgal biofuels. Nature Biotechnology 28(2), 126-8, ISSN: 1087-0156

[82] Tarwadi S. J. and Chauhan V. D. (1987). Seaweed biomass as a source of energy. Energy 12(5), 375-8, ISSN: 0360-5442

[83] Templier J., Largeau C., Casadevall E. and Berkaloff C. (1992). Chemical inhibition of resistant biopolymers in outer walls of the A and B races of Botryococcus braunii. Phytochemistry 31(12), 4097-104, ISSN: 0031-9422

[84] Travieso L., Sanchez E. P., Benitez F. and Conde J. L. (1993). Arthrospira sp. intensive cultures for food and biogas purification. Biotechnology Letters 15(10), 1091-4, ISSN: 1573-6776

[85] Vergara-Fernández A., Vargas G., Alarcón N. and Velasco A. (2008). Evaluation of marine algae as a source of biogas in a two-stage anaerobic reactor system. Biomass and Bioenergy 32(4), 338-44, ISSN: 0961-9534

[86] Wang B., Li Y., Wu N. and Lan C. Q. (2008). CO2 bio-mitigation using microalgae. Applied Microbiology and Biotechnology 79(5), 707-18, ISSN: 1432-0614

[87] Wittmann C., Zeng A. P. and Deckwer W. D. (1995). Growth inhibition by ammonia and use of a pH-controlled feeding strategy for the effective cultivation of Mycobacterium chloropheolicum. Applied Microbiology and Biotechnology 44(3-4), 519-25, ISSN: 1432-0614

[88] Xiao Y. Y. and Roberts D. J. (2010). A review of anaerobic treatment of saline wastewater. Environmental Technology 31(8-9), 1025-43, ISSN: 0959-3330

[89] Yen H.-W. and Brune D. E. (2007). Anaerobic co-digestion of algal sludge and waste paper to produce methane. Bioresource Technology 98(1), 130-4, ISSN: 0960-8524

[90] Yerkes D. W., Boonyakitsombut S. and Speece R. E. (1997). Antagonism of sodium toxicity by the compatible solute betaine in anaerobic methanogenic systems. Water Science and Technology 36 (6-7), 15-24, ISSN: 0273-1223

[91] Zamalloa C., Vulsteke E., Albrecht J. and Verstraete W. (2011). The techno-economic potential of renewable energy through the anaerobic digestion of microalgae. Bioresource Technology 102(2), 1149-58, ISSN: 0960-8524

Chapter 7

Advanced Monitoring and Control of Anaerobic Digestion in Bioreactor Landfills

Mohamed Abdallah and Kevin Kennedy

Additional information is available at the end of the chapter

1. Introduction

Despite recent increases in recycling, composting, and incineration, the sanitary landfill remains the predominant and most economical municipal solid waste (MSW) management alternative. Modern MSW landfills strive to optimize the design, construction, and operation processes in order to mitigate many of the potentially negative impacts, and improve the profitability. The bioreactor landfill (BL) is considered one of the promising developments that have recently gained significant attention. This waste-to-energy technology requires specific management activities and operational procedures that enhance the microbial decomposition processes inside the landfill resulting in higher production of landfill gas [1]. The recirculation of leachate, which is conducted by recycling the water passing through and collected from the landfill, is considered the main operational characteristic in the BL to increase moisture, and consequently stimulate the biodegradation process (Figure 1). The potential benefits of the BL include increased waste settlement rates and airspace utilization, decreased costs for leachate treatment, more rapid gas production (which improves the economics of gas recovery), and more rapid waste stabilization (which may reduce the post-closure maintenance period). These potential benefits have led to many full-scale BL applications in the last decade, mostly in the United States, resulting in the generation of design and operation data. In 2004, the Solid Waste Association of North America conducted an inventory that identified over 70 BLs in North America [2]. Many of these experiences revealed scale-up issues and technical limitations that merit further research and development [3-5].

One of the most critical, yet little studied, issues in the operation of BLs is process control. In field applications, unsupervised operational procedures can disturb the dynamics of the landfill biological processes causing serious consequences on the overall evolution of the ecosystem, i.e., unstable and sometimes unsuccessful transition from one operational phase to

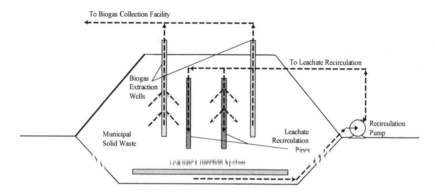

Figure 1. Schematic of an anaerobic bioreactor landfill

another. Dealing with the BL as a dynamic and evolving biological system could solve many of the BL control issues especially those pertinent to daily operation such as leachate recirculation. For example, one of the main operational issues, which are addressed in the present work, is the large variation in the characteristics of the collected leachate, which sometimes makes the leachate (as produced) unsuitable for recirculation. At the same time, the physical, chemical, and biochemical growth requirements of the bacterial consortia inside the BL change significantly during the different operational phases. *It is therefore necessary to manipulate the collected leachate before recirculation in order to suit the prevailing reactions and conditions inside the BL.* Several techniques have been tested in laboratory studies to enhance the performance of BLs either directly or indirectly through the manipulation of the recirculated leachate: pH adjustment, nutrients addition, and biosolids addition [1, 6-8]. However, these techniques are rarely, if ever, used in field applications due to lack of well-defined methodologies and the huge cost if applied excessively in an uncontrolled fashion. Applying advanced process control techniques offers an alternative solution for this problem. *Developing a control system that optimizes the leachate recirculation and manipulation processes based on real-time conditions of the controlled BL can provide a flexible engineered solution that is applicable to any typical landfill site..*

The proposed Sensor-based Monitoring and Remote-control Technology (SMART) features an expert controller that manipulates the controllable variables of the bioreactor process based on online monitoring of key system parameters. The objective of this control framework is to provide the optimal operational conditions for the biodegradation of MSW, and also, to enhance the performance of the BL in terms of biogas production. A comprehensive analysis of the process control of BLs is presented, followed by the conceptual framework of SMART including its structure, components, and instrumentation. In conclusion, a pilot-scale implementation of the control system is discussed.

1.1. Bioreactor landfill ecosystem

Controlling the BL requires a good understanding of the system and its dataflow including inputs, outputs, and interconnecting processes. The basic principles and mechanisms of the BL are well documented in the literature [9-11]. A simplified dataflow diagram for an anaerobic BL is shown in Figure 2. The BL can be considered as an anaerobic fixed-bed reactor in which the biodegradable organic fraction of the solid waste is the substrate. The factors affecting the biological processes in landfills can be grouped to: (1) factors related to the microbial environment (e.g., moisture, temperature, nutrients availability, and toxicity), and (2) factors related to the landfill site including: climate conditions (e.g., air temperature and precipitation), waste characteristics (e.g., particle size and composition), and site-specific settings (e.g., collection and injection systems). The BL concept is based on employing specific operational activities to control the influencing factors in a positive manner, e.g., applying leachate recirculation to optimize waste moisture. From the process control point of view, i.e., based on the feasibility of real-time manipulation, the first group of factors can be considered *controllable* inputs to the BL process, while the second group of factors is *uncontrollable*. The management techniques through which the *controllable* factors can be controlled are discussed below.

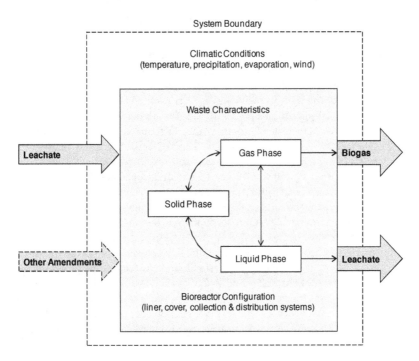

Figure 2. Data flow diagram of the bioreactor landfill ecosystem

1.1.1. Leachate recirculation

Moisture addition has been proved repeatedly to stimulate the methanogenic population in the landfill waste matrix. Leachate recirculation is considered the most effective method to increase moisture content of waste in a controlled fashion, which could reduce the time required for landfill stabilization from several decades to two to three years [12]. Leachate recirculation has been proven to achieve better BL performance in terms of biogas production by several lab-, pilot-, and full-scale studies [13-17]. In full-scale applications, leachate recirculation at Trail Road landfill enhanced waste settlement and resulted in 30% airspace recovery, which was used for landfilling more waste [4]. In another full-scale study by [18], leachate recirculation achieved more rapid biogas production, increased settlement rates, and accelerated decreases in the concentration of certain contaminants in leachate. According to [19], moisture increase alone does not enhance methane production. It is the nutrients, inocula and buffers, which in addition to moisture, enhances biodegradation to the greatest extent. It was shown in [8] that added alkalinity, dissolved oxygen level, and presence of methanogenic bacteria in the recirculated liquid considerably influenced the hydrolysis rate and onset of methanogenesis. *Therefore, it is suggested that, not only moisture addition, but also the quality of the leachate affects the impact/outcome of recirculation significantly.* Hence, there are two main aspects of the recirculation process that can be controlled: the **quantity** and **quality** of the recirculated leachate.

The **quantity** of the leachate generated is site-specific and a function of water availability, weather conditions, characteristics of the waste, as well as the liner and cover design [10]. In order to achieve the benefits of leachate recirculation, leachate has to be recycled at optimal rates that achieve sufficient contact with waste. The effect of varying leachate recirculation rates was studied in lab simulations [13, 16, 17, 20]. These studies demonstrated that higher recirculation rates result in better BL performance in terms of biogas production. It was suggested that leachate recirculation should be adjusted according to the phases of waste stabilization [21]. This practice was applied successfully in [13] as well as [22] who varied the leachate recirculation rates in lab scale BLs based on 7 and 4 operational stages, respectively. Unsupervised high rate of recirculation may result in: (1) washout of large amounts of organic matter before the methanogenic phase, thereby reducing the biological methane potential, (2) production of leachate containing high concentrations of short chain fatty acids which either inhibits methanogenesis directly or by lowering the pH, (3) excessive accumulation of leachate within the landfill, which may breakout from landfill slopes, (4) increase of pore water pressure and decrease of the shear strength of the waste matrix which affect the geotechnical slope stability, (5) increase in the hydrostatic head on the base liner, leading to higher risk for ground water contamination, and (6) drop in the internal temperature of the landfill especially in cold regions. *Therefore, leachate recirculation rate has to be selected such that the desired moisture content levels, moisture movement, and supplements distribution are provided, and at the same time, the pre-mentioned issues are monitored and incorporated in the decision-making process.*

The **quality** of leachate is highly dependent on waste composition and operational phase [10]. Leachate has been reported to contain a wide range of inorganic and organic compounds including toxicants such as aliphatic/aromatic hydrocarbons and halogenated organics [23]. Typically, the concentration of constituents, including pollutants, in leachate decreases with the waste age. The large temporal variation in the biochemical characteristics of leachate - as produced - makes it sometimes unsuitable for recirculation. For example, the concentrations of dissolved organic substances in young leachate are usually much higher than in older leachate. Continuous recirculation of young leachate in early phases of operation will increase the concentration of short chain fatty acids inside the BL which either inhibits methanogenesis directly or indirectly by lowering the pH of the system. Recently, researchers examined the use of different leachate (e.g., mature leachate from older landfill cells) for recirculation [16, 17, 21]. Alternatively, young and mature leachates were used interchangeably over four operational stages along the BL lifespan [22]. They used young leachate in phase I, then mature leachate in phase II and when the characteristics of produced leachate became suitable, they switched back to young leachate in phases III and IV. The same concept was applied by [20] who rotated the recirculated leachate between fresh waste and stabilised waste reactors until a balanced microbial population was established. Other studies combined leachate with water, as simulated rainfall, which simulated field conditions and diluted the leachate [13, 24]. The addition of supplemental water to the recirculated leachate in early operational phases could promote dilution of inhibitory substances and reduce leachate strength resulting in more favourable methanogenic conditions [25]. *Therefore, supplemental water can be used in combination with other leachate manipulation techniques – shown below - to correct certain process deviations, reduce the impact of detrimental substances, and/or enrich the concentration of other beneficial compounds.*

1.1.2. pH Adjustment

Methanogenic bacteria are sensitive to pH, with an optimal range between 6.8 and 7.2, and could be inhibited by acidic conditions at pH less than 6.7. Therefore, pH of recycled leachate can have a significant effect on waste stabilization and methane production. This understanding of microbial ecology has promoted the addition of buffer to adjust the pH of leachate prior to recycling it back to landfill. Buffering as a control option may be best used in response to changes in leachate characteristics (i.e., a drop in pH or increase in volatile acids' concentration). Leachate recirculation with a buffering system to control the pH has been found to result in shorter acidogenic stage leading to earlier initiation of the methanogenic stage, and concomitant higher gas production [7, 8, 25].

1.1.3. Bioaugmentation

Bioaugmentation or inoculation of the landfill has been investigated, usually through the addition of bio-solids from wastewater treatment facilities [1]. The optimal inoculum should provide suitable consortia and concentration of microorganisms, as well as nutrients such as nitrogen and phosphorus. It was stated in [23] that initiating fermentation in BLs can be promoted by addition of large amounts of methanogenic microorganisms in the form of effluent and sludge from an anaerobic sewage digester since the population of such microorganisms in fresh MSW is typically low. In [6], moisture saturation conditions was examined with digested sewage sludge, with fertilizer, and without additives. It was found that moisture and sewage sludge additions resulted in the shortest acidogenic phase and highest gas production. However, it has been suggested that any measured beneficial effects associated with the addition of biosolids may be due to buffering or moisture addition rather than inoculation [26]. Moreover, generic conclusions regarding the effect of sludge addition cannot be drawn, since different types and percentages of sludge might have been used in different experiments.

1.1.4. Nutrients addition

Nutrients required for anaerobic degradation of waste are generally low, and therefore, nutrients are expected to be available especially during early phases of biodegradation [7]. It was found that all the necessary nutrients and trace heavy metals are available in most landfills, but insufficient mixing and heterogeneity of the wastes may result in nutrient-limited zones [23, 27]. Experimentally, it was proven that the addition of nitrogen and phosphorous stimulated methane production rapidly decreased organic concentration in leachate, and shortened the initial phase before methane generation commenced [1, 28].

1.2. Identification of control problem

While most studies reported process improvements associated with leachate recirculation and manipulation processes, other studies found the contrary, such as toxicity and souring conditions. The results reported in many studies are different, and sometimes contradicting, since the same substance can be useful or harmful depending on its dose. This can be explained by the general effect of increasing salt concentration in anaerobic systems shown in Figure 3. A substance which is essential to a biological process can stimulate the bacterial growth at low concentrations. However, as concentrations increase above optimal, the rate of microbial activity decreases until the process is inhibited. Similarly, this trend can describe the effects of adding leachate and other amendments on the BL performance. In addition to the dose, other factors may affect the results: (1) operational factors, such as the type and characteristics of amendments, and (2) operational phase and progressive evolution of the BL.

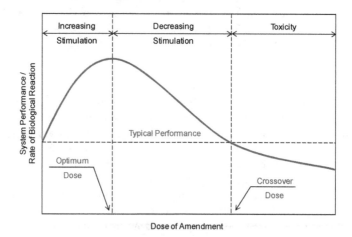

Figure 3. Effect of adding amendments on BL performance (modified from [29])

In conclusion, specific growth needs of the BL bacterial consortia changes with time, concomitantly the required leachate characteristics are continuously changing such that leachate as produced in its original form may not always be ideal for recirculation. The goal of the present research is the development of a real-time monitoring and expert decision making system that can adjust both, leachate characteristics and rates of recirculation according to the ecological requirements of each operational phase to provide the optimum conditions for waste biodegradation in BLs.

2. The proposed control system

The main real-time control tool in an anaerobic BL is leachate recirculation combined with amendment addition to provide both optimal moisture content and distribution of essential additives. The pH of the recirculated leachate can be adjusted by adding buffer, while inoculum in the form of anaerobic digested sludge, can be used both as a buffer and a rich source of methanogenic bacteria. At later BL operational phases, nutrients can be added as needed to supply the nutritional needs of the bacterial consortia. Supplemental water can be added to dilute concentrated leachate (as a remedy for toxicity) and to account for any shortage in available recyclable leachate for moisture control. The rate of application of any of these amendments can be decided based on measurable parameters in the leachate as well as the specific requirements of each BL operational phase. In conjunction with recirculation, certain parameters such as pore water pressure, landfill internal temperature, and hydrostatic head on the liner must be monitored and considered as they are influenced by recirculated leachate, and can affect BL operation.

2.1. Control scheme

The biological processes occurring in the landfill are largely anaerobic. Similar to anaerobic digesters, the landfill ecosystem is sensitive to environmental conditions such as pH, temperature, moisture, toxic compounds, and presence of oxygen. In fact, much of what is known or assumed concerning processes in landfills has primarily come from experiences with anaerobic digesters [10]. For this reason, the required control for an anaerobic BL is analogous to that of an anaerobic digester, with the latter more easily to control being a well-mixed reactor [7]. There are various control schemes that can be applied in managing biochemical systems. The most widespread control schemes are: feedback, feed-forward, and open-loop. Feedback control is a control mechanism that uses information from measurements to manipulate a variable so that the desired result is achieved. Alternatively, feed-forward control mechanism predicts the effects of measured disturbances and takes corrective action to achieve the desired result. On the other hand, the open-loop controller does not utilize feedback to determine whether the input achieved the desired goal or not, and can neither engage in machine learning nor correct any errors that it could make. Thus far in landfill sites, process control is accomplished, if ever, based on a non-feedback scheme. Therefore, the present study aims at applying feedback control in the management of BLs.

In feedback control, the variable being controlled is measured and compared with a target value. The difference between the measured and desired value is called the *error*. Feedback control manipulates inputs of the system to minimize this error. Figure 4 shows a generic component block diagram of an elementary feedback controller. The output of the system is measured by a *sensor* and the *control element* represents an actuator or control device. The *error* in this system would be the *Measured Output - Desired Output*.

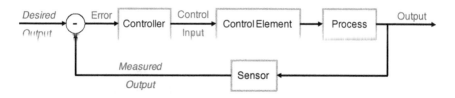

Figure 4. Block diagram of a basic feedback control loop

The potential advantages of feedback control lie in the fact that it obtains and utilizes data at the process output [30]. Therefore, the controller takes into account unforeseen disturbances in the process. Feedback control architecture ensures the desired performance by altering the inputs immediately once deviations are observed regardless of their reason. Thus, it reduces operator workload by eliminating the need for human adjustment of the control variable. An additional advantage is that by analyzing the output of a system, unstable processes may be stabilized. Feedback controls do not require detailed knowledge of the system and, in particular, do not require a mathematical model of the process. The controller can be easily duplicated from one system to another.

On the other hand, the time lag in the system is potentially the main disadvantage of feedback control. A process deviation occurring near the beginning of the process will not be recognized until the process output. The feedback control will then have to adjust the process inputs in order to correct this deviation. This results in the possibility of substantial deviation throughout the entire process [30]. The system could possibly miss process output disturbances and the error could continue without adjustment resulting in a steady state error. When the feedback controller proves unable to maintain stable closed-loop control, operator intervention is then required. Finally, feedback control does not take predictive control action towards the effects of known disturbances, and depends entirely on the accuracy with which the controlled output is measured.

2.2. Control framework

The proposed Sensor-based Monitoring and Remote-control Technology (SMART) system features software and hardware interacting components that provide real-time monitoring and expert control of BLs. Figure 5 shows a general diagram of the control system. The dashed lines indicate the sensory data, while the dot-dashed lines represent the commands.

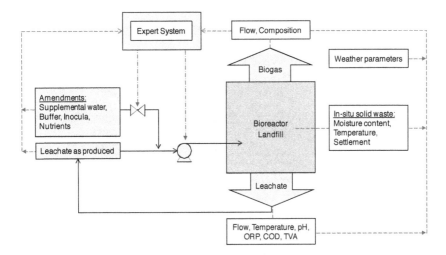

Figure 5. Schematic of the SMART control system

The control system has a geographically and functionally distributed architecture in which the BL is divided into basic blocks. Each block has its own local sensory data acquisition and control units. In addition, global sensory units are to provide measurements for the landfill body altogether as one block. All these local and global components are connected and remotely controlled by a global data processing and decision making unit. The controller continuously monitors two types of sensory data: process parameters (such as moisture and temperature), and returned feedback from performance indicators (such as biogas production and settlement). The decision made by the control algorithm is transmitted to the actuators, after authorization from the site operator, to inject the computed volumes of the selected amendments in order to manipulate the characteristics of the recirculated leachate. This batch control process runs continuously along the lifetime of the BL cell.

2.3. System components

The SMART system incorporates six interacting components: (1) Local Sensory Unit, (2) Global Sensory Unit, (3) Primary Sensory Data Processor, (4) Main Controller Unit, (5) Primary Driving Controller, and (6) Local Driving Unit. The main components of the system are shown in Figure 6, and described in detail below.

Local Sensory Unit (LSU)

The LSU is placed in each block, i.e., n sensory units for the n blocks. Each unit includes a set of analog sensors which quantify the values of different system parameters, such as temperature and moisture content, in the corresponding block. The installed units form a three dimensional grid in order to show the spatial dynamic status of the main parameters within the BL. All LSUs are designed to send the measured data to the Primary Sensory Data Processor.

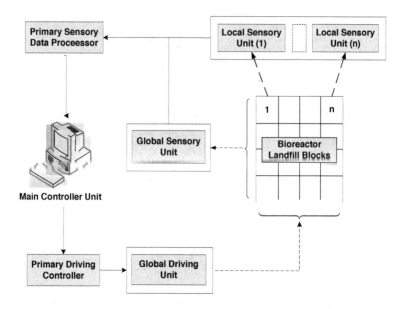

Figure 6. Main components of the SMART control system

Global Sensory Unit (GSU)

The GSU provides global measurements for the landfill body altogether as one block. These measurements include the parameters that are impractical to be determined for each block individually such as leachate characteristics, settlement, hydrostatic head on the liner, as well as biogas quantity and quality. Other examples of global measurements are the weather condition parameters such as air temperature, wind speed and direction, humidity, solar radiation, precipitation, and evaporation. All GSUs are connected to the Main Controller Unit through the Primary Sensory Data Processor.

Primary Sensory Data Processor (PSDP)

The PSDP is responsible for analyzing the acquired data from the Local and Global Sensory Units, and arranging them in a new frame to be delivered to the Main Controller Unit. Although this work could be done by the Main Controller Unit, employing an intermediate device here provides more modularity and flexibility to the system by providing an interface between the software of the Main Controller Unit from one side, and the LSUs from the other side.

Main Controller Unit (MCU)

The MCU is considered the driving brain of the control system. It receives the measured data (inputs), processes them within the developed expert system, and makes the control decision. The operator is prompted with the decision made by the MCU in order to evaluate it, and then approves it to be sent to the Primary Driving Controller in the form of quantified commands.

The operator can overwrite the decision to deal with any unexpected problem or unconsidered scenario in the expert system. The control program was programmed on the LabVIEW™ graphical programming platform (National Instruments, USA). The control program and expert system of MCU are discussed in *Section 2.5*.

Primary Driving Controller (PDC)

The PDC receives the commands from the MCU and distributes it to the different Local Driving Units. Basically, it is a device that de-multiplexes the received data set which holds the commands for all the driving units, and then delivers the commands to each unit separately. This unit combines analog/digital conversion, signal conditioning, and signal connectivity.

Local Driving Unit (LDU)

The LDU receives the commands and performs the required action by driving the corresponding actuator (motorized valves and/or pumps). Similar to the LSU, each of these units is responsible for controlling a single block, i.e., n driving units for the n blocks. Each actuator receives from the PDC the exact quantity required of the amendment it controls.

2.4. Instrumentation

In order to build the on-line monitoring and real-time control system of SMART, all sensors and control elements must be adaptable to automatic operation and because of the aggressive environment of landfills, instruments have to be durable, chemical and corrosion resistant, and robust (especially against overburden pressure). Typical sensor requirements to monitor in-place waste, leachate, and biogas for a generic block in the SMART system are shown in Figure 7. In this instrumentation system, sensors are controlled remotely by the PSDP, whereas the final control elements are controlled by the PDC. The PSDP/PDC unit transmits/receives the input/output signals via standard communication protocols (such as RS-232 or RS-485) to/from the MCU.

In-place waste is monitored by LSU bundles which are evenly distributed in the BL body forming a three-dimensional grid. Each bundle measures moisture content, temperature, and water pressure (Figure 7, objects 10-12, respectively). Electrical resistivity and capacitance (frequency domain) technologies are suitable technologies for moisture measurements, and are compatible with automated monitoring systems. Waste temperature can be measured using thermocouples or thermistors, with the latter built into most commercial moisture and pressure sensors. However, thermocouples are still the preferred stand-alone temperature monitoring devices because they are reliable, inexpensive and the higher accuracy of thermistors is not needed in landfill applications. Thermocouples of types T (-250 to 350°C) or K (-200 to +1350°C) or J (-40 to +750°C) are widely used in landfill applications. Pore water pressure is measured using vibrating wire or solid state piezometers. Settlement is measured using settlement plates, whereas hydrostatic head on the liners is monitored by differential pressure transducers (Figure 7, objects 13 and 14, respectively). Landfill biogas flow is metered and totalized onsite using turbine or thermal dispersion flow meters (Figure 7, object 15). Biogas is analyzed for carbon dioxide and methane with dual wavelength infrared gas analyzers, whereas, oxygen is monitored via a zirconium dioxide sensor (Figure 7, object 16).

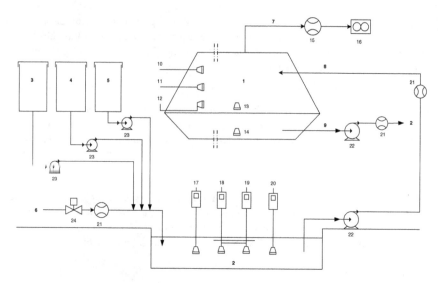

Figure 7. Schematic instrumentation diagram of the SMART system: (1) bioreactor landfill cell; (2) leachate storage facilities; (3) buffer tank; (4) inocula tank; (5) nutrient tank; (6) water supply; (7) collected biogas flow line; (8) recirculated leachate flow line; (9) collected leachate flow line; (10) moisture sensors; (11) thermocouples; (12) piezometers; (13) settlement plates; (14) pressure transducers; (15) gas flow meter; (16) inline gas analyzer; (17) liquid thermistor; (18) pH probe; (19) ORP electrode; (20) ammonia-selective electrode; (21) liquid flow meters; (22) pumps; (23) dosing pumps; and (24) electrically actuated valves.

Collected leachate is analyzed for major parameters such as chemical oxygen demand (COD), volatile fatty acids (VFA), oxygen reduction potential (ORP), and pH. The pH, ORP, and ammonia are measured by inline double-junction temperature-compensated pH, ORP, and ion-selective electrodes, respectively connected to a transmitter (Figure 7, objects 18-20, respectively). Online analyzers for COD and VFA are commercially available, however due to their high capital and maintenance costs as well as the slow reaction time in landfill processes, determination of these parameters by standard offline analytical methods is still the most economic and practical approach, and therefore is used in SMART. Leachate flow rate and cumulative flow are measured via Coriolis mass flow sensors equipped with totalizers (Figure 7, object 21). On the control side, GDU units include electrically actuated double-diaphragm or peristaltic pumps, and diaphragm valves that can safely handle particulate-laden and corrosive liquids (Figure 7, objects 22-24, respectively).

2.5. Expert system

The control program receives the measured data (inputs), processes them within the MCU expert system, makes the control decision, and sends it to the LDUs in the form of quantified commands. The expert system is designed to determine the required volumes of leachate, make-up water as well as bioaugmentation and nutritional amendments necessary to provide the BL microbial consortia with their optimum growth requirements. It was assumed that the chemical/biochemical characteristics of the effluent leachate are representative of the conditions within the whole BL waste matrix. Regulating the characteristics of the recirculated leachate alters the characteristics of the waste matrix through which it percolates, in a gradual stepwise manner, over a number of cycle times. It is the premise of the system to identify the current operational phase of the controlled bioreactor, and accordingly determines quantities of leachate, buffer, supplemental water, and inoculum/nutrition amendments required to provide the landfill microbial consortia with their growth needs.

The data flow diagram and hierarchy of the developed control program are shown in Figure 8. The structure of the program is composed of multiple cascading mathematical calculations (MCs 1-5) based on a main logic controller (LC). The control sequence in Figure 8 is repeated every operational cycle. The LC is discussed below (why a logic controller is needed? which method should be used? how the model is developed?), and then the mathematical calculations are presented.

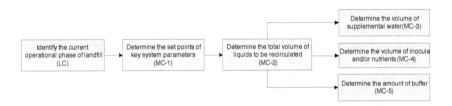

Figure 8. Dataflow diagram of the control program

2.5.1. Logic controller

Bioreactor landfills undergo the typical waste decomposition phases of sanitary landfills (in the order of: *initial/aerobic, transition, acid formation, methane generation,* and *final maturation* phases) but in a shorter time frame [7, 9, 31]. The determination of the current operational phase of the BL is vital because the bacterial consortia change significantly throughout the BL lifetime, and accordingly so do the conditions for their optimal growth. In order to stimulate the decomposition process and consequently biogas generation, those requirements have to be adequately provided. Practically, the identification of the dominant operational phase of the BL at a given time is challenging especially because of factors such as the heterogeneity of the waste which may cause system parameters not to follow their normal expected trends. Moreover, since landfills receive waste continually over several years, these progressive phases occur simultaneously, but in different neighbouring locales. The temporal and spatial dimensions of each phase depends on many factors such as waste characteristics, landfill design, operational strategy, and environmental conditions, that can be characterized by changes in various physical and biochemical indicator parameters.

In recent years, intelligent control of large-scale industrial processes has brought about a revolution in the field of advanced process control [32]. Knowledge-based techniques, such as fuzzy logic which uses linguistic control rules capturing the know-how of the experienced human operators, proved to be robust and reliable solutions for dealing with complex and ill-defined processes, such as those encountered in the operation of a BL. In fact, no conventional controller could efficiently operate such a complex process because it is practically impossible to predict its behaviour especially with the heterogeneity of waste. Fuzzy logic has been applied successfully to control various biological treatment systems such as anaerobic digesters [33], biological reactors [34], and wastewater treatment plants [35].

Therefore, the objective was to employ the modeling capabilities of fuzzy logic in developing a knowledge-based controller that determines the operational phase based on quantifiable input parameters of leachate and biogas, while taking uncertainty issues into consideration. The selected input variables include the leachate's COD, total volatile acids (TVA), pH, ORP, and methane content ($\%CH_4$) in biogas, whereas, the single output variable is an index that defines the current operational phase of the BL, hereafter named the *Phase Index*.

Model development

The **first step** in the design of a Fuzzy Logic Controller (FLC) is to build the *data base* which contains the membership functions defined for each input and output variable. Each variable is expressed by linguistic terms (fuzzy sets) within its predefined range (universe of discourse). The degree of truth of a fuzzy set A is defined by a membership function μ_A, which is represented by a real number in the interval [0, 1] depending on the degree at which it belongs to the set. This is different from conventional numerical sets where an element either belongs or does not belong to a particular set (membership = 0 or 1). This distinctive feature is advantageous for controlling biological ecosystems, like the BL, where the change in input variable does not cause the controlled process to shift abruptly from one state to another. Instead, as the variable changes, it loses its membership in one fuzzy set while gaining membership in the next. This is a logical approach to account for the fact that a part of the BL may be in a particular operational phase, while adjacent parts may be in other phases.

Membership functions (MFs) can have different shapes such as triangular, trapezoidal, bell-shaped (Gaussian), or wave-shaped (Sigmoid). In the present FLC, fuzzy sets were defined by trapezoidal and/or triangular (special case of the trapezoidal shape) MFs where the uncertainty in each variable is represented by the most likely interval (i.e., the range at membership degree = 1.0) and the largest likely interval (i.e., the range at membership degree = 0.0) as shown in Figure 9. These intervals facilitate the interpretation of overlapping and disagreement in the compiled data ranges. The membership value is constant in the most likely interval [b, c], and increasing linearly from 0 to 1 between (a & b) and decreasing linearly from 1 to 0 between (c & d), thus providing the trapezoidal shape. For the special case of the triangular MF, the only difference to the trapezoidal MF is that the most likely interval [b, c] is a single point.

Figure 10 shows the MFs defined for a sample input (ORP) and the single output (*Phase Index*). The linguistic labels (fuzzy sets) used to describe the ORP values are *positive* (P), *zero* (Z), *negative* (N), and *very negative* (VN). The '*Phase Index*' variable was defined by the basic phases that typically characterize the BL lifespan; *aerobic* (A), *transition* (T), *acid formation* (AF), and *methane generation* (MG).

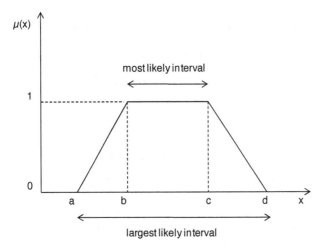

Figure 9. Typical trapezoidal membership function

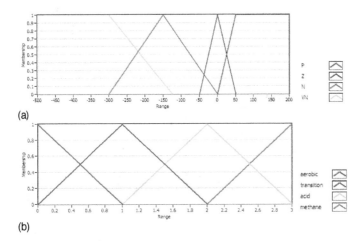

Figure 10. Membership functions for: a) ORP, and b) Phase Index

The **second step** in the design of FLC is developing the *rule base* for the controlled process. The *rule base* consists of fuzzy rules which are stated as IF–THEN statements that define the system behavior and predict the output variable. A typical fuzzy rule can include several variables in the antecedent (IF part) and consequent (THEN part) of the rule. If a rule has more than one antecedent, a fuzzy operator such as AND, OR, or NOT, is used to connect them, and to determine how to calculate the truth value of the aggregated rule antecedent. In the present

FLC, five basic statements (rules) were created to define the expected operational phase based on different quantifiable parameters. The probabilistic-type of the OR operator, which uses the probabilistic sum of the degrees of membership of the antecedents, was applied in the formulated rules. The following is an example of the developed rule base statements:

IF 'ORP' is 'VN' OR 'pH' is 'HN' OR 'COD' is 'H' OR 'TVA' is 'I' OR '%CH$_4$' is 'H'

THEN 'Phase Index' is 'MG'

In the above rule, VN, HN, H, I, H, and MG are fuzzy sets that denote *very negative, high neutral, high, intermediate, high,* and *methane generation,* respectively. The complete fuzzy rules as well as parameters of membership functions defined in the FLC are presented in [36].

Example: Based on the compiled knowledge base, when the ORP of the leachate is -250 mV, it has a 0.3 membership in the *"negative"* fuzzy set, and a 0.7 membership in the *"very negative"* fuzzy set (see Figure 10a). This allows the single input (-250 mV) to be processed with multiple rules, i.e., the fuzzy rules that include *"negative"* and *"very negative"* ORP in their antecedents. Although all the invoked rules influence the output, the rules with higher truth values (*"very negative"* in this case) have the greatest effect. This weighing system helps in dealing with the uncertainties in the landfill ecosystem, as well as simplifying the complexity of the controlled process.

The data base and rule base represent the knowledge components based on which the FLC makes the decision. The knowledge was compiled from information presented in [7, 37-39]. Table 1 shows the reported ranges of the input system parameters in the compiled studies.

Parameter	Study	Phase II Transition	Phase III Acid Formation	Phase IV Methane Generation	Phase V Maturation
COD, mg/l	[7]	20 - 20,000	11,600 - 34,550	1,800 - 17,000	770 - 1,000
	[38]	-	15,000 - 41,000	1,000 - 41,000	-
TVA, mg/l	[7]	200 - 2,700	1 - 30,730	0 - 3,900	0
	[38]	-	7,000 - 15,000	10,000	0
pH	[7]	5.4 - 8.1	5.7 - 7.4	5.9 - 8.6	7.4 - 8.3
	[38]	-	5 - 6	5.6 - 7.1	-
	[37]	-	5.8 - 6	6 - 7.8	7.1
%CH$_4$	[38]	-	0	0 - 50	40
	[37]	-	-	23 - 62	-
ORP, mV	[38]	50 - (-50)	50 - 0	0 - (-125)	-
	[39]	-	(-100)	(-300)	-

Table 1. Ranges of selected system parameters at the main operational phases

The data base and rule base are incorporated in the typical FLC components, shown in Figure 11, which includes: (1) *fuzzification unit,* (2) *inference engine,* and (3) *defuzzification unit.* The *fuzzification unit* converts the input variables into fuzzy sets using the predefined membership functions. The *inference engine* then processes the fuzzy inputs based on their relevant fuzzy rules, and determines the fuzzy output(s). As mentioned above, the *inference engine* invokes more than one rule, which results in having different memberships in multiple output fuzzy sets. In the present LC, the *inference engine* uses the product implication method in which each output MF is scaled down at the truth value of the corresponding aggregated rule antecedent. The output from this step is an irregular area under the scaled-down membership functions. Finally, the *defuzzification unit* incorporates a number of fuzzy sets in a calculation that gives a single numeric value for each output.

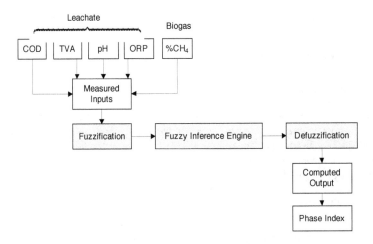

Figure 11. Typical structure of a fuzzy logic controller

In order to help visualize the non-linear characteristics of the *Phase Index*, surface plots were generated by varying two variables while the other variables remained constant. This can generate an infinite number of response surface, however if grouped for each pair of inputs, the number of possible groups of response surfaces becomes equal to the combination $C\,(n,\,2)$ $= n! \,/\, 2! \,(n - 2)!$ where n is the number of input variables. In the present FLC, 10 groups of response surfaces can be established for the 10 possible pairs of input variables. Figure 12 shows the response of the output variable '*Phase Index*' to changes in two pairs of the input variables, namely ORP and COD as well as TVA and pH, at the average defined value for the other input variables. The non-linear variation of the response intensity for the different values of input variables is considered one of the main advantages of the fuzzy logic system. Moreover, SMART's numeric representation for the operational phase offers a unique feature being able to obtain the transitional stage of the controlled BL. For example, when the '*Phase Index*' is equal to 2.7, this means that the bioreactor is transitioning from the acid formation phase (2.0) to the methane generation phase (3.0). The value (2.7) indicates also that the BL microbial ecosystem is closer to the methanogenic stage.

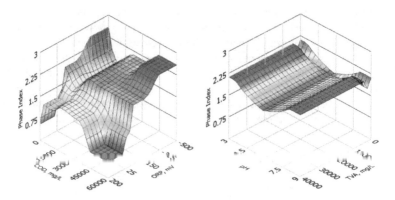

Figure 12. Response surfaces for two pairs of inputs: 1. COD/ORP (left), and 2. TVA/pH (right)

2.5.2. Mathematical calculations

As shown in Figure 8, the program sequence starts with the logic controller (LC) which identifies the current operational phase of the BL based on quantifiable characteristics of the generated leachate and biogas. The output of LC is a real number in the interval [0, 3] that expresses the BL operational phase, where 0 is the aerobic phase and 3 is the methanogenic phase. The output from LC is the input to the first mathematical step (MC-1).

Target set points

In MC-1, set points of pH (leachate), Carbon/Nitrogen (C/N) ratio (leachate), and moisture content (solid waste matrix) are computed based on the BL operational phase determined from LC. Table 2 shows default set points used in the present study for the two main BL operational phases. It should be noted that these set points may vary depending on several site-specific factors such as holding capacity of waste matrix, degree of compaction, and waste composition.

Parameter	Medium	Phase III Acid Formation	Phase IV Methane Generation
pH	Leachate	5.5-6.5	6.8-7.2
C/N ratio	Leachate	10	15
Moisture content, %	Waste Matrix	50	60

Table 2. Set points of process parameters at the Acid Formation and Methane Generation phases

MC-1 applies linear interpolation between the predefined parameter values (shown in Table 2). The parameter setpoint (S) at a given phase (P) can be calculated as follows:

$$S_P = S_i + [(S_{i+1} - S_i) \times (P - i)] \tag{1}$$

Where P is the computed phase from LC, i is the integer part from the computed *Phase Index* P, S_i is the setpoint at phase i, and S_{i+1} is the setpoint at phase $i+1$.

Recirculation volume

MC-2 computes the total required volume of recirculated liquids to raise the water content of the waste matrix from its current level to the desired setpoint. The liquid volume is calculated as follows:

$$V_{liquid} = \left(\frac{S_{mc} \times w}{\rho_{water}} \right) - \left(\theta \times V_{waste} \right) \tag{2}$$

Where V liquid is the total required volume of liquids to be added in a cycle (m³), S_{mc} is the setpoint for the gravimetric water content (calculated in MC-1), θ is the measured volumetric water content, ρ water is the water density (t/m³), and w is the bulk weight of the waste (t).

Supplemental water addition

One of the main benefits of supplemental water addition is to dilute elevated concentrations of pollutants in leachate which may inhibit the microbial consortia in the waste matrix. The primary inhibitors in MC-3 can include, but are not limited to, ammonia-nitrogen, VFA, and their free unionized fractions, as well as alkali cations. The concentrations of selected inhibitors are used to compute a factor (D) for the required dilution (i.e., dilution water as a fraction of the liquid recirculated). D is calculated as the greatest of individually calculated dilution indices required to bring each of the potential inhibitors, if any, to its nontoxic range, as follows:

$$D = Max \left(1 - \frac{C_{target}}{C_{inhibitor}} \right) \tag{3}$$

Where $C_{inhibitor}$ is the concentration of an inhibitor in leachate (g/m³), and C_{target} is the nontoxic concentration of that inhibitor (g/m³). The required supplemental water volume can then be calculated by multiplying the volume of leachate produced in previous operational cycle by the dilution factor.

Nutritional requirements

Next, MC-4 determines additional nutrient requirements using the set point for C/N ratio as well as the concentrations of TOC and TN of the generated leachate. The addition of a nitrogen source to the BL is controlled according to the C/N ratio. The volume of nutritional source is calculated as:

$$V_{nutrients} = \frac{\left(\dfrac{TOC}{S_{C/N}} \right) - TN}{TN_{nutrients}} \times V_{liquid} \tag{4}$$

Where $V_{nutrients}$ is the required volume of the nutritional source (m³), $S_{C/N}$ is the setpoint calculated for the C/N ratio, V_{liquid} is the volume of liquid calculated in MC-2 (m³), and TN, TOC and TN nutrients are the concentrations of total nitrogen of diluted leachate, total organic carbon of diluted leachate, and total nitrogen of the nutritional source to be used, respectively (g/m³).

Buffering requirements

Next, the required amount of buffer is calculated in MC-5. The buffer salt is used to adjust the pH and provide external source of alkalinity to the system. MC-5 calculates the required bicarbonate alkalinity to be added to the leachate regardless of the resultant pH. The buffer is added to provide the difference between the required alkalinity (CO_2/water buffering system) and the available alkalinity in the system. The available bicarbonate alkalinity can be calculated as:

$$BA = ALK - 0.83 \times f \times VFA \tag{5}$$

Where BA is the bicarbonate alkalinity (mg $CaCO_3$/L), ALK is the total alkalinity (mg $CaCO_3$/L), VFA is the concentration of the volatile fatty acids (mg/L), 0.83 is a unit conversion factor (Equivalent weight of $CaCO_3$/Equivalent weight of VFA), and f is a factor for the percentage of VFA titrated at the pH endpoint of the alkalinity test. On the other hand, the required alkalinity (RA) for the CO_2/water buffering system can be calculated as:

$$RA = K_1 \times K_H \times P_{CO_2} \times E_{CaCO_3} \times 10^{S_{pH}} \tag{6}$$

Where RA is the required concentration of bicarbonate ion for CO_2 neutralization (g $CaCO_3$/L), K_1 is the ionization constant for carbonic acid, K_H is the hydration equilibrium constant, P_{CO_2} is the partial pressure of CO_2 in the system (fraction of CO_2 in the composition of air), S_{pH} is the target pH as computed in MC-1, and E_{CaCO3} is the equivalent weight of $CaCO_3$. The added alkalinity is the difference between the required and available alkalinity in the system. The volume of buffer to provide the required alkalinity can be calculated as:

$$V_{buffer} = \frac{\left[RA - BA\right] \times E_{buffer} \times V_{liquid}}{C_{buffer}} \tag{7}$$

Where V_{buffer} is the required volume of buffer, E_{buffer} is the equivalent weight of buffer salt, C_{buffer} is the concentration of buffer salt in solution, and V_{liquid} is the volume of recirculated liquid. The amount of buffer to be added should be equal or greater than the amount required to bring the pH up to the setpoint calculated from MC-1.

3. Application and evaluation of SMART

The new concepts proposed and incorporated in SMART were demonstrated in a real operational prototype. Specifically, the concept of temporal determination of the BL operational phase as the starting step for initiating the other subsequent computations to determine the various amendments to be added to manipulate the leachate recirculated. Concomitantly, the main objectives of this research phase were to: (1) implement the software and hardware components of SMART on a pilot-scale prototype, and (2) evaluate the system viability to control the BL versus a conventional open-loop leachate control (OLLC) scheme, in which recirculation rate is fixed and the leachate quality is not changed.

3.1. Experimental setup

Experimental work was conducted on two bioreactor setups; Cell-1 and Cell-2. Figure 13 shows the configuration of a single bioreactor cell (675 litres volume) with its leachate collection and recycling tanks. An equal mixture of residential and food wastes were thoroughly mixed while loaded to the bioreactor cells. The average total organic fraction and water content of the mixed waste was 73%, and 48%, respectively. Each bioreactor cell was equipped with three type-T thermocouples measuring temperature in different radial positions at three equidistant vertical levels in the waste matrix. In addition, three moisture sensors were placed at the same monitoring spots in order to measure the volumetric water content using frequency domain technology. The biogas generated went through a micro-turbine wheel flow meter, followed by an inline infrared methane analyzer. Leachate was collected by gravity from a lower outlet port connected to a collection tank with a mechanical mixer. This tank also received the flow from the amendments' tanks through tube lines with actuated solenoid valves (SMART-controlled). The recirculated leachate was manipulated by adding amendments such as inoculum (anaerobic digester sludge), nutritional source (plant fertilizer), buffer (sodium bicarbonate), and supplemental water. After mixing with amendments, leachate was recycled in a cyclic batch mode using a submersible pump (SMART-controlled).

After loading the bioreactor cells, the first nine months were used to examine the communication and synchronization between system components, as well as test run of the system. By the end of this period, Cell-2 has already started producing methane and surpassed Cell-1 in terms of all performance and evolution parameters. In order to effectively assess the system, SMART was applied on Cell-1 (the inadequately performing cell) for four months so as to evaluate the performance. In parallel, Cell-2 was running according to an OLLC scheme, at a constant rate of leachate recirculation equal to a predetermined percentage (8%) of the initial volume of waste matrix. The discussion is presented in two main sections: (1) assessment of the control actions made by SMART, and (2) evaluation of the system performance through its effect on leachate and biogas.

3.2. Evaluation of SMART control decisions

There has been no consensus in the literature on the optimal leachate recirculation rates in BLs, and the reported rates are extremely diverse to over 400 fold [17]. It was also found

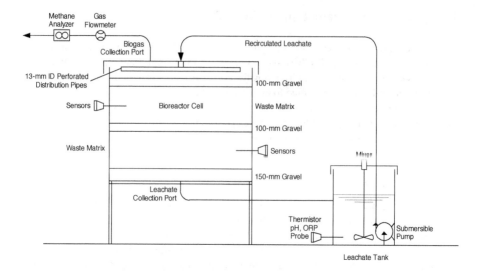

Figure 13. Configuration and instrumentation of the prototype bioreactor cell

that higher recirculation rates do not necessarily achieve better performance of the BL [1, 24]. Alternatively in SMART, recirculation rates vary based on the site-specific and real-time conditions, and so every BL is controlled according to its own evolution. Figure 14 shows the different recirculated volumes of leachate as determined by SMART for Cell-1, as well as the various fractions of leachate, water, buffer, and sludge in the recirculated liquid in each cycle. It can be observed that the calculated volumes of leachate and other amendments did not follow a predictable trend, and they varied significantly over time (34±7 L/cycle). However, the volumes of amendments followed a decreasing trend that seemed to restart every four operational cycles (1-4 & 5-8).

3.2.1. System evolution

The *Phase Index*, determined by the logic controller, for the two cells is shown in Figure 15. The progress of Cell-1 surpassed that of Cell-2 which was also evolving but at slower rate. It can be seen that, while at the beginning of this test, Cell-2 was ahead of Cell-1 with a PI of 1.6 (Cell-1) versus 2.0 (Cell-2), the SMART-controlled Cell-1 was able to catch up and actually surpassed Cell-2 in four operational cycles. It is clear that since Cell-2 was running with an open-loop control scheme, the improvement in the evolution pattern of Cell-1 can be mostly attributed to the implementation of SMART. The fuzzy logic controller was able to track the BL evolution by identifying the operational phase at any time based on multiple parameters of leachate and biogas. The computed *Phase Index* descri-bed the transitioning progress between the main phases of BL, which enabled the interpo-lation of the evolving growth requirements for the bacterial population inside the BL, and led to successful transition from one phase to another.

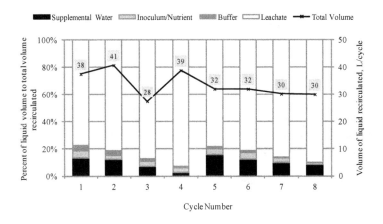

Figure 14. Cyclic recirculated liquid volumes and amendment fractions in Cell-1

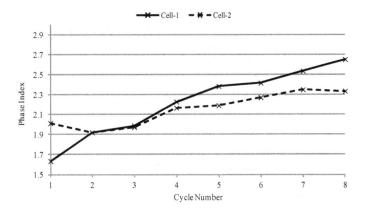

Figure 15. Progress of the Phase Index of Cell-1 and Cell-2

3.2.2. Control strategy

During the operation period, the operator had to interfere occasionally so as to insure the control actions address all potential problems. This man-computer interaction was crucial due to: (1) the instability and unexpected behavior of the BL system, in part due to its complexity and nonlinear responses, and (2) the fact that the reasoning of the fuzzy logic is limited to its knowledge base. Therefore, applying a semi-automated control strategy, rather than a fully automated one, was found to achieve more stable performance of the system. In this control strategy, SMART collects and analyzes the data, performs the computational effort to determine the optimum operational strategy, and then aids the site operator to apply the final operational decision through the computer interface.

3.2.3. Feedback control scheme

The control actions determined by SMART were based on multiple leachate and biogas parameters acquired from previous cycles. The response time of the BL ecosystem, i.e., time from changing a system parameter to when its effect (feedback) on system performance is detected, was found to be sufficient to facilitate the application of the feedback control scheme. The BL performance was significantly improved with the application of closed-loop control (in Cell-1) as opposed to an open-loop strategy (in Cell 2).

3.3. Evaluation of process parameters

3.3.1. Organic matter

The development of oxidizable organic concentration in the leachate produced is plotted in terms of COD and VFAs in Figure 16. The average degradation rate of COD in Cell-1 (controlled by SMART) was 330 mg/L.d compared to 110 mg/L.d in Cell-2. The COD concentrations in leachate from Cell-2 were fluctuating and the final COD was about 10% less than the initial concentration (from 116 to 105 g/L). After 40 days, COD concentration in leachate from Cell-1 was consistently less than that of Cell-2 which shows that the implementation of SMART had a positive effect on the degradation of organic matter.

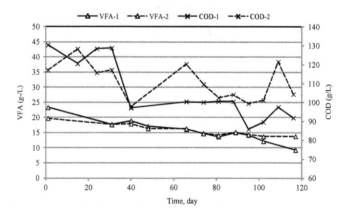

Figure 16. Evolution of organic concentration of leachate from Cell-1 and Cell-2

As shown in Figure 16, the conversion of VFAs to methane was increasing slowly resulting in lower and mostly similar concentrations of VFA in leachate from both cells. However at day 95, the VFA concentration in Cell-1 started to drop, leading to an overall conversion rate of 120 mg/L.d compared to 50 mg/L.d in Cell-2. The last recorded VFA concentration was less than

10 g/L in leachate from Cell-1 compared to 14 g/L in Cell-2. It is therefore clear that the SMART control system stimulated the methanogenic activity which gradually consumed the produced VFAs, until the conversion rate of VFA became greater than the production rate (starting from the 95th day).

3.3.2. Biogas production

The CH_4 fractions of the biogas produced from both cells are shown in Figure 17. The performance of Cell-1 in terms of the rate of increase of the CH_4 fraction was improved. SMART was successful in leading the cell through the transitional stage from acid formation to methane generation. The CH_4 content increased from 10 to 62% in Cell-1 in a four-month period, while Cell-2 continued to increase but at slower rate going from 40 to 58%. Based on the equations of the trend lines fitted to the actual data of cumulative production in Figure 17, the rate of increase in methane production in Cell-1 was 1.7 fold higher than that of Cell-2. By the end of operation, the cumulative biogas production reached 23 and 14 m³ which corresponds to a specific production of 61 and 35 L/kg of waste in Cell-1 and Cell-2, respectively.

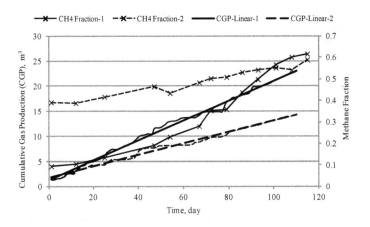

Figure 17. Development of methane production and methane content in the biogas produced

3.4. Future aspects and potential implications

The implementation of a sensor-based control strategy in full-scale BL faces two main issues: (1) instrumentation of the system, and (2) the heterogeneity of the waste matrix which affects the degree to which the measurements are representative. While in-situ measurements of leachate and biogas are well established, the main instrumentation problem is the subsurface monitoring for in-place waste, such as: moisture content and temperature. The difficulty arises from the following issues: (1) instrument failure is most likely to occur since no specialized sensing technologies and installation procedures exist for landfill application, (2) installation techniques are very challenging and obstruct daily site operations, (3) cables are subject to physical damage due to heavy equipments, differential settlement, and aggressive environment, and (4) cable conduits create pathways for lateral breakout of leachate and gas. It is clear that, with all these operational issues, current monitoring techniques are neither robust nor efficient. The solution for these issues can be realized via two approaches: (1) using non-intrusive surface methods for subsurface monitoring; e.g., for moisture measurements: seismic waves [40], ground penetrating radar [41], and fiber optics [42], or (2) using wireless communication techniques to eliminate the huge capital cost and operational problems associated with conventional wired techniques [43]. Both approaches can also solve the heterogeneity problem in a way that: in the first approach, a three-dimensional image of moisture distribution can be produced, and in the second approach, more wireless sensors can be used to give higher resolution data. In addition, soft computing methods can be used to deal with the uncertainty in measurements, and by using adaptive systems, monitoring and control programs can learn and adapt to the controlled BL.

Given the rapid development in both instrumentation and full-scale applications of BLs, it is expected that robust subsurface monitoring techniques will appear in the near future. However, research in the area of advanced BL process control like the present research, has to move in-parallel and not to wait until a flawless method to measure subsurface parameters is ready. In fact, process control research can motivate the search for robust and reliable sensory equipment. Therefore, SMART can be currently applied in full-scale BLs if some technical modifications of in-situ monitoring are considered, e.g., monitoring in/out liquid to/from the BL can effectively replace the in-situ measurements of moisture content by means of continuously conducting a real-time water balance.

4. Conclusions

The present work developed a control framework in which an expert system is responsible for the operation of BLs. The main control objective of the system was to optimize the performance of the BL by manipulating the quantity and quality of leachate recirculated so as to supply the microbial consortia inside the BL with their optimal growth requirements. The proposed control framework and guidelines were described, and an assessment was conducted for a SMART-controlled pilot-scale BL in order to examine the applicability, feasibility, and effectiveness of the technology. The following conclusions were drawn:

1. The control system successfully determined the quantity and quality of recirculated liquid based on the BL operational stage and multiple process leachate and biogas parameters.

2. The performance of the BL was significantly improved with the application of closed-loop control as opposed to an open-loop strategy.

3. Leachate manipulation techniques, such as buffering, bioaugmentation, and supplemental water addition, were proven to be potentially effective control tools that are able to adjust/optimize the leachate characteristics.

4. Recirculating variable calculation-based amounts of leachate and other amendments resulted in a positive influence on the overall performance of the BL system.

5. The pilot-scale implementation of SMART demonstrated the feasibility of the system. Since all the incorporated hardware components are commercially available, the system can be readily scaled-up to a larger scale application.

Author details

Mohamed Abdallah and Kevin Kennedy

Department of Civil Engineering, University of Ottawa, Ottawa, Canada

References

[1] Warith M. Bioreactor Landfills: Experimental and Field Results. Waste Management. 2002;22(1):7-17.

[2] SWANA. Summary List of North American Bioreactor Landfill Projects 2004 [cited 2012 December 1st]. Available from: www.swana.org/pdf/swana_pdf_295.pdf.

[3] Reinhart DR, McCreanor PT, Townsend T. The Bioreactor Landfill: Its Status and Future. Waste Management & Research. 2002;20(2):172-86.

[4] Warith M, Smolkin P, Caldwell J. Effect of Leachate Recirculation on Enhancement of Biological Degradation of Solid Waste: Case Study. Practice Periodical of Hazardous, Toxic, and Radioactive Waste Management. 2001;5(1):40-6.

[5] Zhao XD, Musleh R, Maher S, Khire MV, Voice TC, Hashsham SA. Start-up Performance of a Full-scale Bioreactor Landfill Cell under Cold-Climate Conditions. Waste Management. 2008;28(12):2623-34.

[6] Cossu R, Blakey N, Trapani P, editors. Degradation of mixed solid waste in conditions of moisture saturation. International Sanitary Landfill Symposium; 1987; Cagliari, Italy.

[7] Reinhart DR, Townsend TG. Landfill Bioreactor Design and Operation. Boca Raton, FL: Lewis Publishers; 1998.

[8] He PJ, Shao LM, Qu X, Li GJ, Lee DJ. Effects of Feed Solutions on Refuse Hydrolysis and Landfill Leachate Characteristics. Chemosphere. 2005;59(6):837-44.

[9] Pohland FG, Alyousfi B. Design and Operation of Landfills for Optimum Stabilization and Biogas Production. Water Science and Technology. 1994;30(12):117-24.

[10] ElFadel M, Findikakis AN, Leckie JO. Environmental Impacts of Solid Waste Land-filling. Journal of Environmental Management. 1997;50(1):1-25.

[11] Hilger HH, Barlaz MA. Anaerobic decomposition of refuse in landfills and methane oxidation in landfill covers. 2007 (Ed.3):818-42.

[12] Pohland FG, Behavior and Assimilation of Organic and Inorganic Priority Pollutants Codisposed with Municipal Refuse: Project Summary: U.S. Environmental Protection Agency, Center for Environmental Research Information; 1993.

[13] San I, Onay TT. Impact of Various Leachate Recirculation Regimes on Municipal Sol-lil Wasle Liegradation journal of Hazardous Materials, 2001,87(1-3).259-71.

[14] Mehta R, Barlaz MA, Yazdani R, Augenstein D, Bryars M, Sinderson L. Refuse Decomposition in the Presence and Absence of Leachate Recirculation. Journal of Environmental Engineering-ASCE. 2002;128(3):228-36.

[15] Bilgili MS, Demir A, Ozkaya B. Influence of leachate recirculation on aerobic and anaerobic decomposition of solid wastes. Journal of Hazardous Materials. 2007;143(1-2):177-83.

[16] Filipkowska U. Effect of Recirculation Method on Quality of Landfill Leachate and Effectiveness of Biogas Production. Polish Journal of Environmental Studies. 2008;17(2):199-207.

[17] Benbelkacem H, Bayard R, Abdelhay A, Zhang Y, Gourdon R. Effect of Leachate Injection Modes on Municipal Solid Waste Degradation in Anaerobic Bioreactor. Bioresource Technology. 2010;101(14):5206-12.

[18] Morris JWF, Vasuki NC, Baker JA, Pendleton CH. Findings from Long-Term Monitoring Studies at MSW Landfill Facilities with Leachate Recirculation. Waste Management. 2003;23(7):653-66.

[19] Buivid MG, Wise DL, Blanchet MJ, Remedios EC, Jenkins BM, Boyd WF, et al. Fuel Gas Enhancement by Controlled Landfilling of Municipal Solid-Waste. Resources and Conservation. 1981;6(1):3-20.

[20] Chugh S, Clarke W, Pullammanappallil P, Rudolph V. Effect of Recirculated Leachate Volume on MSW Degradation. Waste Management & Research. 1998;16(6): 564-73.

[21] Jiang JG, Yang GD, Deng Z, Huang YF, Huang ZL, Feng XM, et al. Pilot-scale Experiment on Anaerobic Bioreactor Landfills in China. Waste Management. 2007;27(7): 893-901.

[22] Suna Erses A, Onay TT. Accelerated Landfill Waste Decomposition by External Leachate Recirculation from an Old Landfill Cell. Water science and technology. 2003;47(12):215-22.

[23] Lisk DJ. Environmental Effects of Landfills. Science of the Total Environment. 1991;100:415-68.

[24] Sponza DT, Agdag ON. Impact of Leachate Recirculation and Recirculation Volume on Stabilization of Municipal Solid Wastes in Simulated Anaerobic Bioreactors. Process Biochemistry. 2004;39(12):2157-65.

[25] Sanphoti N, Towprayoon S, Chaiprasert P, Nopharatana A. The Effects of Leachate Recirculation with Supplemental Water Addition on Methane Production and Waste Decomposition in a Simulated Tropical Landfill. Journal of Environmental Management. 2006;81(1):27-35.

[26] Stegmann R, Spendlin HH. Enhancement of biochemical processes in sanitary landfills. International Sanitary Landfill Symposium; Cagliari, Italy1987. p. 1-16.

[27] Rees JF. Optimization of Methane Production and Refuse Decomposition in Landfills by Temperature Control. Journal of Chemical Technology and Biotechnology. 1980;30(8):458-65.

[28] Pohland F. Anaerobic treatment: fundamental concepts, applications, and new horizons. In: Malina J, Pohland F, editors. Design of Anaerobic Processes for the Treatment of Industrial and Municipal Wastes. Water Quality Management Library. Lancaster, USA: Technomic; 1992.

[29] McCarty P. Anaerobic Waste Treatment Fundamentals, Part Three: Toxic Materials and their Control. Public Works. 1964;92(11).

[30] Kim E, Liu B, Roehm T, Tout S. Feedback Control 2007 [cited 2012 December 1st]. Available from: https://controls.engin.umich.edu/wiki/index.php/Feedback_control.

[31] Kim J, Pohland FG. Process Enhancement in Anaerobic Bioreactor Landfills. Water Science and Technology. 2003;48(4):29-36.

[32] Manesis SA, Sapidis DJ, King RE. Intelligent Control of Wastewater Treatment Plants. Artificial Intelligence in Engineering. 1998;12(3):275-81.

[33] Estaben M, Polit M, Steyer JP. Fuzzy Control for an Anaerobic Digester. Control Engineering Practice. 1997;5(9):1303-10.

[34] Bae H, Seo H, Kim S. Knowledge-based Control and Case-based Diagnosis based upon Empirical Knowledge and Fuzzy Logic for the SBR Plant. Water Science and Technology. 2006 2006;53(1):217-24.

[35] Garcia C, Molina F, Roca E, Lema JM. Fuzzy-based Control of an Anaerobic Reactor Treating Wastewaters Containing Ethanol and Carbohydrates. Industrial & Engineering Chemistry Research. 2007;46(21):6707-15.

[36] Abdallah M. A novel computational approach for the management of bioreactor landfills. Ottawa, Canada: University of Ottawa; 2011.

[37] Barlaz MA, Schaefer DM, Ham RK. Bacterial Population Development and Chemical Characteristics of Refuse Decomposition in a Simulated Sanitary Landfill. Applied and Environmental Microbiology. 1989;55(1):55-65.

[38] Pohland FG, Kim J. Insitu Anaerobic Treatment of Landfills for Optimum Stabilization and Biogas Production. Water Science and Technology. 1999;40(8).

[39] Gerardi MH. The Microbiology of Anaerobic Digesters. Hoboken, N.J.: Wiley-Interscience; 2003.

[40] Catley AJ, Seismic Velocity analysis to determine moisture distribution in a bioreactor landfill: Carleton University (Canada); 2007.

[41] Cassiani G, Fusi N, Susanni D, Deiana R. Vertical Radar Profiling for the Assessment of Landfill Capping Effectiveness. Near Surface Geophysics. 2008;6(2):133-42.

[42] Sayde C, Gregory C, Gil-Rodriguez M, Tufillaro N, Tyler S, van de Giesen N, et al. Feasibility of Soil Moisture Monitoring with Heated Fiber Optics. Water Resources Research. 2010;46:8.

[43] Nasipuri A, Subramanian KR, Ogunro V, Daniels JL, Hilger HA. Development of a wireless sensor network for monitoring a bioreactor landfill. GeoCongress 2006: Geotechnical Engineering in the Information Technology Age; 2006.

Chapter 8

Determination of Anaerobic and Anoxic Biodegradation Capacity of Sulfamethoxasole and the Effects on Mixed Microbial Culture

Zeynep Cetecioglu, Bahar Ince, Samet Azman,
Nazli Gokcek, Nese Coskun and Orhan Ince

Additional information is available at the end of the chapter

1. Introduction

During last decades, concentration of human and veterinarian antibiotics in the environment, natural and engineered systems have been increased because of high amount production and consumption. This situation has aroused great concern due to the possibility of harmful effects on human, animals and plants [1,2]. Occurrence and fate of these compounds are one of the main issues because of their unknown potential risks and their effects on the environment. Approximately 500 tonnes of them are produced and consumed every year in the worldwide. Antibiotics are resistant to conventional biological treatment process and the wastewaters including these compounds are directly discharged to the receiving water bodies without efficient treatment. Hospitals and pharmaceutical industries are the main sources of high antibiotic concentration release to the environment [3]. Also sewage systems can transport these molecules and/or their metabolites since metabolization of them by humans and animals cannot be achieved completely [4]. During the transportation of antibiotics throughout treatment plants, elimination of these compounds can occur via biodegradation, photolysis and sorption to sludge but ultimate degradation of these compounds cannot be achieved in conventional treatment plants [4, 5, 6]. As a result of the introduction of metabolized and/or active antibiotics to the receiving water bodies caused an increase in the ratio of multi-antibacterial resistant pathogens [7].

Sulfamethoxasole (SMX) is a sulfonamide bacteriostatic antibiotic that is used to treat urinary tract infections. SMX inhibits the multiplication of bacteria, since they are competitive inhibitors of *p-amino benzoic acid* in the folic acid metabolism cycle [8]. Sulfonamide antibiotics, including SMX, have been found in the activated sludge processes and digested sludges in varying concentration from ng/L to µg/L levels [5, 8]. Behaviour of sulfonamide antibiotics has been reported as recalcitrant molecules thus sorption and desorption are the main pathways on antibiotic elimination from aquatic phases [9, 10]. Biodegradability of SMX has not been widely studied for anaerobic systems. There are few studies about anaerobic biodegradability characteristics of SMX in the literature [11, 12].

In this chapter, the aim is to reveal the anoxic and anaerobic biodegradability characteristics of SMX and the effects of this compound on microbial community. In this scope, biodegradation capacity and the effects on the microorganisms were investigated by destructive batch tests based on a modified version of Anaerobic Biodegradability of Organic Compounds OECD 311 protocol [13] under three different electron acceptor conditions; nitrate reducing, sulfate reducing, and methanogenic conditions. Quantification of defined microbial groups was also carried out to determine the effects of SMX on abundance of microbial community.

1.1. Antibiotics

Antibiotics are among the most important groups of pharmaceuticals and chemotherapeutic agents that inhibit or terminate the growth of microorganisms, such as bacteria, fungi, or protozoa without affecting host [11, 12]. The term antibiotic used for drugs that block any of these microorganisms. Other terms as chemotherapeutics or antimicrobials are not synonymous because of their scopes; the term of antimicrobial is used for the medicine which is also effective against viruses and the expression "chemotherapeutical" referring to compounds used for the treatment of disease which kill cells, specifically microorganisms or cancer cells. The term "chemotherapeutical" may also refer to antibiotics (antibacterial chemotherapy).

The expression of antibiotic is originally used to describe any agent with biological activity against living organisms; however, "antibiotic" now refers to substances with antibacterial, anti-fungal, or anti-parasitical activity. During the years, this definition has been changed and now it includes also synthetic and semi-synthetic products. There are approximately 250 different compounds registered for use in medicine and veterinary application [16].

In this chapter, the term "antibiotic" refers only to drugs that kill or inhibit bacteria. Antibiotics that are sufficiently nontoxic to the host are used as chemotherapeutic agents in the treatment of infectious diseases of humans, animals and plants. They are extensively used for prevention and treatment of diseases caused by microorganisms in human and veterinary medicine as well as in aquaculture nowadays. Also, they are being still used as growth factor in livestock farming. Some compounds may be used for different purposes such as in growing fruit and in bee keeping other than human or veterinary medicine. The application purposes may vary from country to country.

Antibiotics are classified as their chemical structures and the mechanism of inhibition of microorganisms and they can be divided into subgroups such as β-lactams, quinolones, tetracyclines, macrolides, sulfonamides and others. The active compounds of antibiotics are often complex molecules, which may have different functionalities. In the environment, these molecules could be found as neutral, cationic, anionic, or zwitterionic forms. Because of the different functionalities within one molecule, their physicochemical and biological properties may change with pH levels [17].

1.2. Sulfamethoxazole

In this chapter, sulfamethoxazole (SMX) is selected as model compound. The systematic name of this compound is 4-amino-N-(5-methylisoxazol-3-yl)-benzenesulfonamide. Sulfamethoxazole and other sulfonamides have a similar structure to p-aminobenzoic acid and inhibit to the synthesis of nucleic acids in sensitive microorganisms by blocking the conversion of p-aminobenzoic acid to the coenzyme dihydrofolic acid, a reduced form of folic acid; dihydrofolic acid is obtained from dietary folic acid so sulfanomides do not have any influence on human cells. Their action is primarily bacteriostatic, although they may be bactericidal where concentrations of thymine are low in surrounding medium. The sulfonamides have a broad spectrum of action, but the development of widespread resistance has greatly reduced their usefulness, and susceptibility often varies widely even among nominally sensitive pathogens like Gram-positive and Gram-negative cocci.

There are several mechanisms of resistance including alteration of dyhydropteroate synthetase, the enzyme inhibited by sulfonamides, to a less sensitive form, or an alteration in folate biosynthesis to an alternative pathway; increased production of p-aminobenzoic acid; or decreased uptake or enhanced metabolism of sulfonamides.

Resistance may result from chromosomal alteration, or may be plasmid-mediated and transferable, as in many resistant strains of enterobacteria. High-level resistance is usually permanent and irreversible. There is complete cross-resistance between the different sulfonamides [18].

1.3. Consumption and occurrence

The yearly consumption of antibiotics worldwide is estimated between 500 tons [19]. Approximately 90% of the consumed antibiotics are excreted via urinary or fecal pathways from the human body after partial or no metabolism and they are transferred to the domestic sewage plants or directly to the environment. Conventional biological treatment of domestic sewage provides very low or no reduction for these compounds, which usually by-pass treatment and accumulate in the receiving waters.

Antibiotic consumption changes depending on the country and/or region however the situation is scarce and heterogenous. Country specific consumption for groups of antibiotics in DDDs can be found for Europe on the ESAC homepage [20]. Using patterns of different regions and countries are given Table 1. The relative importance of the different use patterns in different countries is still not known.

An increasing number of studies have been done to determine the source, occurrence, fate, and effects on the ecosystem of antibiotics. However, there is still a lack of understanding and knowledge of these compounds. So studies maybe focus on the strategies about stream segregation and at-source treatment of the concentrated streams appears.

Region/ Country	Total volume used in human medicine (ton/year)	Volume used in human medicine (gram per capita)	Thereof in hospitals (%)	Unuse medicaments	Measured in sewage up to (µg/L)	Measured in surface water up to (µg/L)	Reference
World wide	100000-200 000	N.D.	N.D.	N.D.	N.D.	N.D.	[19]
EU + Switzerland	8367	22.4	N.D.	N.D.	N.D.	N.D.	[21]
USA	4800	17	70	N.D.	1.9	0.73	[22, 23]
Canada	N.D.	N.D.	N.D.	N.D.	N.D.	0.87	[24]
Switzerland	34.2	4.75	20-40	N.D.	0.57	0.2	[25]
Germany	411	4.95	25	20-40	6	1.7	[16, 26]
Denmark	40	7.4	N.D.b	N.D.	5N.D.	N.D.	[27]
Austria	38	4.7	N.D.	20-30	N.D.	N.D.	[28]
Netherlands	40.9	3.9	20	N.D.	4.4	0.11-0.85	[29]
Italy	283	4.88	N.D.	N.D.	0.85-	0.25	[30]
Turkey	N.D.	31.4	N.D.	N.D.	N.D.	N.D.	[31]

Table 1. Country specific antibiotics consumption and occurrence data (N.D.: not defined)

1.4. Production and manufacturing

Pharmaceutical industries have minor importance on the sewage treatment plants. Only in some Asian countries, wastewaters from this industry contributes to the sewage and cause an increase in the concentration of single compound up to mg/L level [32, 33, 34]. Also in developed countries, manufacturing plants increases the total antibiotic concentrations in the domestical wastewater [35].

The main problem for this industry is that they still use the physicochemical treatment technologies in the plant to remove the compounds from their wastewater. However, this approach is expensive.

1.5. Elimination and treatment

In the literature, there are lots of studies focused on the fate of these compounds in conventional domestical wastewater treatment plants and also lab-scale applications in the innovative treatment methods. Elimination and/or treatment of these organic compounds are the results of biotic and abiotic processes. While biotic process is the biodegradation by microorganisms, abiotic processes are sorption, hydrolysis, oxidation-reduction, and photolysis.

1.5.1. Sorption

Before to assess the sorption characteristics of antibiotics, it is necessary to consider their physical and chemical parameters. Tolls [36] investigated the sorption behavior of these compounds in soil and the results showed that sorption mechanism of antibiotics could be very complex and difficult.

Additionally, binding to particles or the formation of complexes may prevent their detection. For example, tetracyclines are able to form complexes with double cations such as magnesiumor calcium [37]. Also humic substances cause the change in the surface properties and sites available for sorption and reactions. Gu and Karthikeyan [38] reported that there is a strong interaction between humic acids, hydrous Al oxide and tetracycline. Some studies showed that antibiotics used in medicine such as fluoroquinolones and macrolides can reach the terresrial environment by sewage sludge [38, 39].

Also sorption mechanism is a significant process for sulphonamides [36]. However, knowledge about the interaction of antibiotics with sludge and of sediments with sludge in activated sludge plants as well as the subsequent potential for their release back into the environment is still too sparse.

1.5.2. Photolysis

Photochemical process can be important in the surface waters and treatment plant effluents as another elimination process [40-43]. In the environment, photolysis process is not effective in turbid water or river and lakes, which are shadowed. So, the in the lab-scale experiments cannot reflect the photochemical process in the nature. Also, effectiveness of depletion process can differ under different environmental conditions such as pH, temperature, water hardness [44] and depends on type of matrix, location, season, latitude [45].

One of the problems about this type of process is that incomplete photo-transformation and photo-degradation can cause to more or less stable or toxic compounds although this does not necessarily have to happen [46-48].

The significance and extent of direct and indirect photolysis of antibiotics in the aquatic environment are different for each compound because some of them are light sensitive (e.g. quinolones, tetracyclines, sulphonamides, tylosin, nitrofuran antibiotics). However, not all compounds are photo-degradable [49]. Tetracyclines are senstive to photo-degradation. Samuelsen [50] investigated the sensitivity of oxytetracycline towards light in seawater as well as in sediments. The antibiotics proved to be stable in sediments rather than in seawater. As no mechanism of decomposition other than photolysis is known for them [51], the substance remains in the sediment for a long period, as shown by [52]. Boree et al. [53] showed that sulphanilic acid was found as a degradation product common to most of the sulpha drugs.

1.5.3. Hydrolysis and thermolysis

Another important pathway for the non-biotic decomposition of organic substances in the environment is hydrolysis. Some instability in water could be demonstrated for some tetracycylines [54]. In general, the hydrolysis rates for oxytetracycline increase with reascept to temperature at pH 7. The half-lives of oxytetracycline under investigation changed by differences in temperature, light intensity and flow rate from one test tank to another. However sulphonamides and quinolones are known as resistant antibiotic to hydrolysis.

1.5.4. Oxidation

Pharmaceutical industry wastewaters including antibiotic are well known for the difficulty of their elimination by conventional biological treatment methods and their important contribution to environmental pollution is due to their fluctuating and recalcitrant nature. For this reason, oxidation processes are usually applied.

The presence of carbon–carbon double bonds, aromatic bonds or nitrogen is a necessary essential for this application. However, the presence of these structural elements does not provided the fast and full degradation or even the complete degradation.

The effect of ozonation on the degradation of oxytetracycline in aqueous solution at different pH values (3, 7 and 11) was reported by Li *et al.* [55]. The study was designed that ozonation as a partial step in a combined treatment concept is a potential technique for biodegradability enhancement. It has been shown that COD removal rates increase with increasing pH as a consequence of enhanced ozone decomposition rates at elevated pH values. The results of bioluminescence data indicate that the initial by-products after partial ozonation (5–30 min) of oxytetracycline were more toxic than the parent compound [55].

Sulfamethoxazole was also efficiently degraded by ozonation [56]. An improvement in biodegradability by the increasing of BOD5/COD ratio from 0 to 0.28 was observed by the authors after 60 min of ozonation. The acute toxicity of the intermediates was checked and a slight acute toxicity increment in the first stage of ozonation was found. pH variation was found as important parameter on TOC and COD removal efficiencies. The complete sulfamethoxazole removal was achieved for an in photo-Fenton process [57]. Toxicity and inhibition tests pointed in the same direction: no toxic effect of oxidized intermediates was determined and also no inhibition was detected on activated sludge activity.

1.5.5. Biodegradation

Biodegradability of most antibiotics has been checked and it was found that they are not biodegradable under aerobic conditions until today [3, 11, 55, 58, 59]. Biodegradability characteristics have been weak for most of the compounds investigated in laboratory tests such as the OECD test series (301–303, 308) – even for some of the ß-lactams (Alexy *et al.*, 2004). Out of 16 antibiotics tested, only benzyl penicillin (penicillin G) was completely mineralized in a combination test (combination of the OECD 302 B and OECD 301 B tests; [11]).

Biodegradation for tetracycline was not observed during a biodegradability test (sequence batch reactor), and sorption was found to be the principal removal mechanism for tetracycline in activated sludge [61].

Some antibiotics occurring in soil and sediment proved to be quite persistent in laboratory testing as well as in field studies. Some of them were not biodegradable also under anaerobic conditions [12] others did [62]. Substances extensively applied in fish farming had long half-lives in soil and sediment, as reported in several investigations [63]; [64]; [65]; [66]; [67]; [68]; [69]). However, some substances were at least partly degradable ([70]; [71]; [72], [66]; [68]; [73]). Maki *et al.* [62] found that ampicillin, doxycycline, oxytetracycline, and thiamphenicol were significantly degraded, while josamycin remained at initial levels. Tylosin was biodegraded [42].

1.6. Problem definition and aim

The yearly consumption of antibiotics is 500 tons throughout the world according to the data of 2001. Approximately 90% of the consumed antibiotics after being partially metabolized or not being metabolized are excreted by the help of urea or feces from the body and transferred to the domestic sewage plants. These antibiotics are discharged into the receiving environment with no or low elimination after being treated in conventionally operated domestic sewage plants. While the concentration of these materials in domestic wastewaters and surface waters are in µg/l level, in pharmaceutical wastewater they are in 100-1000 mg/L level [74, 75, 76, 77]. As this low concentration in the surface wastewaters cause important problems in the ecosystem, it necessitates the removal of high antibiotic amount that are found in the pharmaceutical wastewaters. However, because the chemical removals of these materials are costly, biological treatment is essential. Antibiotics are the one of these compounds and the most often discussed pharmaceuticals because of their potential role in the spread and maintenance of (multi)resistance of bacterial pathogens. There are lots of studies that have been done in Europe and North America on the detection and removal of antibiotics in the receiving environment and the treatment plant [4, 5, 23, 24, 78-84]. However, the studies on the treatability of these antibiotics biologically are quite few [61, 85]. Also the scope of the studies done on the biodegradability potential of these materials is limited [11, 12, 86, 87]. Additionally, the studies on the microbial groups and species that are responsible for degradation have not been done, yet.

In this scope, determination of biodegradation characteristics of the refractory compounds and their toxic/inhibition effects on microbial community is substantial for environmental engineering. For this aim, the biodegradability of these sulfamethoxazole under anoxic and anaerobic conditions and also changes in microbial groups under the different conditions are explained in this chapter.

2. Materials and methods

2.1. Experimental approach

This study involves setting-up batch biodegradation test to investigate biodegradation characteristics of sulfamethoxasole (SMX) under anoxic and anaerobic conditions. The biodegradation test bottles were set up under nitrate reducing conditions (NRC), sulfate reducing conditions (SRC) and methanogenic conditions (MC). Experiment was carried out for 120 days. During the experiment, gas production was monitored daily. Destructive sampling was done in four different times (at 0th, 20th, 60th and 120th day). Wet chemical analysis (dissolved organic carbon [DOC], SMX measurements and electron acceptor measurements) and microbiological analysis (quantitative real-time PCR [Q-PCR]) were carried for four sampling times.

2.2. Set–up of batch biodegradation test bottles

In this study, two different seed sludges were used for setting-up of the batch tests. For NRC, the seed was taken from anoxic part of a domestic wastewater treatment plant in Istanbul whereas; test tubes for the SRC and MC were inoculated by anaerobic sludge from a full-scale UASB reactor treating alcohol distillery effluents.

The batch tests were constructed in 120 mL serum bottles, 100 mL of active volume, according to modified OECD 311 protocol [13]. The constituents of each experimental set for NRC, SRC and MC conditions are given in Table 2, 3 and 4, respectively. Also chemicals of the trace element solution and their amounts are given in Table 5. SMX was chosen as the model carbon source. The test tubes were set up as duplicates including positive and negative controls. Phenol was chosen as slowly biodegradable carbon source for positive control set. Negative control sets were constructed without any carbon source to determine endogenous decay. All sets were set-up in an anaerobic cabinet (Coy Laboratory Products, U.S.).

Experimental sets were destructed in 4 different sampling times. The first set was destructed immediately after all the test tubes were set-up, the other three sets were spoiled in day 20, day 60 and day 120. In each test tube, after inoculation 2000 mg/L TVS was maintained. Phenol and SMX concentrations were adjusted to 80±4.5 mg DOC/L and 280±1.0 mg DOC/L within the all experimental groups. The dissolved organic carbon (DOC) value of negative control bottles was 18.6±1.5 mg/L. All solutions were deoxygenated and adjusted to pH 7. Biodegradation test bottles were incubated at 20 °C and 35 °C for NRC and MC/SRC, respectively. All test bottles were stored at dark chambers to ensure occurring only biodegradation and sorption mechanisms during the experiment. The test tubes were shaken daily by hand.

CONSTITUENT	AMOUNT (g)
Anhydrous potassium dihydrogen phosphate (KH2PO4)	0,27
Disodium hydrogen phosphate dodecahydrate (Na2HPO4.12H2O)	1,12
Ammonium chloride (NH4Cl)	0,53
Potassium Nitrate (KNO3)	1
Calcium chloride dihydrate (CaCl2.2H2O)	0,075
Magnesium chloride hexahydrate (MgCl2.6H2O)	0,1
Iron (II) chloride tetrahydrate (FeCl2.4H2O)	0,02
Resazurin (oxygen indicator)	0,001
Sodium sulphide nonahydrate (Na2S.9H2O)	0,1
Stock solution of trace elements	10 ml
Add de-oxygenated water	to 1 liter

Table 2. Medium for nitrate reducing conditions (10 mM potassium nitrate)

CONSTITUENT	AMOUNT (g)
Anhydrous potassium dihydrogen phosphate (KH2PO4)	0,27
Disodium hydrogen phosphate dodecahydrate (Na2HPO4.12H2O)	1,12
Ammonium chloride (NH4Cl)	0,53
Potassium Sulfate (K2SO4)	1,8
Calcium chloride dihydrate (CaCl2.2H2O)	0,075
Magnesium chloride hexahydrate (MgCl2.6H2O)	0,1
Iron (II) chloride tetrahydrate (FeCl2.4H2O)	0,02
Resazurin (oxygen indicator)	0,001
Sodium sulphide nonahydrate (Na2S.9H2O)	0,1
Stock solution of trace elements	10 ml
Add de-oxygenated water	to 1 liter

Table 3. Medium for sulfate reducing conditions (10 mM potassium sulfate)

CONSTITUENT	AMOUNT (g)
Anhydrous potassium dihydrogen phosphate (KH2PO4)	0,27
Disodium hydrogen phosphate dodecahydrate (Na2HPO4.12H2O)	1,12
Ammonium chloride (NH4Cl)	0,53
Calcium chloride dihydrate (CaCl2.2H2O)	0,075
Magnesium chloride hexahydrate (MgCl2.6H2O)	0,1
Iron (II) chloride tetrahydrate (FeCl2.4H2O)	0,02
Resazurin (oxygen indicator)	0,001
Sodium sulphide nonahydrate (Na2S.9H2O)	0,1
Stock solution of trace elements	10 ml
Add de-oxygenated water	to 1 liter

Table 4. Medium for methanogenic conditions

CONSTITUENT	AMOUNT
Manganese chloride tetrahydrate (MnCl2.4H2O)50 mg	50 mg
Boric acid (H3BO3)	5 mg
Zinc chloride (ZnCl2)	5 mg
Copper (II) chloride (CuCl2)	3 mg
Disodium molybdate dihydrate (Na2MoO4.2H2O)	1 mg
Cobalt chloride hexahydrate (CoCl2.6H2O)	100 mg
Nickel chloride hexahydrate (NiCl2.6H2O)	10 mg
Disodium selenite (Na2SeO3)	5 mg
Add de-oxygenated water	to 1 liter

Table 5. Stock solution of trace elements

Headspace pressure was measured by hand-held pressure transducer (Lutron PM-9107, U.S.A.) every day. At each sampling time, biogas composition of the samples was determined via gas chromatography (Perichrom, France). DOC concentration of each sample was measured by Shimadzu ASI-V TOC analyser (Japan). Nitrate and sulfate concentrations were measured by DIONEX ICS 1500 ion chromatograph (U.S.A.). SMX measurements within the solid and liquid phase were carried by the protocol that is proposed previously by Karcı and Balcıoglu [88].

2.3. Calculation of mass balances

Theoretical CO_2 (Th CO_2) and Theoretical Biogas (Th biogas), which were used for evaluation of biodegradation, were calculated according DOC, gas and ion chromatography results. Mass balances were calculated by the assumptions, which were described by Ritmann and McCarty [89]. Simplified mass balances were given in Equation 1-3 for NRC, SRC and MC, respectively.

$$SMX + NO_3^- \rightarrow Biomass + CO_2 + N_2 + H_2O \qquad (1)$$

$$SMX + SO_4^{2-} \rightarrow Biomass + CO_2 + H_2S + HS^- + H_2O \qquad (2)$$

$$SMX + H_2O \rightarrow Biomass + CO_2 + CH_4 \qquad (3)$$

Ultimate biodegradation ratios were estimated by comparison of $ThCO_2$ and Th biogas production (which were assumed to be produced as a result of 100% biodegradation of tetracycline) were compared to actual CO_2 and biogas production within the batch tests, DOC elimination and SMX measurements.

2.4. Microbiological analyses

Genomic DNA (GDNA) was extracted from 0.5 g sludge using the Fast DNA Spin Kit for Soil (Qbiogene Inc., U.K.) following the manufacturer's instructions.

Q-PCR procedure recommended by Roche was followed and a Light Cycler Master Kit (Roche, Applied Science, Switzerland) was used to set up the reaction (2.0 µl master mix, 1.6 µl $MgCl_2$ 1.0 µl Primer F and R, 13.4 µl H2O, 1 µl sample). Absolute quantification analysis of the GDNA was carried out with a Light Cycler 480 Instrument (Roche Applied Science, Switzerland). Primers used in the quantification are given in Table 6.

Significant differences were determined according to independent sample t-test. Pearson correlation was used for the interactions between variables. All the statistical analyses were conducted by using SPSS (IBM, U.S.A) and $p < 0.05$ level was used for significance.

Primer	Target Gene	Target Microorganisms	Respiration Conditions	References
Bac519f	16S rRNA	Bacteria	All	90
Bac907r				
Arc349f	16S rRNA	Archaea	All	91
Arc806r				
Met348f	16S rRNA	Methanogens	All	92
Met786r				
DSRp2060F	Sulfites reductase beta sub-unit (dsrB)	Sulfate Reducing Bacteria	Only sulfate reducing condition	93

Table 6. Primers and target groups for Q-PCR analysis

3. Results and discussion

3.1. Methane generation

Biogas generation in the test bottles operated under sulphate reducing and methanogenic conditions was observed daily. However methane content of the biogas in the test bottles were determined in each sampling time before the destruction of the test bottles as 0th, 20th, 60th and 120th days. Produced methane volume in sulfamethoxazole (SMX) and reference item (REF) fed bottles with non-carbon source (NC) fed bottles under sulphate reducing and methanogenic conditions were given in Figure 1 and 2, respectively.

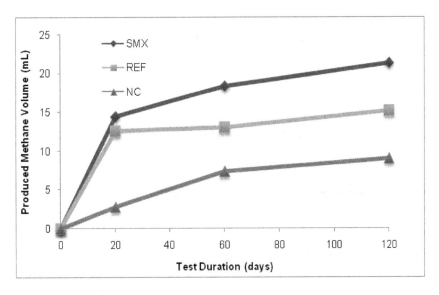

Figure 1. Methane production under sulfate reducing conditions

As seen in Figure 1, the maximum methane production associated with SMX was 21 mL while the maximum values were determined as 15 L and 9 mL in REF and NC test bottles, respectively, under sulphate reducing conditions. These values increased to 132 mL, 41 mL and 23 mL, respectively. This wide difference is expected as a result of sulphate inhibitory effect on methanogens. Another point to show the inhibition that most of the methane was produced during first 20 days under methanogenic conditions while methane production was slower under sulphate reducing conditions. Also it was known that sulphate reducers are much more versatile than methanogens and in environments where sulfate is present, sulfate-reducing bacteria compete with methanogenic consortia for common substrates. Compounds like propionate and butyrate, which require syntrophic consortia in methanogenic environments, are degraded directly by single species of sulfate reducing bacteria [94].

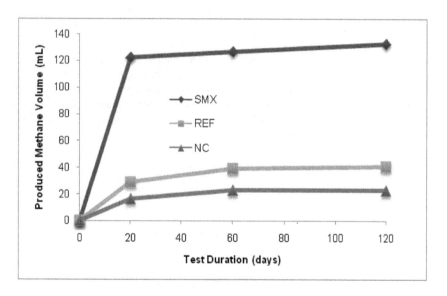

Figure 2. Methane production under methanogenic conditions

Positive and negative control groups were used to increase the reliability of the experiment. For positive control groups phenol was used as a carbon source. For all three electron-accepting conditions, phenol was biodegraded at the ratios between 74- 78% in 120 days, which indicated the ultimate biodegradation according to OECD protocol [13]. Measured CO_2 and biogas production within the negative control groups subtracted as blanks to reveal the actual biodegradation ratios. The CO_2 productions in the negative control test bottles reached a total of 4-12 mL in 120 days corresponding to 70 -100% of the theoretical CO_2 (Th CO_2) production while biogas production reached 40 mL corresponding to 100% of the Th biogas occurred via degradation of biomass completely.

3.2. Removal of dissolved organic carbon

Total organic carbon parameter was used to compare the biodegradation capacity of the antibiotic and reference item under nitrate reducing, sulphate reducing and methanogenic conditions. Also electron acceptors were measured in the test bottles. DOC removal was higher in the first 60 days in all electron-accepting condition. The removal between 60th-120th days, any significant changes were not observed.

In Figure 3, DOC and nitrate concentration changes in respect to time are given. in the beginning of the experiment, nitrate concentration in each bottle was 250 mg/L. This concentration decreased to less than 10 mg/L in the first 20-day period of the experiment. Also decrease in DOC values was parallel to nitrate concentration except of SMX test bottles. The decrease in the DOC continued first 60 days while nitrate concentration was 1 mg/L.

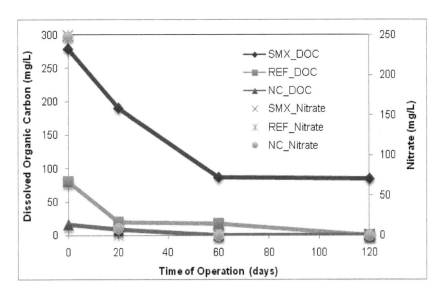

Figure 3. DOC and nitrate concentration in SMX, REF and NC bottles under nitrate reducing conditions

As seen in Figure 4, most of the DOC in the SMX bottle was consumed in the first 60 days. Also sulphate concentration decreased from 480 mg/L to 59 mg/L during same period.

In Figure 5, changes in DOC concentrations under methanogenic conditions are given. The results indicated that the removal of DOC mechanism was more quickly in the first 20 days. This pattern was also similar with the other respiration conditions. Also reference item was consumed in the same period. DOC concentration in SMX bottles decreased from 280 mg/L to 88 mg/L in the first 20 days. After this day, only 18 mg/L DOC was consumed and the final DOC concentration in SMX test bottles was determined as 70 mg/L.

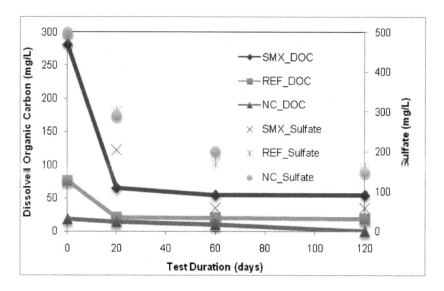

Figure 4. DOC and sulphate concentration in SMX, REF and NC bottles under sulphate reducing conditions

The most efficient DOC removal in SMX test bottles was observed under SRC and MC as 78/ and 74%, respectively. Under NRC, DOC removal was detected as 71%.

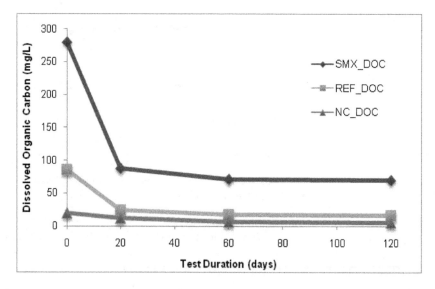

Figure 5. DOC concentration in SMX, REF and NC bottles under methanogenic conditions

In another study showed that SMX affected the propionic acid degradation and acetic acid utilization pathways in the higher concentrations [95]. Sponza and Demirden [96] also showed while sulfamerazine, which is another antibiotic from sulfonamid group, was being fed to the anaerobic system, an increase in VFA accumulation was observed with respect to rising of antibiotic concentration. Decreased utilization of butyrate and propionate is consistent with the fact that these substrates are used directly by bacteria, homoacetogens. SMX also has a bacteriostatic inhibition effect on folic acid production of especially gram positive and negative cocci [18]. VFAs are not directly used by methanogens, however different groups of syntrophic bacteria use specific VFAs.

3.3. Biodegradability of sulfamethoxazole and mass balance

SMX measurement was done for water and sludge matrix. The recovery was found as 92% after solid-phase extraction (SPE). Antibiotic measurement in the sludge showed that the SMX concentration in the sludge did not change in respect to time and it is found as 50,4±3 mg/L. This result indicated that velocity of biodegradation and sorption mechanisms are similar during the test. Antibiotic concentrations in water samples are given in Figure 6.

Antibiotic removal for three electro accepting conditions was same and it was detected as approx. 98% in water matrix. If the sorption mechanism takes into the consideration, the removal decreased to 70%. Most of the antibiotic was removed in the first 60 days. There is no significant change between 60th-120th days. Also the decrease in electron acceptor concentration under nitrate reducing conditions may be caused a negative impact on microbial activity [97]. However, it was clear that antibiotic removal was faster under methanogenic conditions. The results showed that 68% of SMX was removed under methanogenic conditions while ultimate SMX removal was 70%.

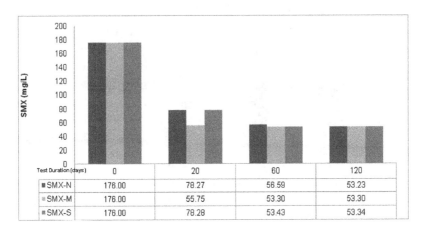

Figure 6. Sulfamethoxazole concentration under nitrate reducing (SMX-N), sulphate reducing (SMX-S) and methanogenic (SMX-M) conditions

Figure 7 shows ultimate biodegradation ratios (evaluated according to gas production only derived from SMX biodegradation), sorption ratios according to SMX measurements within sludge and soluble microbial products (SP) and/or transformation products (TP) ratios that were calculated via DOC removal ratio compared with SMX biodegradation for each electron accepting condition throughout the operating period. SMX showed non-biodegradable behavior under SRC, NRC and MC according to OECD protocol [13].

SMX measurements within the sludge samples of the all experimental groups showed that 29% of the SMX sorbed to the solid media throughout the experiment time. Sorbed part of the SMX did not change for four sampling time. Stabile results indicated that sorption processes are more dominant rather than desorption processes since all serum bottles were shaken daily in order to increase the bioavailability of the carbon source. Yang et al. also confirmed the rapid sorption processes rather than biodegradation [10].

Under MC, biogas production showed that 23% of the SMX was mineralized. However, according to SMX and DOC measurements 40% of the SMX were removed from the liquid phase. This result indicated that parent compound transformed to SP and/or TP. 17% of the SMX was remained in liquid phase as its potential SP and/or TP. Gartiser et al. reported SMX as non-biodegradable compound (2.3%) as well [11]. Different results of two studies mainly emanated by the application of different methods and duration time of the experiments.

Under SRC, 32% of the SMX was ultimately biodegraded whereas; 8% of the parent compound transformed to SP and/or TP. Under NRC, 38% of the SMX was mineralized to CO_2 and 2% of the SMX converted to residual SP and/or TP. Biodegradation ratios within the conventional treatment plants which is reported by Hong et al. [98] complies with our results. In their study 40% of the SMX removed from liquid phase. Also in our study, anoxic biodegradation rate was the highest removal rate among the experimental groups. Overall elimination within three electron-accepting conditions was calculated as 69 %.

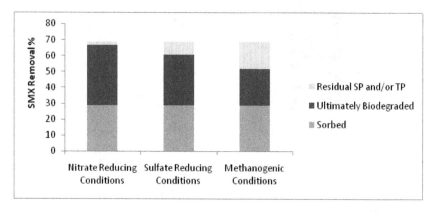

Figure 7. Biodegradation of SMX under different e-accepting conditions

3.4. Microbiological analyses

Q-PCR analyses were carried out for four sampling times. Four different taxonomic groups were quantified. These were; Bacteria, Archaea, methanogenic Archaea and Sulfate Reducing Bacteria (SRB). There was no significant change in the amount of these populations during 120 days (data not shown). However, under methanogenic conditions, biogas, antibiotic concentration and microbial quantification data indicate that there was a strong correlation between antibiotic concentration and amount of bacterial and methanogenic species. This correlation was a strong proof of the usability of SMX and showed that the bacterial and archaeal community continued to work together while this compound was only carbon source. Figure 8 and Table 7 show the changes within these groups under methanogenic conditions and their correlations with each other, respectively.

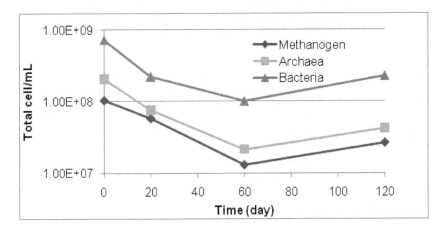

Figure 8. Microbial cell counts under methanogenic conditions

Table 7. Correlation analyses for methanogenic conditions (n=3, p<0.05)

As it can be seen in Table 7 for methanogenic conditions, there was a strong positive correlation between methanogenic count and bacterial count; also, they had a positive correlation with DOC and SMX concentration change over time. This observation shows that there may be syntrophic relationship between bacteria and methanogens in methanogenic conditions. In addition, SMX concentration is strongly correlated with bacterial and methanogenic count.

Insignificant change in microbial groups can be explained with two approaches. 1- SMX may have an inhibition on bacterial growth so SMX biodegradation delayed and microbial cells have faced with starvation as described by Gartiser et al. [12]. 2- Multi antibacterial resistant bacteria have been described by many authors [7, 99]. Based on that knowledge, bacterial populations in the batch tests might have gained resistance to SMX with time but would not able to use SMX efficiently. In this case, bacterial populations lowered their metabolic functions to survive rather than grow. Also changes in the population dynamics can be derived from the microbial interactions. Thus methanogens were not affected by SMX. Otherwise biogas production wouldn't have occurred or would have been inhibited because of their suscepti- bility to toxic compounds [100].

More detailed microbiological approach was applied in a parallel study [95]. The author tested the inhibition effect and biodegradability characteristic of same compound in long-term semi- continuous operation under anaerobic conditions. In that study, *Clostiridum spp.* was found in the system independently of operation time and SMX concentration in the system according to 16S rDNA clone library and denaturing gradient gel electrophoresis (DGGE) studies. It was also expected due to they are responsible to fermentation and some species especially produce the ethanol. Additionally, *Clostridium spp.* have the role on the starch degradation by exo- enzymes. Other OTUs, which were detected in the system, almost belong to the uncultured clones or unclassified bacterial cultured species. In addition to bacterial results, archaeal studies showed that acetoclastic methanogenic species disappeared in the last phases of operation in which SMX concentration increased. However the abundance of hydrogenotro- phic methanogenswere higher than acetoclastic species and they were dominant during the operation.

Looking at this point, more detailed microbiological approach was needed to give the answers of two main questions: Which microbial groups are directly affected by SMX and which ones utilize this compound as a substrate? Next generation sequencing (NGS) based on DNA and also cDNA produced from total RNA represent more details about microbial community in each operation period. NGS is a novel sequencing technology for metagenomic studies. The main advantage of this technique is to sequence the mix GDNA directly without any prelimi- nary study. By this means, the process does not cover the bias coming from polymerase chain reaction (PCR) and cloning. Additionally stable isotope probing (SIP) technique may be a good option to find the answer about which microbial groups utilize directly the SMX. In this technique, a labeled compound is given as substrate and then the produced GDNAs are monitored by labeled elements coming from utilized compound.

4. Conclusion

In the light of evaluations presented above, the significant findings of the study on the biodegradability characteristics of sulfamethoxazole under different electron accepting conditions may be outlined as follows:

The results suggested that the nature of the biodegradability characteristic of SMX are similar under nitrate reducing, sulphate reducing and methanogenic conditions and it was clear that biological treatment is suitable for this compound to remove from the wastewater during long retention times. However, methanogenic conditions should be selected because of obtaining biogas to use as energy source.

Microbial studies showed a syntrophic relationship between bacteria and methanogens in methanogenic conditions. Quantification of the main microbial groups has given general idea about the effect of SMX and showed the next step to clarify this mechanism: To focus microbial kinetics in terms of metabolic expressions on mRNA level and also quantification of antibiotic resistance genes in the system, which is operated under methanogenic conditions, give more information about the removal mechanism of this compound. Also for detailed information about microbial community and changes in the community, NGS is a good option. SIP is the direct method to observe which microbial species utilize the SMX and its transformation products.

Acknowledgements

This study was funded by The Scientific and Technological Research Council of Turkey (TUBITAK), Project No: 109Y012.

Author details

Zeynep Cetecioglu[1], Bahar Ince[2], Samet Azman[1], Nazli Gokcek[1], Nese Coskun[1] and Orhan Ince

*Address all correspondence to: cetecioglu@itu.edu.tr

1 Istanbul Technical University, Environmental Engineering Department, Istanbul, Turkey

2 Bogazici University, Institute of Environmental Sciences, Istanbul, Turkey

References

[1] Li, B, & Zhang, T. Biodegradation and adsorption of antibiotics in the activated sludge process. Environmental Science and Technology. (2011). , 44, 3458-3473.

[2] Xiao, Y, Chang, H, Jia, A, & Hu, J Y. Trace analysis of quinolone and fluoroquinolone antibiotics from wastewaters by liquid chromatography electrospray tandem massspectrometry. Journal of Chromatography A, (2008).

[3] Kummerer, K. Pharmaceuticals in the Environment: Sources, Fate, Effects and Risks. 1st ed. Berlin, Springer-Verlag; (2001).

[4] Hirsch, R, Ternes, T, Haberer, K, & Kratz, K L. Occurrence of antibiotics in the aquatic environment. Science of the Total Environment. (1999).

[5] Golet, E M, Strehler, A, Alder, A. C, & Giger, W. Determination of fluoroquinolone antibacterial agents in sewage sludge and sludge-treated soil using accelerated sol-

vent extraction followed by solid-phase extraction. *Analytical Chemistry*. (2002). , 74(21), 5455-5462.

[6] Andreozzi, R, Caprio, V, Ciniglia, C, & De Champdore, M. Lo Giudice R, Marotta R, Zuccato E. Antibiotics in the environment occurrence in Italian STPs, fate, and preliminary assessment on algal toxicity of amoxicillin. *Environmental Science and Technology*. (2004). , 38(24), 6832-6838.

[7] Munir, M, Wong, K, & Xagoraraki, I. Release of antibiotic resistant bacteria and genes in the effluent and biosolids of five wastewater utilities in Michigan. *Water Science and Technology*. (2011). , 45, 681-693,

[8] Nieto, A, Borrull, F, Pocurull, E, & Marce, R M. Occurrence of pharmaceuticals and hormones in sewage sludge.*Environmental Toxicological Chemistry*.(2010). , 29(7), 1484-1489.

[9] Drillia, P, Dokianakis, S N, Fountoulakis, M S, Kornaros, M, Stamatelatou, K, & Lyberatos, G. (2005). *Journal of Hazardous Materials.*, 122, 259-265.

[10] Yang, S, Lin, C F, Lin, A Y, & Hong, P A. Sorption and biodegradation of sulfonamide antibiotics by activated sludge: Experimental assessment using batch data obtained under aerobic conditions. *Water Research.*(2011). , 45, 3389-3397.

[11] Gartiser, S, Urich, E, Alexy, R, & Kummerer, K. Anaerobic inhibition and biodegradation of antibiotics in ISO test schemes. *Chemosphere*. (2007). , 66, 1839-1848.

[12] Gartiser, S, Urich, E, Alexy, R, & Kümmerer, K. Ultimate biodegradation and elimination of antibiotics in inherent tests. *Chemosphere*. (2007). , 67, 604-613.

[13] Anaerobic biodegradability of organic compounds in digested sludge- method by measurement of gas production (OECD 311)France; (2006). OECD- Organization for Economic Co-operation and Development.

[14] Korolkovas, A. (1976). Essentials of Medicine Chemistry. Wiley, New York.

[15] Foye, W. O, Lemke, T. L, & Williams, D. A. (1995). Principles of Medicinal Chemistry. MD. Williams&Wilkins, Baltimore.

[16] Kummerer, K, & Henninger, A. (2003). Promoting resistance by the emission of antibiotics from hospitals and households into effluents. Clin. Microbiol. Infec. , 9, 1203-1214.

[17] Cunningham, V. (2008). Special characteristics of pharmaceuticals related to environmental fate. In. Kummerer, K. (Ed.), Pharmaceuticals in the Environment. Sources, Fate, Effects and Risk, third ed. Springer, Berlin Heidelberg, , 23-34.

[18] Sweetman, S. C. Ed.), (2009). Martindale. The complete drug reference, 36th edition. Pharmaceutical Press, London UK.

[19] Wise, R. (2002). Antimicrobial resistance. priorities for action. J. Antimicrob. Chemoth., , 49, 585-586.

[20] http://www.esac.ua.ac.be/main.aspx?c=*ESAC2&n=10661

[21] European Federation of Animal Health (FEDESA)(2001). Antibiotic use in farm animals does not threaten human health. FEDESA/FEFANA Press release, Brussels, 13 July.

[22] Union of Concerned Scientists(2001). Percent of all Antibiotics Given to Healthy Livestock. Press release, 8 January, Cambridge, MA

[23] Kolpin, D. W. Fur Kolpin long, E.T., Meyer, M.T., Thurman, E.M., Zaugg, S.D., Barber, L.B., Buxton, H.T., (2002). Pharmaceuticals, hormones, and others organic wastewater contaminants in U.S. streams, a national reconnaissance. Environ Sci Technol 36.1202-1211., 1999-2000.

[24] Miao, X. S, Bishay, F, Chen, M, & Metcalfe, C. D. (2004). Occurrence of Antimicrobials in the final effluents of waste water treatment plants in Canada. Environ Sci Technol 38., 3533-3541.

[25] Kummerer, K. (2004). Pharmaceuticals in the Environment, Ed. Kummerer, K., 2nd Ed. Springer, Verlag.

[26] Rönnefahrt, I. (2005). Verbrauchsmengen in der Bewertung des Umweltrisikos von Humanarzneimitteln, In. Umweltbundesamt (Hrsg) Arzneimittel in der Umwelt- Zu Risiken und Nebenwirkungen fragen Sie das Umweltbundesamt. Dessau, UBA texte 29/05.

[27] Kummerer, K. (2008). Antibiotics in the Environment. Pharmaceuticals in the Environment, Ed. Kummerer, K., 3rd Ed. Springer, Verlag., 3-35.

[28] Sattelberger, S. (1999). Arzneimittelruckstande in der Umwelt, Bestandsaufnahme und Problemstellung. Report des Umweltbundesamtes Österreich, Wien.

[29] Verbrugh, H. A, & De Neeling, A. J. Eds, (2003). Consumption of antimicrobial agents and antimicrobial resistance among medically important bacteria in the Netherlands. SWAB NETHMAP.

[30] Calamari, D, Zuccato, E, Castiglioni, S, Bagnati, R, & Fanelli, R. (2003). Strategic survey of therapeutic drugs in the rivers Po and Lambro in northern Italy, Environ Sci Technol., , 37, 1241-1248.

[31] Karabay, O. (2009). Türkiye'de antibiyotik kullanımı ve direnç nereye gidiyor?, AN-KEM Dergisi, 23(2), 116-120.

[32] Larsson, D. G, De Pedro, C, & Paxeus, N. (2007). Effluent from drug manufactures contains extremely high levels of pharmaceuticals. J. Hazard. Mater. , 148, 751-755.

[33] Li, D, Yang, M, Hu, J, Ren, L, Zhang, Y, Chang, H, & Li, K. (2008). Determination and fate of oxytetracycline and related compounds in oxyteracycline production wastewater and the receiving river. Environ. Toxicol. Chem., , 27, 80-86.

[34] Li, D, Yang, M, Hu, J, Zhang, Y, Chang, H, & Jin, F. (2008). Determination of penicillin G and its degradation products in a penicillin production wastewater treatment plant and the receiving river. Water Res., , 42, 307-317.

[35] Thomas, K. V. (2008). The relevance of different point sources. Lecture given at the "First International Conference on Sustainable Pharmacy", April 2008, Osnabruck, Germany <http //www dbu.de/550artikel27007 1771html> //- h

[36] Tolls, J. (2001). Sorption of veterinary pharmaceuticals in soils. a review. Environ. Sci. Technol., , 35, 3397-3406.

[37] Christian, T, Schneider, R. J, Färber, H. A, Skutlarek, D, Meyer, M. T, & Goldbach, H. E. (2003). Determination of antibiotic residues in manure, soil, and surface waters, Acta Hydroch. Hydrob., , 31, 36-44.

[38] Gu, C, & Karthikeyan, K. G. (2008). Sorption of the antibiotic tetracycline to humic-mineral complexes. J. Environ. Qual. , 37, 704-711.

[39] Trivedi, P, & Vasudevan, D. (2007). Spectroscopic investigation of ciprofloxacin speciation at the goethite-water interface. Environ. Sci. Technol., , 4, 3153-3158.

[40] Viola, G, Facciolo, L, Canton, M, & Vedaldi, D. Dall'Acqua, F., Aloisi, G.G., Amelia, M., Barbafina, A., Elisei, F., Latterini, L., (2004). Photophysical and phototoxic properties of the antibacterial fluoroquinolones levofloxacin and moxifloxacin. Chem. Biodivers., , 1, 782-801.

[41] Werner, J. J, Chintapalli, M, Lundeen, R. A, Wammer, K. H, Arnold, W. A, & Mcneill, K. (2007). Environmental photochemistry of tylosin. efficient, reversible photoisomerization to a less-active isomer, followed by photolysis. J. Agric. Food Chem., , 55, 7062-7068.

[42] Hu, D, & Coats, J. R. (2007). Aerobic degradation and photolysis of tylosin in water and soil. Environ. Toxicol. Chem. , 26, 884-889.

[43] Hu, Z, Liu, Y, Chen, G, Gui, X, Chen, T, & Zhan, X. (2011). Characterization of organic matter degradation during composting of manure-straw mixtures spiked with tetracyclines. Bioresource Technology, doi.j.biortech.2011.05.003.

[44] Werner, J. J, Arnold, W. A, & Mcneill, K. (2006). Water hardness as a

[45] Kallenborn, R, Fick, J, Lindberg, R, Moe, M, Nielsen, K. M, Tysklind, M, & Vasskog, T. (2008). Pharmaceutical residues in Northern European environments. consequences and perspectives. In. Kummerer, K. (Ed.), Pharmaceuticals in the Environment. Sources, Fate, Effects and Risk, third ed. Springer, Berlin-Heidelberg, , 61-74.

[46] Arslan-alaton, I, & Caglayan, A. E. (2006). Toxicity and biodegradability assessment of raw and ozonated procaine penicillin G formulation effluent, Ecotoxicol. Environ. Safe, , 63, 131-140.

[47] Gonzalez, O, Sans, C, & Esplugas, S. (2007). Sulfamethoxazole abatement by photo-Fenton toxicity, inhibition and biodegradability assessment of intermediates. J. Hazard. Mater. , 146, 456-459.

[48] Paul, T, Miller, P. L, & Strathmann, T. J. (2007). Visible-light-mediated TiO2 photocatalysis of fluoroquinolone antibacterial agents. Environ. Sci. Technol., , 41, 4720-4727.

[49] Turiel, E, Bordin, G, & Rodríguez, A. R. (2005). Study of the evolution and degradation products of ciprofloxacin and oxolinic acid in river water samples by HPLC-UV/MS/MS-MS. J. Environ. Monitor., , 7, 189-195.

[50] Samuelsen, O. B. (1989). Degradation of oxytetracycline in seawater at two different temperatures and light intensities, and the persistence of oxytetracycline in the sediment from a fish farm. Aquaculture, , 83, 7-16.

[51] Oka, H, Ikai, Y, Kawamura, N, Yamada, M, Harada, K, Ito, S, & Suzuki, M. (1989). Photodecomposition products of tetracycline in aqueous solution.J. Agric. Food Chem., , 37, 226-231.

[52] Lunestad, B. T, & Goksøyr, J. (1990). Reduction in the antibacterial effect of oxytetracycline in sea water by complex formation with magnesium and calcium. Dis. Aquat. Organ., , 9, 67-72.

[53] Boree, A. L, Arnold, W. A, & Mcneill, K. (2004). Photochemical fate of sulfa drugs in the aquatic environment. sulfa drugs containing five-membered heterocyclic groups, Environ. Sci. Technol., , 38, 3933-3940.

[54] Halling-sørensen, B, Lykkeberg, A, Ingerslev, F, Blackwell, P, & Tjørnelund, J. (2003). Characterisation of the abiotic degradation pathways of oxytetracyclines in soil interstitial water using LC-MS-MS. Chemosphere 50 (10), 1331-1342.

[55] Li, K, Yediler, A, Yang, M, Schulte-hostede, S, & Wong, M. H. (2008). Ozonation of oxytetracycline and toxicological assessment of its oxidation by-products. Chemosphere , 72, 473-478.

[56] Dantas, R. F, Contreras, S, Sans, C, & Esplugas, S. (2007). Sulfamethoxazole abatement by means of ozonation. J. Hazard Mater. , 150, 790-794.

[57] Gonzalez, O, Sans, C, & Esplugas, S. (2007). Sulfamethoxazole abatement by photo-Fenton toxicity, inhibition and biodegradability assessment of intermediatesJ. Hazard. Mater. , 146, 456-459.

[58] Ingerslev, F, & Halling-sørensen, B. (2001). Biodegradability of metronidazole, olaqiondox, and tylosin, and formation of tylosin degradation products in aerobic soil/manure slurriesEcotoxicol. Environ. Safe , 48, 311-320.

[59] Ingerslev, F, Torang, L, Loke, M. L, Halling-sørensen, B, & Nyholm, N. (2001). Primary biodegradation of veterinary antibiotics in aerobic and anaerobic surface water simulation systems.*Chemosphere* , 44, 865-872.

[60] Gartiser, S, Urich, E, Alexy, R, & Kuummerer, K. (2007). Ultimate biodegradation and elimination of antibiotics in inherent tests.*Chemosphere*, , 67, 604-613.

[61] Kim, S, Eichhorn, P, Jensen, J. N, Weber, A. S, & Aga, D. S. (2005). Removal of antibiotics in wastewater. effect of hydraulic and solid retention timeson the fate of tetracycline in the activated sludge process. *Environ. Sci. Technol., 39*, 5816-5823.

[62] Maki, T, Hasegawa, H, Kitami, H, Fumoto, K, Munekage, Y, & Ueda, K. (2006). Bacterial degradation of antibiotic residues in marine fish farm sediments of Uranouchi Bay and phylogenetic analysis of antibiotic-degrading bacteria using 16S rDNA sequences. *Fisheries Sci.* , 72, 811-820.

[63] Jacobsen, P, & Berglind, L. (1988). Persistence of oxytetracyline in sediment from fish farms. *Aquaculture 70*, 365-370.

[64] Hansen, P. K, Lunestad, B. T, & Samuelsen, O. B. (1992). Effects of oxytetracycline, oxolinic acid and flumequine on bacteria in an artificial marine fish farm sediment. *Can. J. Microbiol.* , 38, 307-1312.

[65] Samuelsen, O. B, Torsvik, V, & Ervik, A. (1992). Long-range changes in oxytetracycline concentration and bacterial resistance towards oxytetracycline in fish farm sediment after medication. *Sci. Total Environ.,* , 114, 25-36.

[66] Samuelsen, O. B, Lunestad, B. T, & Fjelde, S. (1994). Stability of antibacterial agents in an artificial marine aquaculture sediment studied under laboratory conditions. *Aquaculture*, , 126, 183-290.

[67] Hektoen, H, Berge, J. A, Hormazabal, V, & Yndestad, M. (1995). Persistence of antibacterial agents in marine sediments. Aquaculture , 133, 175-184.

[68] Capone, D. G, Weston, D. P, Miller, V, & Shoemaker, C. (1996). Antibacterial residues in marine sediments and invertebrates following chemotherapy in aquaculture, *Aquaculture*, , 145, 55-75.

[69] Marengo, J. R, Brian, O, Velagaleti, R. A, & Stamm, R. R. J.M., (1997). Aerobic biodegradation of (14C)-sarafloxacin hydrochloride in soil. *Environ. Toxicol. Chem.,* , 16, 462-471.

[70] Donoho, A. L. (1984). Biochemical studies on the fate of monensin in animals and in the environment. *J. Anim. Sci.* , 58, 1528-1539.

[71] Gilbertson, T. J, Hornish, R. E, Jaglan, P. S, Koshy, K. T, Nappier, J. L, Stahl, G. L, Cazers, A. R, Napplier, J. M, Kubicek, M. J, Hoffman, G. A, & Hamlow, P. J. (1990). Environmental fate of ceftiofur sodium, a cephalosporin antibiotic. Role of animal excreta in its decomposition. *J. Agric. Food Chem.* , 38, 890-894.

[72] Samuelsen, O. B, Solheim, E, & Lunestad, B. T. (1991). Fate and microbiological effects of furazolidone in a marine aquaculture sediment. *Sci. Total Environ.,* , 108, 275-283.

[73] Thiele-bruhn, S. (2003). Pharmaceutical antibiotic compounds in soils- a review. *J. Plant. Nutr. Soil Sci.,* , 166, 145-167.

[74] Daughton, C. G, & Ternes, T. A. (1999). Pharmaceuticals and personal care products in the environment. agents of subtle change? *Environmental Health Perspectives,* , 107, 907-938.

[75] Stuer-lauridsen, F, Birkved, M, Hansen, L. P, Lutzhoft, H. C. H, & Halling-sorenson, B. (2000). Environmental risk assessment of human pharmaceuticals in Denmark after normal therapeutic use. *Chemosphere,* , 40, 783-793.

[76] Chelliapan, S, Wilby, T, & Sallis, P. J. (2006). Performance of an upflow anaerobic stage reactor (UASR) in the treatment of pharmaceutical wastewater containing macrolide antibiotics, *Water Res.,* 40 (3). , 507-516.

[77] Amin, M. M, Zilles, J. L, Greiner, J, Charbonneau, S, Raskin, L, & Morgenroth, E. (2006). Influence of the antibiotic erythromycin on anaerobic treatment of a pharmaceutical wastewater, *Environ. Sci. Technol.,* , 40, 3971-3977.

[78] Hartig, C, Storm, T, & Jekel, M. (1999). Detection and identification of sulphonamide drugs in municipal wastewater by liquid chromatography coupled with electrospray ionization tandem mass spectrometry. *J. Chromatogr. A,* , 854, 163-173.

[79] Alder, A. C, Mcardell, C. S, Golet, E. M, Ibrics, S, Molnar, E, Nipales, N. S, & Giger, W. (2001). Occurrence and fate fluoroquinolone, macrolide and sulfonamide antibiotics during wastewater treatment and in ambient waters in Switzerland. *In Pharmaceuticals and Personal Care Products in the Environment. Scientific and Regulatory Issues.* Daughton, C.G. and Jones-Lepp, T. (eds). Washington D.C.. Am. Chem. Soc., , 56-69.

[80] Golet, E. M, Alder, A. C, Hartmann, A, Ternes, T. A, & Giger, W. (2001). Trace determination of fluoroquinolone antibacterial agents in solid-phase extraction urban wastewater by and liquid chromatography with fluorescence detection. *Anal. Chem.* 73, 3632-3638.

[81] Lindsey, M. E, Meyer, M, & Thurman, E. M. (2001). Analysis of trace levels of sulfonamide and tetracycline antimicrobials, in groundwater and surface water using solid-phase extraction and liquid chromatography/mass spectrometry. *Anal. Chem.,* , 73, 4640-4646.

[82] Golet, E. M, Xifra, I, Siegrist, H, Alder, A. C, & Giger, W. (2003). Environmental exposure assessment of fluoroquinolone antibacterial agentsfrom sewage to soil. *Environ. Sci. Technol.* , 37, 3243-3249.

[83] Mcardell, C. S, Molnar, E, Suter, M. J. F, & Giger, W. (2003). Occurrence and fate of macrolide antibiotics in wastewater treatment plants and in the Glatt Valley Watershed, Switzerland. *Environ. Sci. Technol.*, , 37, 5479-5486.

[84] Gobel, A, Thomsen, A, Mcardell, C. S, Alder, A. C, Giger, W, Thesis, N, Löffler, D, & Ternes, T. (2005). Extraction and determination of sulfonamide and macrolide antimicrobials and trimethoprim in sewage sludge. *J. Chromatogr. A* , 1085, 179-189.

[85] Drillia, P, Dokianakis, S. N, Fountoulakis, M. S, Kornaros, M, Stamatelatou, K, & Lyberatos, K. G. (2005). On the occasional biodegradation of pharmaceuticals in the activated sludge process. The example of the antibiotic sulfamethoxazole J Haz. Mat. , 122, 259-265.

[86] Alexy, R, Scholl, A, & Kummerer, K. (2004). Elimination and degradability of 18 antibiotics studied with simple tests, *Chemosphere*, , 57, 505-512.

[87] Matamoros, V, Caselles-osorio, A, Garcia, J, & Bayona, J. M. (2008). *Sci. Tot. Env.*, 394. , 171-176.

[88] Karci, A, & Balciogli, A I. Investigation of the tetracycline, sulfonamide, and fluoroquinolone antimicrobial compounds in animal manure and agricultural soils in Turkey, *Science of the Total Environment*. (2009). , 407, 4652-4664.

[89] Ritmann, B E, & Mc Carty, P L. Environmental Biotechnology: Principles and Applications, 1st ed.,U.S.A., Mc GrawHill; (2001).

[90] Lane, D J. *S/23S rRNA sequencing, Nucleic acid techniques in bacterial systematics.* England, Wiley;(1991). , 1991, 205-248.

[91] Takai, K, & Horikoshi, K. Rapid detection and quantification of members of the archaeal community by quantitative PCR using fluorogenic probes. *Applied Environmental Microbiology*. (2000). , 66, 5066-5072.

[92] Sawayama, S, Tsukahara, K, & Yagishita, T. Phylogenetic description of immobilized methanogenic community using real-time PCR in a fixed-bed anaerobic digester. *Bioresource Technology*.(2006). , 97, 69-76.

[93] Geets, J, Borremans, B, Diels, L, Springael, D, Vangronsveld, J, Van Der Lelie, D, & Vanbroekhoven, K. DsrB gene-based DGGE for community and diversity surveys of sulfate-reducing bacteria. *Journal of Microbiological Methods*. (2005). , 66, 194-205.

[94] Stams, A. J. M. Oude Elferink, S. J.W.H., Westermann, P.Metabolic Interactions Between Methanogenic Consortia and Anaerobic Respiring Bacteria. In: Scheper, T. (ed.) Advances in Biochemical Engineering Biotechnology, (2003). , 31-56.

[95] Cetecioglu, Z. Evaluation of Anaerobic Biodegradability Characteristics of Antibiotics and Toxic/Inhibitory Effect on Mixed Microbial Culture. PhD Thesis. Istanbul Technical University, (2011).

[96] Sponza, D. T, & Demirden, P. (2007). Treatability of sulfamerazine in sequential up-
      flow anaerobic sludge blanket reactor (UASB)/completely stirred tank reactor (CSTR)
      processes. Separation and Purification Technology, 56. , 108-117.

[97] Thomas, K. L, Lloyda, D, & Boddya, L. Effects of oxygen, pH and nitrate concentra-
      tion on denitrification by Pseudomonasspecies, FEMS Microbiology Letters, (1994).

[98] Hong, C, & Ying, H J. LeZheng W and Bing S. Occurrence of sulfonamide antibiotics
      in sewage treatment plants. Chinese Science Bulletin. (2008). , 53(4), 514-520.

[99] Iwane, T, Urase, T, & Yamamoto, K. Possible impact of treated wastewater discharge
      on incidence of antibiotic resistant bacteria in river water. Water Science and Technolo-
      gy. (2001). , 43, 91-99.

[100] Dworkin, M, Falkow, S, Rosenberg, E, Schleifer, K H, & Stackebrandt, E. The Prokar-
      yotes, 3nd ed., U.S.A, Springer; (2006).

Chapter 9

Sustainable Post Treatment Options of Anaerobic Effluent

Abid Ali Khan, Rubia Zahid Gaur,
Absar Ahmad Kazmi and Beni Lew

Additional information is available at the end of the chapter

1. Introduction

The strategy of treating sewage by the common and known aerobic process has been shifted back to anaerobic processes in the recent years with the advent of high rate anaerobic systems such as up-flow anaerobic sludge blanket reactor (UASB), anaerobic contact process, anaerobic filter (AF) or fixed film reactors and fluidized bed reactors. The high rate anaerobic processes, like UASB have several advantages such as low capital, operation and maintenance costs, energy recovery in the form of biogas, operational simplicity, low energy consumption, and low production of digested sludge (van Haandel & Lettinga, 1994; Gomec, 2010; Khan et al., 2011a).

During early 1970s, due to the energy crisis and the above advantages, the UASB process was recognized as one of the most feasible method for the treatment of sewage in developing tropical and sub-tropical countries like India, Brazil and Colombia; where financial resources are generally scarce. However, the quality of UASB effluent rarely meet the discharge standards despite several modifications; such as settlers at the top of gas-liquid-solid-separator, addition of AF, two UASB reactors placed in series and even the incorporation of an external sludge digester (Lew et al., 2003; Khan et al., 2011a).

Since early 1980, the discussion on the applicability of UASB process for the treatment of sewage has been presented by Lettinga and co-researchers (Lettinga et al., 1980; Lettinga et al., 1993; Lettinga, 2008; Seghezzo et al., 2002; von Sperling and Chernicharo, 2005) and the results indicated that about 70% chemical oxygen demand (COD) removal can be achieved in warm climates countries (Schellinkhout and Collazos, 1992; Souza and Foresti, 1996; Khan et al., 2011a). Since its inception a lot of research has been done on this process and technology has been given wider publicity. Presently about 30 UASB based sewage treatment plants (STPs) are in operation in India and more than 20 are under construction (MoEF, 2005 and 2006). In total, about 200 UASB reactors are used for municipal and industrial applications (Khan, 2012).

The UASB reactor treating domestic wastewater can produce two main valuable resources, which can be recovered and utilized: methane and the effluent. The methane gas, which is produced during the COD removal can be recovered (from 28% to 75%) and transformed into energy (Mendoza et al., 2009). In energy terms, $1m^3$ of biogas with 75% methane content is equivalent to 1.4 kWh electricity. The biogas can be used to run dual fuel generators or street lighting (Arceivala and Asolkar, 2007). According to Arceivala and Asolkar (2007) approximately 23% methane gas was observed dissolved in UASB effluents, therefore, the recovery of dissolved methane gas is discretionary and may not be acceptable in case of sewage treatment due to high expenditure costs and complexity. However, the methane gas evolved to the headspace (gas phase) can be of much importance and easily collected. For high strength industrial wastewaters the recovery of dissolved methane gas is favoured in view of the global warming and its fuel value. Moreover, at high temperature the solubility of gaseous compounds decreases. Therefore, the issue of gas recovery especially dissolved methane gas must be carefully reviewed for each individual case in terms of economics and desirability.

The produced effluent can be used in agriculture irrigation or disposed. However, the inability of UASB process to meet international disposal standards, owing to its anaerobic nature has given enough impetus for the subsequent post treatment. Furthermore, the growing concern over the impact of sewage contamination on rivers and lakes and the increasing scarcity of water in the world along with rapid population increase in urban areas give reasons to consider appropriate technologies for the post treatment of anaerobic effluent in order to achieve the desired effluent quality and save receiving water bodies.

A variety of post treatment configurations based on various combinations with UASB have been studied, such as aerobic suspended growth, aerobic attached growth, combined suspended and attached growth aerobic processes, anaerobic processes, natural treatment processes, physical processes and physico - chemical processes. UASB followed by final polishing units (FPU) or polishing pond (PP) is a common process used at several STPs in India, Colombia and Brazil, since the technology is simple in operation (von Sperling and Mascarenhas, 2005; von Sperling et al., 2005; Chernicharo, 2006; Khan et al., 2011a). However, still the final effluent is generally devoid of dissolved oxygen (DO) and rich in nutrient. Moreover, polishing ponds operate at long hydraulic retention time (HRT), around 1 day, leading to a high land requirement (Khan, 2012).

Other post treatment options widely used in India are activated sludge process (ASP) and aeration-polishing pond. A demonstration scale Down-flow hanging sponge reactor is also in operation (Tandukar et al., 2005 and 2006). Several other options such as plain aeration i.e. without using biomass, are the next technology option for the post treatment of anaerobic effluents but, limited studies have been performed. A bench scale batch aeration investigated by Khan (2012) has demonstrated that aeration systems operating at 1 to 2 h HRT are able to reduce the BOD of UASB effluent to discharge standards but, unable to remove nutrient. In the same study, continuous aeration of UASB effluent with and without activated sludge could remove nutrient.

Similarly, sequencing batch reactor (SBR), moving bed bio-film reactor, sand filtration, dissolved air flotation, rotating biological contactors (RBC), wetlands and others are still under investigations at bench and pilot scale. Results are promising; however, more studies are needed at pilot or demonstration level with actual environmental conditions in order to scale-up these technologies for best treatment concept. If stringent disposal standards need to follow, aeration with biomass can effectively reduce the organics, nutrients and odour causing substances like sulfides. Some of these processes are exclusively discussed in subsequent section

Recently two different aerobic biomass based processes viz. continuous fill intermittent decant (CFID) type SBR and intermittent fill and intermittent decant type SBR were investigated by Khan (2012). Several researchers investigated the CFID and SBR and results revealed that the CFID can reduce the nitrogen to less than 10 mg/L as nitrogen. SBR is highly efficient to remove the nitrogen and phosphorous. Detailed studies were carried out on different aerobic treatment processes by Khan (2012).

Another latest concept of treatment is the 'Natural Biological Mineralization Route' (NBMR), which can be applied for the treatment of anaerobic effluents as suggested by Lettinga (2008) and elucidated in detail, by Khan (2012). This treatment concept enables conserving or recovery of useful by-products in the form of fertilizers, soil conditioners and renewable energy. The whole concept consists of treatment units of several micro aerobic and aerobic systems and dealt in subsequent section.

The objective of this chapter is to summarize different post treatment options for anaerobic effluent in general and specifically effluent of UASB reactor treating sewage. Natural biological mineralization route (NBMR) concept is also explained for an economical and efficient treatment.

2. Anaerobic effluent/ UASB effluent characteristics

The effluent characteristics in terms of biological oxygen demand (BOD), COD, suspended solids (SS), nutrients (N & P), microbial pathogens and reduced species such as sulfides explained as follows:

2.1. Organics and suspended solids

The BOD, COD and SS of various anaerobic treatment systems anaerobic ponds, UASB reactors, Imhoff tank and septic tanks treating sewage without any post treatment system has been reported to vary from 60 to 150; 100 to 200 and 50 to 100 mg/L, respectively (Chernicharo, 2006; Foresti et al., 2006). The process efficiency depends on different factors like strength and composition e.g. fraction of industrial wastewater infiltrated, temperature and diurnal fluctuations. The dissolved mineralized compounds such as ammonia, phosphate and sulfides in the effluent also varied with these factors. The performance of these treatment systems highly depends on temperature and decreases with a decrease in temperature (Lew et al., 2003 and 2004; Elmitwalli et al., 2001). The performance of UASB reactors (COD, BOD and TSS influent, effluent and removal) treating sewage at different temperatures is summarized in Table 1.1.

2.2. Nutrients (N and P)

Insignificant or negligible removal of nutrient may be expected in anaerobic systems treating sewage (Foresti et al., 2006; Moawad et al., 2009). The primary reason of poor removal of nutrients in anaerobic process is organic nitrogen and phosphorous hydrolyzed to ammonia and phosphate, respectively, which are not removed by anaerobic processes and in consequence, their concentration increases in the liquid phase. The concentration of ammonia nitrogen and phosphorous in anaerobically treated sewage has been reported to be from 30-50 and 10-17 mg/L, respectively (Foresti et al., 2006).

2.3. Highly mineralized or reduced compounds

Sulfur compounds exist as sulfides in anaerobic systems effluent treating sewage. The effluent total sulfides concentration to the highest degree depends on concentration of sulfates in influent and sulfate reducing bacterial activity present in the reactor. Generally, sulfide concentrations around 7-20 mg/L have been observed in the UASB effluent treating sewage, which increases the effluent oxygen demand (Khan, 2012). Moreover, the chemical and biochemical oxidation also depends on sulfides concentration along with other reduced species such as Fe^{2+}, mercaptans etc. although low ferrous ion concentration has been observed in the anaerobic effluent of systems treating sewage. However, Vlyssides et al. (2007) investigated the effect of ferrous ions addition to influent to enhance COD removal. The addition of ferrous ion induces a stable and outstanding conversion rate of COD and was proved to enhance the biological activity of UASB reactor; otherwise ferrous ions results by reduced environment if sewage is treated by UASB reactor.

2.4. Indicators of microbial pathogens

The reduction of fecal coliforms is around one order of magnitude (from around 10^8 to 10^7) in UASB systems although they are not designed for pathogenic removal, while helminth eggs removal efficiency has been reported to be 60–90% (Chernicharo et al., 2001; von Sperling et al., 2002; Chernicharo, 2006; von Sperling and Mascarenhas, 2005).

Hence, for ideal and sustainable treatment the high rate anaerobic treatment systems especially UASB rector must be integrated with novel and innovative post treatment systems based on NMBR sequence. Numerous post treatment system or combination of anaerobic pre-treatment (i.e. UASB reactor) followed by aerobic systems were investigated at laboratory and pilot scale levels for the treatment of sewage. Most of these combinations were found viable option for the treatment of effluent of UASB reactor.

Country	Capacity	Temp. (°C)	HRT (h)	Influent (mg/L)			Effluent (mg/L)			Removal Efficiency (%)		
				COD	BOD	TSS	COD	BOD	TSS	COD	BOD	TSS
Japan	-	-	6	600	291	333	222	153	-	63	53	-
Japan	1148 L	-	6	532	240	-	197	79	-	63	67	-
India	5 MLD	25	10	590	167	-	201	60	-	66	64	-
-	-	-	8	463	214	174	125	39	47	73	82	73
India	5 MLD	20-31	6	560	210	420	140	53	105	74-78	75-85	75-89
Brazil	106 L	21-25	4 7	265	150	123	143	60	33	30	61	73
-	110 L	12-18	18	465	-	154	163	-	42	65	-	73
Colombia	35m³	23-24	5.2	430-520	-	200-250	170	-	65	66	80	69
-	3.7m³	24-26	10-18	660	300	-	178	66	-	73	78	-
Brazil	106 L	20	4	424	195	188	170	61	59	60	69	69
Netherlands	120 L	8-20	12	500	-	-	225	-	-	60-90	-	65-90
Netherlands	6 m³	20	18	550	-	-	165	-	-	70	-	-

Table 1. Treatment Performances of Lab and Full Scale UASB Reactors Treating Sewage *(adopted from Khan et al., 2011a)*

The discussion for the selection of the sustainable technology for the policymakers, engineers, contractors, consultants and authorities of the public sanitation (PuSan sector) has been presented in discussion/ summary part of this chapter.

In addition, more sustainability and treatment performance of these treatment system can be improved if these systems/ combinations were categorized based on their application to remove the suspended solids with or without chemical coagulants, soluble organic and inorganic matter, and removal of reduced compounds such as ferrous ions, sulfides etc. and recovery of methane.

The foremost categories are:

i.   Conventional settling systems and flotation methods with or without chemical coagulants for the removal of suspended solids and soluble organic and inorganic compounds like phosphate or termed as primary post treatment options;

ii.   Application of physical chemical methods to remove and recover dissolved methane from the effluent, which is very important issue for the researchers, engineers and scientists;

iii.   Biological micro-aerobic methods for the removal of highly reduced (malodours) compounds like sulfides and volatile organic S compounds, Fe^{2+} and colloidal matter;

iv.   High rate aerobic systems for nitrification, when combined with denitrification step;

v.   Polishing methods for high rate removals of pathogens and further polishing of the
         secondary treated effluent. The post treatment systems thus, categorized can either
         be used singly or sequentially.

3. Post treatment systems

3.1. Low rate natural settling systems

The highly stabilized suspended matter present in the UASB effluent can be removed by micro-
aeration and settling process. Therefore, proper methods of removal of suspended solids are
needed. Currently, natural settling processes are widely used at full scale STPs. The natural
settling method is often slow and inefficient and sometimes enhanced by addition of chemical
which could easily remove the colloidal and finely dissolved solids, which are separated by
physical aeration. Further, the recovery of resources in terms of phosphates and treated
effluent, if used for irrigation purposes makes it ideal as a sustainable option.

3.1.1. Overland Flow System (OFS)

Chernicharo et al. (2001) investigated extensively OFS operated in two phases in Brazil. This
system is a classical example of a full scale natural system in use for UASB effluent post
treatment and characterized by constant and transient hydraulic regime respectively. Three
slopes (physically identical) for wastewater overland flow constituted the post-treatment
system. A very common weed species named *Brachiaria humidicola* was used as vegetative
cover on the slopes. This weed is known for its high rate of nutrient absorption and high
resistance against flooding.

The good performance of OFS can be achieved at low flow rate application ranging from 0.4 -
0.5 m^3/m.h. The final effluent concentration of the combined system (UASB followed by OFS)
showed average values of BOD from 48 to 62 mg/L; COD from 98 to 119 mg/L and SS from 17
to 57 mg/L. The combined system removed 2 to 3 log-units of FC thereby reducing the residual
FC of effluent to around 8.4×10^4 to 2.4×10^5 MPN/100mL. In addition, a significant removal
of helminth eggs was observed with an average effluent concentration of 0.2 Egg/L. However,
the final effluent quality of the overland flow system was interfered by the transient flow
regime and the high concentrations of solids and organic matter in the UASB reactor effluent.
For these situations, the length of the slope was suggested to be kept above 35 meters.

3.1.2. Polishing Ponds (PP)

Cavalcanti et al. (2001) investigated the feasibility of a single flow-through PP for the post-
treatment of effluent of UASB reactor in Brazil. The plug flow regime was maintained in pond
in order to elevate the fecal coliform removal efficiency of the system. Two distinct HRT of 5d
and 15d were maintained in the pond. At 5d HRT, the average BOD, COD and TSS values
were reduced to 68, 188 and 68 mg/L, respectively. At HRT of 15d these concentrations lowered
down to 24, 108 and 18 mg/L, respectively. Removal of pathogenic microbial indicators was

also encouraging, with the complete removal of helminth eggs at 5d HRT. Moreover, at 15d HRT the effluent FC concentration was very close to 1000 MPN/100mL, with conformity to the WHO guideline for unrestricted irrigation.

Again in Brazil, von Sperling and Mascarenhas (2005) investigated the performance of four shallow (0.40 m depth) PP in series for the treatment of UASB effluent at a total HRT of 7.4 d (1.4–2.5 days in each pond). Based on the results, the final effluent average concentration of BOD and COD were 44 and 170 mg/L, respectively. The mean overall FC removal efficiency was remarkably high, 6.42 log units, or 99.99996%. The high FC removal together with total nitrogen concentration of 10 mg/L in the effluent were found compatible with the discharge standards for urban wastewaters from the European Community, 15 mg/l (70% removal). The ammonia nitrogen concentration in effluent from combined system was 7.3 mg/L (67% removal). However, phosphorus removal was only 28% (effluent total phosphorus concentration of 2.8 mg/L). Others studies on integrated anaerobic-aerobic systems carried out in Brazil showed that shallow ponds in series, even at short HRT, are able to produce effluents complying with the WHO guidelines for unrestricted irrigation in respect to coliforms concentration (lower than 1000 MPN/100 mL). All polishing pond systems were able to produce effluents without helminth eggs, what is in compliance with the WHO guidelines for unrestricted and restricted irrigation (≤1 egg/L, arithmetic mean).

Many UASB reactors combined with PP are located in India. Khan (2012) studied short HRT PP, 1d. The treatment performance was insignificant and merely running as settling tanks with a very limited algal activity. The BOD and TSS removal was generally found less than 50%. Due to very limited algal activity, coliform removal was also restricted to generally 1-2 log unit, however, helminth eggs were removed completely.

3.1.3. Constructed Wetland (CW)

The CW system for wastewater treatment is accepted as a technically and economically feasible alternative for small communities (Okurut et al., 1999). The systems used solid medium (sand, soil or gravel) to develop a natural processes under suitable environmental conditions that lead to the treatment of wastewaters. The plants are densely spaced and, together with the shallow water, provide habitats for animal, bird and insect communities. Vegetation in a wetland provides a substrate (roots, stems, and leaves) upon which microorganisms can grow as they break down organic materials. The most important functions of the plants are: (a) utilization of the nutrients and other constituents; (b) oxygen transfer to the solid medium; (c) support medium for bio-films on the roots and rhizomes (Sousa et al., 2001).

Sousa et al. (2001) investigated the demonstration scale wetland system for the treatment of effluent of UASB reactor for the removal of residual organic matter, suspended solids, nutrients (nitrogen and phosphorus) and fecal coliforms. The 1500 liter UASB reactor was operated at varied HRT (3h and 6h) while the effluent of the UASB reactor was treated in four units of CW, each 10 m long and 1.0 m wide, with coarse sand and operated in parallel under different hydraulic and organic loads. Macrophytes (*Juncus sp*) were planted in three CWs, whereas one CW was operated as a control unit without plants. The results revealed that the effluent COD from the four CW units had substantially constant concentration values, indicating that there was no influence of varied hydraulic load applied and presence of plant in CWs on its removal efficiency.

The phosphorous removal was very efficient during entire period of study. The phosphorous removal was mainly due to the utilization by plants and microorganisms as well as adsorption and precipitation. In the CW without plants, the removal was due to precipitation and adsorption as well as assimilation by the bio-film developed on sand grains. The results indicated that there was no adverse affect of varying hydraulic load or retention time on phosphorous removal efficiency.

The nitrogen removal in the four CW units was satisfactory under variable operation conditions. The total nitrogen removal efficiency varied from 59% to 87% in wetlands containing microphytes. The two basic factors for the removal of nitrogen in wetlands containing microphytes were observed to be assimilation by plants and microorganisms present in wetlands and; probably nitrification due to transport of oxygen from atmosphere by plants. The results indicated that the presence of microphytes enhance the nitrogen removal efficiency significantly. The highest removal efficiency occurred in the unit with lowest hydraulic load corresponding to HRT of 10 d. The removal efficiency of fecal coliforms was observed to be very high in wetlands with microphytes. The increase in hydraulic load reduced the removal efficiency.

3.1.4. Duckweed Pond (DP)

The aquatic macrophyte based treatment systems such as DP can be used to recover the nutrient and transformed them into easily harvested protein-rich by-products. The UASB effluents are highly rich in nutrient which should not be removed but, recovered. DP are covered by floating mat of macrophytes, which prevents light penetration into the pond resulting in shading. The high growth rates of the macrophyte permits regular harvesting of the biomass and hence nutrients are removed from the system. The produced biomass has economic value, because it can be applied as fodder for poultry and fish.

El-Shafai et al (2007) evaluated the performance of a combined UASB and DP system (3 ponds in series). The UASB reactor had a volume of 40 liter and run at 6 h HRT while each pond had 1 m^2 surface area and 0.48 m deapth. The HRT in each pond was 5 d providing total HRT of 15 d in all ponds. The DP were inoculated with $L.$ $gibba$, obtained from a local drain, containing 600 g fresh duckweed per m^2. The system removed 93% COD, 96% BOD and 91% TSS during warm season. Residual values of ammonia, total nitrogen and total phosphorus were 0.41 mg N/L, 4.4 mg N/L and 1.1 mg P/L, with removal efficiencies of 98%, 85% and 78%, respectively. The system achieved 99.998% FC removal during the warm season with final effluent containing 4×10^3 cfu/100 mL. During the winters, the system efficiently removal for COD, BOD and TSS was the same, but not nutrients and fecal coliforms. The coliform count in the effluent was 4.7×10^5 cfu/100 mL. The authors reported that the FC removal in DP was affected by the decline in temperature, nutrient availabilies and duckweed harvesting rate.

3.2. High rate physical chemical methods

3.2.1. Chemically Enhanced Primary Treatment (CEPT) & zeolite column (UASB post treatment)

Aiyuk et al. (2004) proposed an integrated Coagulation and Flocculation- UASB- Zeolite column concept for the low-cost treatment of domestic wastewater. In this integrated treatment system, domestic wastewater is initially treated with CEPT using $FeCl_3$ as a coagulant and polymer to remove suspended material and phosphorus, followed by UASB treatment to remove soluble organics. The effluent of UASB reactor was treated by regenerable zeolites to remove total ammonia nitrogen. The CEPT pre-treatment on average removed 73% COD, 85% SS and 80% PO_4^{3-}. The coagulation/flocculation step of this integrated system produced a concentrated sludge (8.4% solids), which can be stabilized in a conventional anaerobic sludge digester and used as fertilizer for agricultural purposes. After coagulation/ flocculation step, UASB reactor consequently received an wastewater with low total COD, approximately 140 mg/L and it was operated with volumetric loading rate of 0.4 g COD/L.day (HRT of 10 h) and 0.7 g COD/L.day (HRT of 5h). For these conditions, the system removed about 55% COD, thus producing an effluent with a low COD of approximately 50 mg/L (53±28 mg/L). The zeolite removed almost 100% NH_4^+. The integrated coagulation / flocculation–UASB-Zeolite system effectively decreased the TSS and COD upto 88% and more than 90%, respectively. The nitrogen and phosphorus were decreased by 99% and 94%, respectively. The column of zeolite proved most beneficial due to very high removal efficiency of ammonia and the oxidation of residual organic matter. Pathogenic indicators (FC) levels were reduced from 10^7 cfu/L to 10^5 cfu/L, indicating a removal of 99%. The final effluent from the system can be used for crop irrigation or be discharged in surface waters.

Percolation of the UASB effluent through the zeolite ion exchange column resulted in an improved effluent quality (average final effluent total COD of 45±6 mg/L). Still it is possible that the overall integrated system effluent characteristics do not meet desired standards. But, the system operates at low costs, making it suitable for developing countries and rural areas. The final effluent can be used at least for crop irrigation. The recycling/ reuse or disposal of the side streams generated should be explored further and evaluated in future research, together with the energy recovering potential of the CEPT sludge.

3.2.2. Dissolved Air Flotation (DAF)

Based on the results observed from the use of physico-chemical processes for sewage treatment DAF stood up to be an attractive alternative for the post treatment of UASB effluent. DAF system clarifies wastewater by removing floating suspended matter such as oil, fats or solids. The removal is achieved by dissolving air in wastewater under pressure and then releasing the air at atmospheric pressure in a flotation tank. The released air forms tiny bubbles which adhere to the suspended matter causing the suspended matter to float to the surface of the wastewater and form a froth layer where it may then be removed by a skimming device. The feed water to the DAF float tank is often (but not always) dosed with a coagulant (such as ferric chloride or aluminum sulfate) to flocculate the suspended matter. Penetra et. al. (1999) studied a lab scale DAF with previously coagulated effluent from a pilot scale

UASB reactor. Ferric chloride ($FeCl_3$) was used as coagulant and dosages ranged from 30 to 110 mg/L with pH in the range of 5.3 to 6.1, varified with addition of lime. Best results were achieved at a $FeCl_3$ dose of 65 mg/L. The DAF system was found efficient to reduce COD up to 91%, total phosphate up to 96% TSS up to 94%, turbidity up to 97% and sulfides more than 96%. The combined UASB-DAF system was observed to reduce 98% COD, 98% TP, 98.4% TSS and 94% Turbidity.

3.2.3. Two Stage Flotation and UV disinfection (TSF-UV)

The $FeCl_3$ coagulant and cationic polymer used in DAF systems presents a faily good removal efficiency of the UASB effluent, but these processes resulted in a significant volume of sludge. Tessele et al. (2005) investigted a pilot scale UASB ($50m^3$/d flow) reactor followed by conventional two stage flotation and UV disinfection system for nutrient recovery. The proposed two stage flotation unit brings the advantage of separating the biomass and sludge that contain the phosphate and hydoxide. The suspended solids were removed by first stage flotation-flocculation (FF) process referred as F1 followed by second stage DAF referred as F2. Phosphate ions were removed by precipitation and coagulation. The removal mechanism in FF was the formation of small bubble and entrapped in flocs and these flocs floats over the water surface. In second flotation stage, both flocs and fine solids were aimed to removed. The concentration of Fe^{+3} and flocculant varied from 0 to 25 mg/L and 0 to 15 mg/L, respectively. The air flow in FF process was 3.0 L/minutes while DAF air flow rate 0.9 to 1.2 L/minute. The hydraulic loading rate was kept around 49 m/h at an HRT of 2 minutes in DAF, which is higher than in conventional DAF (6-10 m/h). After F2, the effluent was disinfected with low pressure UV lamp operated at a theoretical value of 25 mJ/cm^2. The results present that the combined UASB-TSF-UV process is more efficient than UASB-PP system. The final effluent contained low COD, phosphate ion concentration, turbidity and air/ water surface tension is as high as that of tap water while the ammonia removal was insignificant.

3.2.4. Coagulation-flocculation

Feasibility of coagulation and flocculation as a post treatment process for the effluent of UASB reactor treating domestic sewage were studied by Jaya Prakash et al. (2007). Commonly used coagulants (alum, polyaluminum chloride (PAC), ferric chloride, and ferric sulphate) were used in a series of jar tests to determine the optimum coagulant dose. The optimum chemical dosage was 20 mg/L (as Al) for alum, 24 mg/L (as Al) for PAC, 39.6 mg/L (as Fe) for $FeCl_3$ and 17.6 mg/L (as Fe) for $FeSO_4$. All the tested coagulants were found to be effective in reducing the effluent BOD and SS to less than 20 mg/L and 50 mg/L, respectively. However, coagulation–flocculation alone was not found sufficient to reduce the FC concentration to a permissible limit (1000 MPN/100 mL) for unrestricted irrigation. The final concentration of fecal coliform of UASB reactor effluent was 2300 MPN/100 mL using alum and PAC optimum doses. Moreover, the investigators suggested that disinfection by a chlorine dose of 1-2 mg/L with contact time of 30 minutes could reduce the FC concentration to below 1000MPN/100 mL after treating UASB effluent by coagulation-flocculation process. Further, higher doses of chlorine i.e. 3 mg/L removed all the FC from the sample after coagulation together with the above mentioned optimum alum and PAC doses. However, 4 mg/L of chlorine dose was needed after coagulation with iron coagulants to remove all the FC.

3.3. Micro-aerobic methods (Including removal/ or recovery of dissolved gases)

The UASB effluent contains reduced organic and inorganic species and dissolved methane gas which can be removed by micro-aeration. Micro-aeration implies aeration of treated effluent for about 30 min. The role of micro-aeration is to strip off and to oxidize the reduced species such as sulfides, ferrous ions etc. which exert immediate oxygen demand and remaining easily biodegradable organic pollutants and to remove the dissolved methane gas. Generally, these systems have very short HRT and the amount of excess sludge generated is negligible. The simple physical micro-aeration can be sufficiently remove or strip off the dissolved sulfides or methane from the UASB effluent. However the removal of suspended solids is insignificant from this process.

3.3.1. Down-flow Hanging Sponge (DHS)

DHS reactor was developed by Harada and his research group at Nagaoka University of Technology, Japan, for the aerobic post-treatment of the UASB effluent. In DHS, sponge cubes diagonally linked through nylon string have been used to provide a large surface area to accommodate microbial growth under non-submerged conditions. The wastewater trickled through the sponge cubes supplies nutrients to resident microorganisms. Oxygen is supplied through natural draught of air in the downstream without equipment. The system provides for dissolved methane gas to be recovered. Matsuura et al. (2010) investigated a two stage DHS system for the post treatment of UASB effluent in Nagaoka, Japan. Most of the dissolved methane (99%) was recovered by the two stage system, whereas about 76.8% of influent dissolved methane was recovered by the first stage operated at 2h HRT. The second DHS reactor was mainly used for oxidation of the residual methane and polishing of the remaining organic carbons. The removal of COD and BOD in the first stage was insignificant as there was no air supply; however, high removals were expected in the second stage due to sufficient supply of air, which is quickly oxidize the residual dissolve methane in the upper reactor portion before being emitted to the atmosphere as off-gas.

Agrawal et al. (1997) evaluated for the first time the performance of combined UASB reactor and DHS cube process. With post-denitrification and an external carbon source, 84% in average N ($NO_3 + NO_2$) was removed with an HRT of less than 1 hour, for temperature range of 13 to 30 ^{0}C. The effluent contained a negligible amount of SS and total COD was only in the range of 10 to 25 mg/L. The DHS reactor was capable of stabilizing total nitrogen through nitrification, which ranged from 73-78%. In another study Machdar et al. (2000a & b) observed that the combined UASB+DHS system successfully achieved 96–98% of BOD removal, 91–98% of COD removal, and 93–96% of TSS removal, at an overall HRT of 8 h (6 h for UASB and 2 h for DHS unit). The complete system neither requires external aeration input nor withdrawal of excess sludge. The final BOD effluent concentration was 6-9 mg/L. Similarly, FC removal was 3.5 log with a final count of 10^3 to 10^4 MPN/100mL in the effluent. Nitrification and denitrification in DHS accounted for 72% removal of total nitrogen (effluent concentration of 11 mg N/L) and

60% removal of ammonium nitrogen (effluent NH_4-N of 9 mg N/L) over the total operational period.

3.3.2. Trickling Filter (TF)

The TF consists of a fixed bed of rocks, gravel, slag, polyurethane, foam, sphagnum peat moss, or plastic media over which sewage or other wastewater flows downward promoted a layer or film of microbial slime to grow. Aerobic conditions are maintained by splashing, diffusion, and either by forced air flowing through the bed or natural convection of air if the filter medium is porous. The process mechanism involves sorption of organic pollutants by the layer of microbial slime. Diffusion of the wastewater over the media furnishes dissolved oxygen which the slime layer requires for the biochemical oxidation of the organic compounds and releases CO_2 gas, water and other oxidized end products. Chernicharo and Nascimento (2001) studied the applicability of pilot level TF for polishing the effluent of UASB reactor. The volume of UASB reactor was 416 liter operated at an average HRT of 4h and the TF volume was 60 liters with blast furnace slag of 4 to 6 cm in size used as media. The operational conditions in the UASB reactor was kept constant throughout the study period while the TF was operated at three different phases, low, intermediate and high rate. The performance of UASB reactor was consistent, with removals above 70% in terms of BOD and COD. The final effluent quality was produced when the TF was operated as low and/or intermediate rate. Under these operational conditions the average COD, BOD and SS concentrations were 90, 30 and 30 mg/L, respectively and; hence, complying with the discharging standards. The system proved very efficient under low loading conditions. At high rate conditions the system was not efficient to remove the BOD, COD and SS. The results of this study showed that the TF can be used as the post treatment option for the treatment of UASB effluent for low organic and hydraulic rates in tropical countries.

3.3.3. Micro aeration methods i.e. flash aeration

For the last decade progress has been made on the use of high rate micro-aerobic methods for the removal or recovery of dissolved sulfides contained in anaerobic effluents. Besides, sulfide purging into the atmosphere, micro-aeration can also be utilized for biological oxidation of sulfides into elemental sulfur, which offers an excellent potential for reuse and it has been shown to be a cost effective alternative (Vallero et al., 2003; Chuang et al., 2005; Chen et al., 2010; Khan et al., 2011a and 2011c). The process is generally focused on the treatment of biogas, off-gas, natural gas or low strength wastewaters, like in the case of anaerobic effluents. In addition, micro-aeration of anaerobic system may be an option for increase hydrogen sulfide stripping and methane production (van der Zee et al., 2007). Buisman et al. (1990) developed a low-cost, high-rate biotechnological aerobic process for the oxidation of sulfide into elemental sulfur by a group of colorless sulfur bacteria, where the sulfide oxidation rate was dependent on the oxygen level. The biofilm on a reticulated polyurethane was more suitable to produce sulfate than a free cell suspension of biomass, for the same given oxygen and sulfide concentrations. For efficient achievement of elemental sulfur, high sulfide loads or low oxygen concentrations must be applied (Stefess et al., 1996). Vallero et al. (2003) utilized the micro-aerated reactors for the oxidation of sulfides to elemental sulphur from the liquid phase of anaerobically treated sewage. The results were encouraging and partial conversion of soluble sulfides (HS⁻) into colloidal elemental sulphur was observed.

The produced element sulfur forms transparent globules of up to 1 micro-meter in diameter, which is deposited inside or outside the bacteria. An important issue is the recovery of the colloidal sulfur particles. Janssen et al. (1999) studied the properties of the colloidal sulfur particles and developed an up-side down cone expanded-bed bioreactor for spatially separation of the aeration and oxidation phases. After 50 days of operation 90% of the sludge settled at a velocity greater than 25 m/h and could be easily removed. Although the results are very encouraging, more studies on the application of high purity systems for colloidal matter removal are necessary. One of the most promising technologies for sulfide removal from biogases is a two-step process where gaseous sulfide is dissolved into the liquid in the first step, followed by sulfide oxidation to elemental sulfur. Although little research has been conducted on the subject Chuang et al. (2005) treated a sulfate-rich wastewater in a UASB followed by a floated bed micro-aerated reactor. The floated bed was operated at short HRT (2.8 hours) and during long-term steady state operation results showed that almost all sulfides (>96%) was oxidized to elemental sulfur and sulfate. Annachhatre and Suktrakoolvait (2001) observed a sulfide conversion higher than 90% at sulfide loading rates of 0.13-1.6 kg $S/m^3/d$ and at DOs lower than 0.1 mg/L sulfur was the major end product.

The simplest method of sulfide oxidation is the introduction of micro-aerobic conditions in the anaerobic reactor. Despite the toxicity exerted by oxygen against obligatory anaerobes, its moderate introduction is not expected to have a harmful impact to the biomass, mainly to the limited penetration depth of oxygen in biofilm. Van der Zee et al. (2007) determined the air injected to sulfide ratio to be 8-10:1 (O_2: S in mol units), which was sufficient to reduce the biogas H_2S content to undetectable levels. Element sulfur and sulfate were the main products.

3.3.4. Continuous Diffused Aeration (CDA)

CDA system was investigated to treat the effluent of UASB reactor in India by several authors (Walia, 2007; Khan et al., 2011b; Khan, 2012). The treatment of sewage in a 60 L pilot scale UASB reactor followed by a CDA system and a full scale plant (111MLD capacity; UASB +Aeration+FPU) were investigated by Khan et al. (2011a). The HRT of CDA system was maintained at 15, 30 and 60 min HRT. During aeration at each HRT bulk liquid DO of 5-6 (high) and 1-2 (low) mg/L were maintained. The final COD, BOD and TSS effluent concentrations were 40-60, 25-35, 30- 40 mg/L, respectively, for operating under high DO (5-6 mg/L) and 30 minutes HRT and 30- 50, 18-30, 20-30 mg/L, respectively, at 60 minutes HRT. The combined reduction efficiency of the integrated UASB-CDA system at HRT of 30 and 60 min ranged from 80 to 85% COD, 85 to 90% BOD, 65-75% TSS. A conceptual model was developed wherein it demonstrated that the aerobic nature of the effluent depends on dissolved oxygen (DO), ORP and BOD. Anaerobic UASB effluent becomes aerobic if its BOD is reduced to less than 30 mg/L and minimum values of DO and ORP are observed, 4-5 mg/L and 120-135mV, respectively. Based on experimental results empirical correlations between BOD, ORP and DO have been developed and the results indicated a 50% reduction in BOD of the UASB effluent at HRT of ~100 min. The removal of NH_4-N and total-P was insignificant at any of the maintained HRT. The Integrated UASB-CDA for sewage treatment could be recognized as a sustainable and cost effective option as the combined HRT of the system is still short (8 h for UASB + 0.25-1.0h for aeration, with a total HRT of 8.25-9.0 h). Existing UASB based STPs can be upgraded by installing continuous aeration system through fine pore diffuser and the energy produced by UASB reactor in terms of biogas could be used to operate the aeration system.

3.4. High rate aerobic methods (Including nitrification-denitrification steps)

The poorly biodegradable soluble matter, hazardous compounds or micro pollutants including ammonia-nitrogen and phosphorous present in the UASB effluent sometimes are difficult to be remove by micro-aerobic or simple settling. Therefore, secondary post treatment is required, following the micro-aerobic or settling treatment methods. A number of secondary post treatment processes have been categorized into methods responsible for the removal of (i) poorly biodegradable soluble matter including micro pollutant and hazardous material, (ii) finely dispersed organic matter i.e. colloidal and pathogens removal and (iii) ammonia-N and phosphorous. The removal of residual biodegradable carbon, ammonia nitrogen and phosphorous can also be achieved if the effluent of UASB is treated by high rate aerobic biological treatment methods.

3.4.1. Sequential Batch Reactor (SBR)

The SBR is a fill and draw type modified activated sludge process, where four basic steps of fill, aeration, settle and decant take place sequentially in a single batch reactor. The operation of SBR can be adjusted to obtain aerobic, anoxic and anaerobic phases inside the standard cycles (Droste and Masse, 1995; Surampalli et al., 1997). Sousa and Foresti (1996) proposed a combined system composed of anaerobic-aerobic processes consisting a UASB reactor followed by a SBR. The system performance was evaluated through a bench scale set-up comprising of a 4 litre volume UASB reactor followed by two SBRs of 3.6 litres each. The UASB reactor was fed with partially mixed synthetic substrate in sewage while the SBR received effluent of UASB reactor. The HRT of 4 h in UASB was maintained constant throughout the study while the 4 h cycles in the following sequence of fill (0.10h), reaction (1.9h), sedimentation (1.6h), discharge (0.25h); idle (0.15h) were maintained in SBR. The combined system removed ~85% total nitrogen through nitrification. The COD removal in UASB reactor was around 86% while in SBR around 65% of the remaining, thus, combined systems removed 95% (residual effluent COD of 20 mg/L). The performance of combined system was 96% in terms of TSS removal (residual effluent TSS of 9 mg/L) and 98% in terms of BOD removal (residual effluent BOD of 6 mg/L).

Torres and Foresti (2001) studied the effect of aeration on the performance of SBR treating UASB effluent. The UASB reactor was operated at a constant HRT of 6 h while the SBR performance was monitored at four different duration cycles (24, 12, 6 and 4 h) corresponding to aeration times (AT) of 22, 10, 4 and 2 h, respectively. The overall removal efficiencies of COD and TSS were 91% and 84%, respectively and observed independent of aeration time given in the SBR. However, the nutrients removal was found to be dependent on aeration time. Total nitrogen removal of approximately 90% was achieved for AT longer than 10 h; complete nitrification occurred for AT longer than 4 h; significant phosphate removal (72%) occurred only at the AT of 2 h. Moawad et al. (2009) also investigated the performance of the combined UASB-SBR system under different operating conditions for the treatment of domestic wastewater. The retention time in the UASB was changed from 4 h to 3 h and the aeration time in the SBR cycle varied from 2 to 5h, and then to 9 h. The observed average percentage removal for the three runs for COD, BOD and TSS was 94%, 97% and 98%, respectively. The residual COD, BOD, and TSS were 26, 5.8 and 5.0 mg/L, respectively. Complete nitrification of ammonia was achieved after 5 h aeration in the SBR. The average percentage removal of phosphorus reached up to 65%. Increasing the HRT in the SBR from 2 to 9 h caused a significant improve- ment in FC removal as the geometric count of FC was reduced to 7.5×10^2 MPN/100mL in the effluent of the 3rd run (HRT 9 h).

Khan et al. (2011a) investigated the performance of a pilot scale integrated UASB-SBR system for treatment of sewage. Two different variant of SBR Process were investigated: a Continuous Flow-Intermittent Decant Sequencing Batch Reactor (CFID) and Intermittent Fill- Intermittent Decant Sequencing Batch Rector (IFID) for about 18 months in conjunction with UASB reactor at ambient environment. Initially, the UASB-CFID system was operated at an HRT of 8h in the UASB reactor while it varied in CFID (20, 8 and 6 h),which also had different DO regimes, 4.0 to 5.0 and < 0.5 mg/L, 2.5-3.5 and < 0.5 mg/L and 2.5 to 3.5 and <0.5 mg/L, for the respectively HRT. The BOD and TSS removal efficiency of combined UASB-CFID system was up to 90%. The FC reduction was more than 99%. It was observed that average reactor MLVSS concentration reduced to around 30% at DO of 2.5-3.5 mg/L showing high degree of mineralization. Later, an integrated UASB followed by IFID system for the treatment of sewage was evaluated for the removal of organics and nutrient for more than six months at ambient conditions. The HRT in UASB reactor was maintained constant at 8 h. The IFID was operated at 6h HRT at DO concentration ranged between 2.5 to 3.5 mg/L. Results revealed that the removal of BOD, COD and TSS were 90, 95 and 90%, respectively in IFID. During higher organic loading conditions and low SRT, the removal of phosphorous was significantly higher than that of lower organic loadings and higher SRT. The suitable COD: P ratio of 105~160 helped for the effectively removal of phosphate. The total nitrogen removal was sufficiently good ranged from 80 to 95%.

3.4.2. Activated Sludge Process (ASP)

Activated sludge process is the most widely used process for the treatment of sewage and industrial wastewaters. Atmospheric air is bubbled through wastewater combined with organisms to develop biological flocs which reduce the organic content of the sewage. The combination of wastewater and biological mass is commonly known as Mixed Liquor. von Sperling et al. (2001) monitored a pilot-scale plant comprising of an UASB reactor followed by an activated sludge system treating actual municipal wastewater from a large city in Brazil. The UASB reactor removed 69-84% COD, while ASP only removed remaining COD ranging from 43% to 56%, resulting in 85% to 93% removal achieved through the overall system (residual effluent COD of 50 mg/L avg.). The final effluent SS concentration was 13 - 18 mg/L. Therefore, UASB and ASP configuration was suggested to be a better alternative for warm-climate countries than the conventional activated sludge system alone, considering the total low hydraulic detention time of 7.9h (4.0 h UASB; 2.8 h aerobic reactor; 1.1 h final clarifier), offering the advantages in terms of savings in energy consumption, absence of primary sludge and possibility of thickening and digesting the aerobic excess sludge in the UASB reactor itself.

3.4.3. Rotating Biological Contactors (RBC)

A RBC consists of a series of closely spaced circular disks of plastic material such as polystyrene mounted on a shaft that are partially submerged (typically 40%) in wastewater. The microorganisms grow on the surface of circular disks which breakdown and stabilize organic pollutants in presence of oxygen obtained from the atmosphere as the disks rotate. The development of excessive biofilm growth and sloughing problems besides odor and poor performance occurs when oxygen demand has exceeded the oxygen transfer capabilities and is the major drawback of this technology. These rotating biological contactors offer many advantages like the capability of handling a wide range of flows, low power requirements, low sludge production and excellent process control.

Tawfik et al. (2003) examined the removal of organic matter, nitrification and *E. coli* by UASB-RBC system at different operational temperature (11, 20 and 30°C) and at different organic loading rates with constant HRT of 2.5 h in the RBC. The results showed good performance of the system when operated at lower OLRs of 27, 20 and 14.5 g COD/m^2/day at 11, 20 and 30°C, respectively. The residual COD values were 100, 85 and 72 mg/L for the respectively temperatures. Moreover, a high ammonia removal and low residual values of *E. coli* were found for the RBC at operational temperature of 30°C as compared to the situation for treatment of domestic wastewater and UASB effluent at lower temperatures of 11°C and 20°C. The effluent however, did not comply with WHO guidelines for unrestricted irrigation.

Tawfik et al. (2005) investigated the performance of a combined single stage RBC, two stage RBC and an anoxic up-flow submerged bio-filter followed by a segmental two stage aerobic RBC system. This study was carried out in order to assess the impact of biodegradable COD in an UASB effluent applied to the systems on the removal efficiency of different COD fractions, *E. coli*, ammonia and partial nitrate removal. The two (single stage) RBCs were operated at a constant HRT of 2.5 h and temperature of 21 °C but at different OLRs, 10 and 14 g biodegradable COD/m^2/day due to varied UASB effluent qualities. The results clearly show that the residual values of COD, ammonia and *E. coli* in the final effluent are significantly lower at the lower OLR of 10 g biodegradable COD/m^2/day. In view of the results it is recommend to use a single stage RBC system at OLR of 10 g biodegradable COD/m^2/day and at HRT of 2.5 h for post-treatment of the effluent of UASB reactor operated at high temperature of 30 °C, as it generally prevails in tropical countries.

The performance of a single stage versus two stage RBC system for post-treatment of the effluent of an UASB reactor operated at a low temperature of 12 °C was also evaluated. Both systems were operated at the same OLR of 18 g biodegradable COD/m^2/day and at HRT of 2.5 h. The results demonstrated that the COD fractions, ammonia and *E. coli* content in the final effluent of a two stage RBC system were significantly lower than the effluent of the single stage RBC system. Accordingly, results envisaged a two stage RBC system at an HRT of 2.5 h and

OLR of 18 g biodegradable COD/m²/day for post-treatment of the effluent of a conventional UASB reactor operating at a low temperature of 12 °C.

The nitrogen removal from the nitrified effluent was investigated using a biofilm system consisting of three stages, namely an anoxic up-flow submerged bio-filter followed by a segmental two stage aerobic RBC. The nitrified effluent of the second stage RBC was recycled to the anoxic up-flow submerged bio-filter reactor. The results obtained reveal that the introduction of an anoxic reactor as a first stage combined with recirculation of the nitrified effluent of the second stage RBC was accompanied with a conversion of nitrate into ammonia, at least in case the content of biodegradable COD in the UASB effluent was low, In such a situation the ammonia needs to be nitrified two times, which obviously should be avoided. Therefore in such situations of a too high quality anaerobic effluent in terms of biodegradable COD content, the introduction of a separate anoxic reactor for denitrification as final post-treatment step cannot be recommended.

3.4.4. Aerated Fixed Bed Reactor (AFBR)

A sequence of denitrification reactor (DN), UASB, AFB and settling units treating sewage was evaluated by Sumino et al. (2007). The DN and AFB reactors contained sponge sheets media fixed to both the surfaces of the boards oriented vertically. The air was supplied to the AFB reactor from the bottom of the tank. Granular sludge obtained from food waste treatment plant was used as the inoculum sludge in the UASB reactor and activated sludge from a sewage treatment plant was used as the inoculum sludge in the AFB reactor. The SS recirculation from settling tank was made to the denitrification tank and the poly aluminium chloride PAC was injected to ABF for phosphorous removal. The whole system was studied for more than 300 days under constant HRT of 24 h in three different seasons, summer, autumn and winter. The performance of the combined system was satisfactory with final mean effluent values of soluble COD of 54, 66 and 65 mg/L in the summer, autumn and winter, respectively, while the mean total soluble BOD were 11, 18 and 25 mg/L for the corresponding periods. The information on nitrogen and phosphorous removal and indicators of pathogens was not discussed in this study.

3.4.5. Submerged Aerated Bio-Filter (SABF)

The SABF system is composed of floating porous media through which wastewater and air flows from the bottom of the reactor. The airflow in the SABF system is always in upflow mode, while the liquid flow can be in upflow or downflow mode. These biofilters backwashed routinely at least once in 3 days. The development of thin, homogeneous and active biofilm layer is the main mechanism of biofilters to remove the soluble organic compound and suspended solids from the wastewater. Besides serving as support medium for microorganisms, the granular material also works as an effective filter (von Sperling and Chernicharo, 2005). Gonclaves et al. (1998) investigated an UASB reactor (46 L) followed by a SABF (6.3 L) for domestic sewage treatment. The floating and totally submerged granular medium in the SABF was made of S5 type polystyrene spheres with 3 mm diameter, 1200 m²/m³ specific surface area, 0.04 density and 0.50 m height. The air was injected in the SABF bottom, wastewater co-current through an air compressor.

In the study, the UASB reactor was initially operated at 8h hydraulic retention time and subsequently reduced to 6h and 4h. The 4h HRT in UASB reactor was maintained to investigate the performance of reactor under breakdown situation. Several authors recommended that the HRT in the UASB to be shorter than 5h in order to keep an adequate mechanization activity in UASB reator (Vieira and Garcia Jr., 1992; van Haandel and Lettinga, 1994). However, the performance of the UASB reactor was stable and similar at all HRTs studied. The final mean removal efficiency of the combined system in terms of SS, BOD and COD were 94%, 96% and 91% respectively, which amounts to the final effluent concentration of 10 mg/L, 49 mg/L and 10 mg/L respectively.

Goncalves et al. (1999) studied the combined UASB-SABF system and observed similar results. The experiments were conducted with UASB reactor operated at HRT of 6 h without sludge recirculation and the bio-filter at HRT of 0.5 h. The average removal efficiencies of SS, BOD and COD were 95%, 95% and 88%, respectively, with final effluent quality of 10, 10 and 50 mg/L, respectively. Although the efficiency of the UASB-SABF system was satisfactory in terms of organic matter removal, the removal of the pathogenic microorganisms was very low.

Keller et al. (2004) investigated the combined UASB-SABF system followed by conventional and UV system to enhance the efficiency of the system to remove the pathogenic microorganisms. The results revealed that the 84% of COD (residual effluent COD of 78 mg/L), 86 % of BOD (residual effluent BOD of 26 mg/L) and 86% of TSS (residual effluent TSS of 23 mg/L) removal was achieved. The concentration of *E.coli*, *salmonellae* and *colliphases* were reduced to very low in the final effluent of the system. The association of UASB-SABF confirms the viability of the system with excellent final effluent quality of the system.

3.4.6. Moving Bed Bio-film Reactor (MBBR)

Tawfik et al. (2010) investigated a laboratory-scale integrated UASB reactor followed by a MBBR for sewage treatment at three different combined HRTs, 13.3 (8+5.3), 10 (6+4) and 5.0 h (3+2) under temperature range of 22–35 °C for a period of 290d in Egypt. The working volumes of UASB reactor and MBBR were 10 and 8.0 L respectively. A cylindrical carrier media of 1.85 cm diameter and 1.8 cm long made of polyethylene was used in MBBR. Its specific gravity and effective specific surface area were 0.95 and 363 m^2/m^3 respectively. The dissolved oxygen was maintained at 2.0 mg/L throughout the experiment. The performance of the integrated UASB-MBBR system was monitored in terms of COD fractions and FC. At the HRT of 5-10 h an overall reduction of 80–86% for total COD; 51–73% for colloidal COD and 20–55% for soluble COD was achieved. The removal efficiencies were increased up to 92, 89 and 80%, for total, colloidal and soluble COD respectively by increasing the HRT to 13.3 h. However, the removal efficiency of suspended COD in the combined system remained unaffected when increasing the total HRT from 5 to 10 h and from 10 to 13.3 h. This indicated that the removal of suspended COD was independent of the HRT. Final effluent total COD at three different HRTs were 54, 95 and 142 mg/L respectively. The final average FC counts were 8.9×10^4, 4.9×10^5 and 9.4×10^5 MPN/100 mL, corresponding to overall log reduction of 2.3, 1.4 and 0.7 respectively. The main mechanisms observed for the removal of FC were adsorption into the media and predation by higher microbes such as protozoa and metazoa.

The removal of ammonia nitrogen was also investigated in MBBR. The results revealed that the removal of ammonia nitrogen greatly depends on organic loading rate. About 62% of ammonia nitrogen was removed at OLR of 4.6g $COD/m^2/day$ but the removal efficiency decreased by 34 and 43% at the higher OLRs of 7.4 and 17.8g $COD/m^2/day$, respectively. Nitrogen was mainly reduced by assimilation into biomass and denitrification in anoxic zone in the biofilm. The sludge produced by MBBR showed poor settleability, however, the combined system still produced less sludge compared to conventional ASP. The authors reported that the integrated UASB MBBR system at an HRT of 4 and 5.5 h are technically feasible for sewage treatment.

3.5. Final polishing techniques

To achieve nearly complete removal of pathogens, color and hazardous compounds the UASB effluent needs to be polished after the micro aeration first step or secondary post treatment such as high rate aerobic treatment before reusing for intended purpose or discharging it into receiving water bodies.

3.5.1. Membrane technology

Recently large number of membrane technologies was investigated for secondary and tertiary treatment of sewage. Therefore, in order to achieve the quality of treated effluent up to reuse standard from UASB reactor, YingYu et al. (2009) evaluated the pilot scale cross flow membrane filtration system for polishing the UASB effluent treating low strength sewage in Singapore. A pilot scale UASB reactor (34 litres) was coupled with a side stream membrane module having a centrifugal pump to feed the effluent of UASB reactor into the membrane filtration unit. The HRT of UASB reactor was reduced from initial 10h to 5.5h after 119 days of operation and kept constant throughout the study period. The precise and constant holding tank was used prior to membrane filtration module unit in order to feed constant permeate flow rate. Results clearly showed high performance of UASB reactor for total solids removal at HRT of 10h which, however, significantly were reduced from 91.1 to 83.6% at HRT of 5.5h. At steady state conditions in the UASB reactor, the average TOC removal efficiency was 65% (10 h HRT), which increased to 81% by treating the effluent of UASB reactor through membrane filtration. But, the performance of this system in terms of TOC removal further increased to 73 and 85%, respectively at the HRT of 5.5h. This might be due to the increased up-flow velocity which provides better contact and distribution of wastewater with membrane. But fouling of membrane limits its use for the stated purpose. Therefore, extensive studies were required regarding it controlling factors such as membrane tube diameter and cross flow velocity etc.

YingYu et al. (2010) also proposed membrane filtration for the post-treatment of the effluent of UASB reactor in Singapore. The system comprised of UASB reactor and membrane filtration. The UASB reactor with working volume of 30 liter divided into two parts i.e. a sludge zone and a membrane zone. A gas/liquid separator was installed at the top of the sludge zone to separate the biogas from the liquid suspension. Two flat-sheet membrane modules (0.22 μm, PVDF, 0.1 m^2) were directly submerged into the upper membrane zone of the reactor above gas/liquid separator. The modules of flat sheet membrane were submerged into the UASB reactor to as a barrier to retain the suspended solids present in the effluent of UASB reactor at intermittent permeation and air sparging operating conditions. The whole system was operated at a constant HRT of 12 h at a temperature of 35 °C and no sludge was removed from the reactor, except for sampling. The experimental study was conducted in two phases with varied flux. In phase I, Intermittent permeation was studied at three different flux of 15, 20 and 25 $L/m^2/h$ with varied suction pressure while in phase II, air sparging was investigated at four different air flow rates of 0, 1, 2 and 4 L/h with constant flux of 25 $L/m^2/h$.

The average supernatant TOC was 10.88 mg/L with fairly stable TOC removal efficiencies of over 89% during the whole operation. Finally this study influence that intermittent permeation was more effective for membrane fouling control compared with air sparging.

The coupling of membrane filtration with UASB represented as an efficient treatment technology for raw municipal wastewater at the ambient temperature. But limited studied are available on this system therefore, detailed investigations on demonstration scale.

3.5.2. Slow Sand Filtration (SSF) system

Various researchers investigated effect of hydraulic loading and sand size on the effectiveness of SSF for tertiary treatment of sewage at laboratory and pilot scale level and found that the SSF was capable of removing BOD, SS, turbidity and total coliforms up to 86%, 68%, 88% and over 99%, respectively (Ellis, 1987; Suhail, 1987; Sawaf, 1986, Adham, 1989; Gersberg, et al., 1989). However, limited data is available on the applicability of SSF on UASB effluent. Recently, Tyagi et al. (2009) studied the applicability of slow sand filter at lab scale as a post treatment option for the treatment of effluent of UASB reactor. The sand filter column operated at hydraulic loading rate of 0.14 m/h was found to be most effective in removing turbidity (91.6%), TSS (89.1%), COD (77 %), BOD (85%), TC (99.95%) and FC (99.99%). The average values of COD, BOD and SS in SSF effluent were 27 mg/L, 12 mg/L and 20 mg/L, respectively. The FC concentration was found below the standards set by WHO 1989 (1000 MPN/100 mL). It was concluded that slow sand filters can be effectively runs upto 7 days at a hydraulic load of 0.14 m/h as compared to the common hydraulic load of 0.19 m/h and 0.26 m/h. Hence, slow sand filtration could also be an effective technology for the post treatment of UASB reactor effluent, where treated effluent can be reuse safely for irrigation and other non-potable reuse purposes. However, the major drawback of SSF system was the frequent cleaning and maintenance requirement.

4. Discussion

The installation of post-treatment system to treat the effluent of UASB reactor treating sewage is a challenging task as to find a proper, reliable and efficient system, that is easy in operation and maintenance; technically feasible, and economically viable (Chernicharo, 2006). Amongst all post treatment systems, four natural wastewater treatment systems were extensively investigated as the post treatment units. The effluent quality of the polishing ponds in series satisfies the effluent pathogen disposal standards, but it has few disadvantages such as large land requirement, poor nutrient removal, odor related problems and occasionally high BOD and TSS concentrations in the effluent. The combination of polishing pond and duckweed pond, duckweed and algae pond system was reported to be very efficient but, large area requirement, low pathogens removal and high TSS concentration in the effluent were the main drawbacks of this system. The combination of polishing pond and coarse rock filter system give an effluent with high FC and occasionally high in BOD. In overland flow system for the treatment of effluent of UASB reactor under low organic loading rate, the performance was observed to be satisfactory, with low solids and organic matter concentration in the final effluent. However, helminthes eggs removal was insignificant.

The duckweed pond and constructed wetland system are also observed to be satisfactory in their respective performances but these systems are dependent on the temperature, hydraulic load, harvesting of plants, etc. Despite their good nutrient removal efficiencies these systems thought to be unable to bring down the effluent quality below discharging standards.

Four high rate physico-chemical processes were presented including CEPT- Zeolite Column system, DAF, TSF-UV and chemical coagulation-flocculation. These processes are capable to reduce organic pollutants and turbidity of UASB reactor effluent up to the level required to meet the reuse standards, but not the fecal coliforms. The other major drawbacks of these processes are high dose and cost of chemicals used, and large sludge volume generation. Further, these systems have only been evaluated on lab-scale models and no scaling up has been investigated so far.

The post-treatment of anaerobic effluents can be carried out by micro-aerobic processes such as flash aeration, trickling filters and DHS, where sulfides are oxidize back to sulfate, specially at low sulfide concentrations. The partial sulfides oxidation to elemental sulfur was observed from the application of these technologies for the anaerobic effluents containing low sulfides. However, the aeration has not been optimized.

Two broad categories of biological wastewater treatment systems were categorised under a high rate aerobic systems and extensive discussed, suspended and attached growth systems. Almost all suspended growth processes were found to be very promising. The SBR was found as one of the most suitable technology for the treatment of UASB effluent due to its high effluent quality with effluent BOD and SS concentrations lower than 10 mg/L. The nutrient removal was also efficient; besides the low energy consumption for aeration and low excess sludge production are other major advantages as compared to other aerobic suspended growth system. In the activated sludge process the final effluent quality follows the discharging standards but, the system requires relatively high energy and land area and, with no nutrient removal capabilities. The continuous aeration system for the treatment of UASB effluent would be able to reduce the BOD of UASB effluent to 50%, but rarely satisfies the effluent discharge standards.

Few attached growth biological treatment processes were also summarized. Among them DHS was reported as a promising technology which reduces the BOD and coliforms well below the effluent discharging standards. However, this process requires high initial investment (sponge cost), it clogs often and no nitrogen and phosphorous removal are observed. Another important attached growth process, RBC was extensively investigated at pilot scale level. The RBC was studied under different combinations, such as one, two, three stage RBC and combination of one, two stage RBC and anoxic biofilter followed by two stage RBC. The best performance was achieved by the post treatment of UASB effluent by a combined one stage RBC, two stage RBC and anoxic biofilter followed by two stage RBC system. The RBC is not very commonly used due to its wear and tear of mechanical moving parts. Additional pre-anoxic unit is required for nitrogen removal. Similarly submerged aerated biofilter systems were evaluated for the post treatment of UASB effluent resulting in high BOD and SS removal but, with no nutrient removal capabilities. Another attached growth process, trickling filter was also evaluated for the UASB reactor effluent. This system was able to maintain the effluent BOD, COD and TSS concentration in the permissible range, however, only under low loading conditions.

The most common physical process, slow sand filtration and membrane filtration as a post treatment unit were also discussed. The systems are able to reduce the physical, chemical and microbiological pollutants not only to the desired standards but, also to satisfy wastewater reuse criteria. However, there are few drawbacks, such as frequent clogging of the filter and membranes.

The performance and effluent concentration of different parameters of various combinations are summarized in Table 2.

Among all discussed post treatment systems few of the alternatives produce final effluent with low COD, BOD and SS concentrations. Between all aerobic post treatment systems presented the SBR was found to be the most compact method and it allows for the removal of nutrient along with residual COD. Scantly information is available in literature on coupling of the SBR with UASB. The major advantage of SBR over other aerobic systems is the system flexibility for BOD and nutrient removal.

Low cost sewage treatment technologies are generally preferred for developing countries. Therefore, it is most important to evaluate the treatment sequence keeping in view of total investment including capital cost, operation and maintenance cost and land requirement. A comparison has been made among UASB reactors and its few post treatment systems with conventional ASP system based on energy requirement and generation from UASB reactor i.e. energy audit of UASB reactor per MLD:

The basis of energy audit of a MLD UASB:

- Negligible energy requirement ~6 kW-h/MLD (only for initial pumping) (Tassou, 1988).

- Energy production in the form of Biogas (60-70% methane) - 50 m^3 biogas/MLD sewage treated (Arceivala, 1998).

Integrated systems	Effluent Concentration*						
	BOD (mg/L)	COD (mg/L)	TSS (mg/L)	NH₄-N (mg/L)	TN (mg/L)	TP (mg/L)	FC (MPN/ 100mL)
CEPT+UASB+Zeolite	32 (85)	45 (91)	24 88)	0.3 (99)	0.5 (99)	0.5 (94)	1.0×10⁵ (99)
UASB+DAF	-	17 (98)	4 (98.4)	-	-	0.6 (98)	-
UASB+ Coagulation-flocculation	>20 (91)	>50 (87)	>30 (82)	-	-	-	4.3×10³ (99.9)
UASB+SSF	12 (92.6)	27 (91)	20 (91)	-	-	-	1.1×10¹ (99.99)
UASB+ Polishing Ponds	24 (92)	108 (79)	18 (96)	20 (50)	25 (55)	-	5.8×10² (99.999)
UASB+Constructed Wetlands	-	52 (82)	174 (65)	14 (70)	17.5 (70)	0.74 (89)	1.0×10³ (99.99)
UASB+ Duckweed ponds	14 (96)	49 (93)	32 (91)	0.41 (98)	4.4 (85)	1.1 (78)	4.0×10³ (99.998)
UASB+DHS	2 (99)	40 (94)	0 (100)	6 (80)	6 (89)	-	-
	9 (96)	46 (91)	17 (93)	18 (28)	28 (40)	-	3.4×10⁴ (99.95)
UASB+SBR	5.8 (97)	26 (94)	5.0 (98)	0 (100)	12.6 (77)	1.2 (65)	7.5×10²
UASB+ RBC	-	43	-	2.2 (92)	-	-	9.8×10² (99.9)
UASB+ Aerated fixed bed reactor	11 (93)	54 (83)	10 (94)	-	30 (21)	3 (40)	-
UASB+ Submerged aerated bio-filter	9.4 (96)	37.8 (92)	9.8 (94)	-	27 (36)	-	
	26 (86)	78 (84)	23 (86)	-	-	-	4.1×10⁵ (99)
UASB+ Trickling Filter	17-57 (80-94)	60-120 (74-88)	<30 (73-89)	-	-	-	
UASB+ Anaerobic Filters	<40 (85-95)	60-90 (85-95)	<25 (77-94)	-	-	-	-
UASB+ Overland Flow System	48-62 (53-83)	98-119 (77-83)	17-57	14-18	-	-	8.4×10⁴- 2.4×10⁵ (99-99.9)
UASB+ ASP	-	50 (85-93)	13-18 (82)	-	-	-	-
UASB+ Flash Aeration System	22 (89)	57 (86)	47 (83)	-	-	-	5.0×10³ (99)

*% removal efficiency in parentheses.

Table 2. Treatment Performance of various Integrated UASB Post treatment systems Treating Sewage *(adopted from Khan et al. 2011a)*

- The electricity produced from 1.0 m³ of methane gas generated by UASB is 36,846 kJ at standard condition and approx.7.0 kW-h under field conditions, since 3600kJ is approximately 1 kW-h (Arceivala, 1998; Metcalf and Eddy, 2003).

- Energy saving through reduced diesel consumption by more than 70% by feeding methane gas into the Dual-Fuel Mode Diesel Engine (Arceivala, 1998).

The basis of energy audit of a MLD aerobic post treatment system:

- Energy requirement of Aerobic Process as the sole wastewater treatment process, including initial pumping is approximately 195 kW-h/MLD (Tassou, 1988).

- *Salient features of comparative energy consumption:*

- Energy requirement of post treatment aerobic system treating only 35% BOD (as 65% BOD removal takes place in anaerobic system) is 195 kW-h/MLD x 0.35 = 68.25 kW-h/MLD

- Hence Total Energy Consumption of integrated UASB-Aerobic Process is (6 + 68.25) kW-h/MLD = 74.25 kW-h/MLD compared to 195 kW-h/MLD for the aerobic process only.

Based on existing waste and wastewater treatment technologies Lettinga (2008) suggested (i) a Natural Biological Mineralization Route followed by physico-chemical methods for achieving the quality of treated wastewater for reuse/ or intended purpose such as for irrigation, industrial reuse etc. and, (ii) decentralization of the sanitation and resource recovery and reuse, that is, a concept which incorporates environmental protection where the waste and wastewater transportation is kept at minimum level and where pollutants are brought to an acceptable value at the location.

4.1. Solutions for sustainability treatment options

Sustainable technologies must be needed in order to make sustainable lifestyle of the society and to protect environment. It is difficult to understand and to implement it due to lack of proper parameters which leads to ambiguously the targets or proposed actions taken by politicians and/ or policy makers. Moreover, the quantification of sustainability is vague. For instance, if government implementing extremely stringent standards for protecting the aquatic environment from pollution many question arises, like why a single country or region pursuing a paradisiacal natural environment while at the same time little if any money or technology is made available to contribute to the highly needed environmental improvement in less prosperous countries. These potential combinations can be considered as sustainable solutions if adopted based on NBMR (Khan et al., 2011a).

4.2. Sustainable technology concept

The superiority of sequential anaerobic – aerobic treatment systems over conventional aerobic is more profound with increase in sewage concentration. In countries of limited per capita share of water, like in Africa, Middle East and India the treatment of concentrated sewage via conventional aerobic system is highly expensive, especially with respect to operational costs (Khan et al., 2011a).

The advantages of introducing UASB reactor ahead of aerobic system is obvious, mainly in terms of sludge production and energy consumption. In view of the fact that aeration costs increase linearly with increasing organic loads, adopting the activated sludge system for polishing of anaerobic effluents may not be the most sustainable option for concentrated sewage. Other aerobic systems, such as DHS, SBR and CFID type SBR for UASB effluents post treatment reviewed in this paper are promising options for sewage management at low cost, low land requirement and low sludge production. Moreover, the potential of nutrients recovery and pathogens removal in an aerobic post-treatment for UASB effluents is considerable and the effluent discharge standards established by various national and international environmental agencies can be achieved.

5. Conclusions

Numerous anaerobic/ aerobic treatment concepts were evaluated in this chapter. The best option observed for the sewage treatment was integrated UASB-SBR system. The organics, nutrients and pathogenic pollutant removal efficiency of the integrated treatment approach was capable to achieve the effluent with low BOD ($\approx$5mg/L; 98 % removal), COD (<25 mg/L; up to 95% removal) and TSS (<10 mg/L; up to 98% removal) and nutrients (TN=4 mg/L; NH_4-N=Nil; P=1 mg/L). Ammonium nitrogen and phosphorus levels were decreased up to 98% and 90%, respectively. Fecal coliforms levels fell to <1000 MPN/100 mL, indicating a significant removal of pathogenic indicators. Thus the final effluent from the integrated UASB-SBR system can be reused for unrestricted irrigation or be discharged safely into the surface waters. However, no information is available regarding the efficacy of integrated UASB-SBR system at full scale level for sewage treatment. The performance of existing UASB based STPs can be improved by installing any of the post treatment system demonstrated in this chapter. The energy conservation, resources recovery and carbon credit were the gaps that still need to be explored for the above suggested post treatment options so that a natural biological mineralization route or sequence can be utilized to make the integrated system a viable sustainable option for treatment of sewage and anaerobically treated effluents.

Author details

Abid Ali Khan[1,2], Rubia Zahid Gaur[3], Absar Ahmad Kazmi[1] and Beni Lew[4,5]

1 Department of Civil Engineering, IIT Roorkee, India

2 Royal HaskoningDHV, India

3 Water & Sanitation Specialist, Plan Environ, H-273, GK1, New Delhi, India

4 The Volcani Center, Institute of Agriculture Engineering, Bet Dagan, Israel

5 Department of Civil Engineering, Ariel University Center of Judea and Samaria, Ariel, Israel

References

[1] Agrawal, L.K., Okui, H., Ueki, Y., Harada, H., Ohashi, A. (1997) Treatment of Raw Sewage in a Temperate Climate using a UASB Reactor and the Hanging Sponge Cubes Process. Wat. Sci. Technol., 36 (6-7), 433-440.

[2] Annachhatre AP, Suktrakoolvait S. (2001). Biological sulfide oxidation in a fluidized bed reactor. Environ Technol; 22:661–72.

[3] Aiyuk, S., Amoako, J., Raskin, L., van Haandel, A., Verstraete, W. (2004) Removal of Carbon and Nutrients from Domestic Wastewater using a Low Investment, Integrated Treatment Concept. Wat. Res., 38, 3031–3042.

[4] Arceivala SJ, Asolkar SR, (2007). Wastewater treatment for pollution control and reuse. 3rd ed. Tata McGraw-Hill Publishing Co. Ltd. New Delhi, India.

[5] Arceivala SJ. (1998). Wastewater treatment for pollution control. 2nd ed. Tata McGraw Hill, New Delhi, India.

[6] A1-Adham, S. S. (1989). Tertiary Treatment of Municipal Sewage via Slow Sand Filtration. MS Thesis, King Fahd University of Petroleum & Minerals, Dhahran, Saudi Arabia.

[7] A1-Sawaf, M. F. (1986). Tertiary Wastewater Treatment by Direct Filtration. MS Thesis, King Fahd University of Petroleum & Minerals, Dhahran, Saudi Arabia.

[8] Arceivala, S.J. (2001). Wastewater Treatment for Pollution Control, New Delhi, Tata McGraw Hill.

[9] Buisman CJN, Geraats BG, Ijspeert P, Lettinga. (1990). Optimization of sulfur production in a biotechnological sulfide removing reactor. Biotech Bioeng; 35: 50–6.

[10] Chuang SH, Pai TY, Horng RY. (2005). Biotreatment of sulfate-rich wastewater in an anaerobic/micro-aerobic bioreactor system. Environ Technol. 26(9):993–1001.

[11] Chen C, Ren N, Wang A, Liu L, Lee DJ. (2010). Enhanced performance of denitrifying sulfide removal process under micro-aerobic condition. J Hazard Mater; 179:1147–51.

[12] Cavalcanti, P.F.F., van Haandel, A., Lettinga, G. (2001). Polishing Ponds for Post-treatment of Digested Sewage Part 1: Flow-through Ponds. Wat. Sci. Technol., 44 (4), 237–245.

[13] Chernicharo, C.A.L., Nascimento, M.C.P. (2001). Feasibility of a Pilot- Scale UASB/ Trickling Filter System for Domestic Sewage Treatment, Wat.Sci.Technol., 44 (4), 221-228.

[14] Chernicharo, C.A.L., Cota, R.S., Zerbini, A.M., von Sperling, M., Brito, L.H.N.C. (2001). Post-treatment of Anaerobic Effluents in an Overland Flow System. Wat.Sci.Technol. 44 (4), 229–236.

[15] Chernicharo, C.A.L. (2006). Post Treatment Options for the Anaerobic Treatment of Domestic Wastewater. Reviews in Environmental Sciences and Bio/Technology, 5, 73-92.

[16] Droste, R.L., Masse, D.I. (1995). Anaerobic Treatment in Sequencing Batch Reactors. International Symposium on Technology Transfer. Pre-prints. Salvador, Bahia, Brazil, pp. 353–363.

[17] Ellis, K.V. (1987). Slow Sand Filtration as Technique for the Tertiary Treatment of Municipal Sewages. Wat. Res., 21 (4), 403- 410.

[18] El-Shafai, S.A., El-Gohary, F.A., Nasr, F.A., van der Steen, P. Gijzen, H.J. (2007). Nutrient Recovery from Domestic Wastewater using a UASB-Duckweed Ponds System. Biores. Technol., 98, 798–807.

[19] Elmitwalli, T., Zeeman, G., Lettinga, G. (2001). Anaerobic Treatment of Domestic Sewage at Low Temperature. Wat.Sci.Technol. 44 (4), 33–40.

[20] Foresti, E., Zaiat, M., Vallero, M. (2006). Anaerobic processes as the core technology for sustainable domestic wastewater treatment: Consolidated applications, new trends, perspectives, and challenges. Reviews in Environmental Science and Bio/Technology, 5, 3–19.

[21] Gomec, C.Y. (2010). High-rate anaerobic treatment of domestic wastewater at ambient operating temperatures: A review on benefits and drawbacks. J. Environmental Sci Health - A Tox Hazard Subst Environ Eng. 45 (10): 1169 – 84. DOI: 10.1080/10934529.2010.493774.

[22] Gersberg, R.M., Gearheart, R.A., Ives, M. (1989). Pathogen Removal in Constructed Wetlands. In: DA. Hammer, Editor, Constructed Wetlands for Wastewater Treatment: Municipal, Industrial and Agricultural, Lewis Publishers, Inc., Chelsea, MI (1989), Pp. 431–445.

[23] Gnanadipathy, A., Polprasert, C. (1993). Treatment of Domestic Wastewater with UASB Reactor. Wat. Sci. Technol., 27, 195–203.

[24] Goncalves, R.F., Araujo, V.L., Chernicharo, C.A.L. (1998). Association of a UASB Reactor and a Submerged Aerated Biofilter for Domestic Sewage Treatment, Wat. Sci. Technol., 38 (8-9), 189-195.

[25] Goncalves, R.F., de Araujo, V.L., Bof, V.S. (1999). Combining Upflow Anaerobic Sludge Blanket (UASB) Reactors and Submerged Aerated Biofilters for Secondary Domestic Wastewater Treatment. Wat.Sci.Technol. 40 (8), 71-79.

[26] Jaya Prakash K., V.K.Tyagi, A.A.Kazmi & Arwind Kumar, 2007, Post- Treatment of UASB Reactor Effluent by Coagulation and Flocculation Process, AIChE, Environmental Progress, Vol. 26, No.2 pp 164-168.

[27] Janssen AJH, Lettinga G, de Keizer A. (1999). Removal of hydrogen sulphide from wastewater and waste gases by biological conversion to elemental sulphur: colloidal

and interfacial aspects of biologically produced sulphur particles. Colloids Surf A: Physicochem Eng Aspects; 151:389–97.

[28] Khan, AA, Gaur, RZ, Tyagi, VK, Khursheed, A, Lew, B, Kazmi, AA, Mehrotra I (2011a). Sustainable Options of Post Treatment of UASB Effluent Treating Sewage: A Review. Resource, Conservation and Recycling; Vol. 55 (12); 1232-1251.

[29] Khan, AA, Gaur, RZ, Lew, B, Diamantis, V, Mehrotra, I, Kazmi, AA (2011b). UASB/ Flash aeration enable complete treatment of municipal wastewater for reuse. Bioprocess and Biosystem Engineering. Vol. 35(6):907-13.

[30] Khan, AA, Gaur, RZ, Lew, B, Mehrotra, I, Kazmi, AA (2011c). Effect of Aeration on The Quality of Effluent of UASB Reactor Treating Sewage. Journal of Environmental Engineering- ASCE, Vol. 137 (6); 464-472.

[31] Khan, AA (2012). Post treatment of UASB effluent: Aeration and Variant of ASP. PhD Thesis. IIT Roorkee India.

[32] Keller, R., Passamani- Franca, R.F., Passamani, F., Vaz, L., Cassini, S.T., Sherrer, N., Rubim, K., Santa' Ana, T.D. & Goncalves, R.F. (2004). Pathogen Removal Efficiency from UASB+BF Effluent using Conventional and UV Post- Treatment Systems, Wat. Sci. Technol., 50 (1), 1-6.

[33] Khan, Abid A., (2012). Post Treatment of UASB Effluent: Aeration and Variant of ASP. PhD Thesis. IIT Roorkee, India.

[34] Lettinga, G., van Velsen, A. F. M., Hobma S. W., De Zecuw, W., Klapwijk, A. (1980). Use of the Upflow Sludge Blanket (USB) Reactor Concept for Biological Wastewater Treatment, Especially for Anaerobic Treatment. Biotechnol. Bioengg., 22, 699-734.

[35] Lettinga G., deMan A., van der Last, A. R. M., Wiegant, W., van Knippenberg, K., Frijns, J., van Buuren, J. C. L. (1993). Anaerobic Treatment of Domestic Sewage and Wastewater. Wat. Sci. Technol., 27(9), 67-73.

[36] Lettinga G. (2008). Towards feasible and sustainable environmental protection for all. Aquat Ecosyst Health Manage; 11(1):116–24.

[37] Lew B, Belavski M, Admon S, Tarre S, Green M. (2003). Temperature effect on UASB reactor operation for domestic wastewater treatment in temperate climate regions. Water Sci Technol.48 (3):25–30.

[38] Lew B, Tarre S, Belavski M, Green M. (2004). UASB reactor for domestic wastewater treatment at low temperatures: a comparison between a classical UASB and hybrid UASB-filter reactor. Water Sci Technol.49 (11–12):295–301.

[39] Mendoza L, Carballa M, Sitorus B, Pieters J, Verstraete W. (2009). Technical and economical feasibility of gradual concentric chambers reactor for sewage treatment in developing countries. Electron J Biotechnol 2009; 12(2):1–13.

[40] MoEF (2005 and 2006). Management Information System, Technical Report, National River Conservation Directorate, Ministry of Environment and Forests, New Delhi, India.

[41] Metcalf and Eddy. (2003). Wastewater engineering treatment and reuse. 3rd ed. Tata McGraw Hill Co. New Delhi, India.

[42] Machdar, I., Sekiguchi, Y., Sumino, H., Ohashi, A., Harada, H. (2000b). Combination of a UASB Reactor and a Curtain type DHS (Downflow Hanging Sponge) Reactor as a Cost-Effective Sewage Treatment System for Developing Countries. Wat. Sci. Technol., 42 (3–4), 83–88.

[43] Moawad,A., Mahmoud,U.F., El-Khateeb, M.A., El-Molla, E. (2009). Coupling of Sequencing Batch Reactor and UASB Reactor for Domestic Wastewater Treatment. Desalination, 242, 325–335.

[44] Mbuligwe, S.E. (2004). Comparative Effectiveness of Engineered Wetland Systems in the Treatment of Anaerobically Pre-treated Domestic Wastewater, Ecol. Eng., Vol. 23, 269–284.

[45] Okurut, T. O., Rijs, G. B. J., van Bruggen, J. J. A. (1999). Design and Performance of Experimental Constructed Wetlands in Uganda, Planted with Cyperus Papyrus and Phragmites Mauritianus. Wat.Sci.Technol. 40 (3), 265–263.

[46] Penetra, R.G., Reali, M.A.P., Foresti, E., Campos, J.R. (1999). Post-treatment of Effluents from Anaerobic Reactor Treating Domestic Sewage by Dissolved-Air Flotation. Wat.Sci.Technol. 40 (8), 137-143.

[47] Schellinkhout, A., Collazos, C.J. (1992). Full-scale Application of the UASB Technology for Sewage Treatment. Wat.Sci.Technol. 25 (7), 159-166.

[48] Seghezzo, L., Guerra, R.G., Gonza´lez, S.M., Trupiano, A.P., Figueroa, M.E., Cuevas, C.M., Zeeman, G., Lettinga, G. (2002). Removal Efficiency and Methanogenic Activity Profiles in a Pilot-scale UASB Reactor Treating Settled Sewage at Moderate Temperatures. Wat.Sci.Technol. 45 (10), 243–248.

[49] Sousa, J.T., Foresti, E. (1996). Domestic Sewage Treatment in an Up-flow Anaerobic Sludge Blanket – Sequential Batch System. Wat. Sci.Technol. 33 (3), 73-84.

[50] Sousa, J.T., van Haandel, A.C., Guimarães, A.A.V. (2001). Post-treatment of Anaerobic Effluents in Constructed Wetland Systems. Wat.Sci.Technol. 44 (4), 213–219.

[51] Suhail, A. (1987). Tertiary Wastewater Treatment by Sedimentation and Sand Filtration, MS Thesis, King Fahd University of Petroleum & Minerals, Dhahran, Saudi Arabia.

[52] Sumino, H., Takahashi, M., Yamaguchi, T., Abe, K., Araki, N., Yamazaki,S., Shimozaki, S., Nagano, A., Nishio, N. (2007). Feasibility Study of a Pilot-scale Sewage Treatment System Combining an Up-flow Anaerobic Sludge Blanket (UASB) and an

Aerated Fixed Bed (AFB) Reactor at Ambient Temperature. Biores. Technol., 98, 177–182.

[53] Surampalli, R.Y., Tyagi, R.D., Scheible, O.K., Heidman, J.A. (1997). Nitrification, Denitrification and Phosphorus Removal in Sequential Batch Reactors. Biores. Technol., 61, (151–157).

[54] Stefess GC, Torremans RAM, de Schrijver R, Robertson LA, Kuenen JG. (1996). Quantitative measurement of sulfur formation by steady state and transient-state continuous cultures of autotrophic thiobacillus species. Appl Microbiol Biotechnol; 45:169–75.

[55] Tawfik, A. Zeeman, G., Klapwijk, A., Sanders, W., El-Gohary, F., Lettinga, G. (2003). Treatment of Domestic Sewage in a Combined UASB/RBC system. Process optimization for Irrigation purposes. Wat. Sci.Technol. 48 (1), 131-138.

[56] Tawfik, A., Klapwijk, B., El-Gohary, F., Lettinga,G. (2005). Potentials of Using a Rotating Biological Contactors for Post Treatment of Anaerobically Pre- Treated Domestic Wastewater, Biochem. Engg. J., Vol. 25, 89-98.

[57] Tawfik, A., El-Gohary, F., Ohashi, A., Harada, H. (2010). Optimization of the Performance of an Integrated Anaerobic–Aerobic System for Domestic Wastewater Treatment. Wat.Sci.Technol. 58 (1), 185-194.

[58] Torres P., Foresti, E. (2001). Domestic Sewage Treatment in a Pilot System Composed of UASB and SBR Reactors, Wat.Sci.Technol. 44(4), 247-53.

[59] Tyagi VK, Khan AA, Kazmi AA, Mehrotra I, Chopra AK. (2009). Slow sand filtration of UASB reactor effluent: a promising post treatment technique. Desalination; 249:571-6.

[60] Tessele F, Monteggia LO, Rubio J. (2005). Treatment of municipal wastewater UASB reactor effluent by unconventional flotation and UV disinfection. Water Sci Technol. 52(1–2):315–22.

[61] Tassou SA. (1988). Energy conservation and resource utilization in wastewater treatment plants. Appl Energy. 30:2–8.

[62] van Haandel, A.C., Lettinga, G. (1994). Anaerobic Sewage Treatment: a Practical Guide for Regions with a Hot Climate. John Wiley & Sons, Chichester, UK, 226.

[63] von Sperling, M., Freire, V.H., Chernicharo, C.A.L. (2001). Performance Evaluation of a UASB–Activated Sludge System Treating Municipal Wastewater. Wat.Sci.Technol. 43 (11), 323–328.

[64] von Sperling, M., Chernicharo, C.A.L., Soares, A.M.E. and Zerbini, A.M. (2002). Coliform and Helminth Eggs Removal in a Combined UASB Reactor–Baffled Pond System in Brazil: Performance Evaluation and Mathematical Modeling. Wat.Sci. Technol, 45 (10), 237–242

[65] von Sperling, M., Chernicharo, C.A.L. (2005). Biological Wastewater Treatment in Warm Climate Regions. IWA Publishing, London, 1452.

[66] von Sperling,M., Mascarenhas, L.C.A.M. (2005). Performance of Very Shallow Ponds Treating Effluents from UASB Reactors. Wat.Sci. Technol., 51 (12), 83-90.

[67] von Sperling, M., Bastos, R.K.X., Kato, M.T. (2005). Removal of E. coli and Helminthes Eggs in UASB: Polishing Pond Systems in Brazil. Wat.Sci. Technol., 51 (12), 91-97.

[68] Vlyssides A, Barampouti EM, Mai S. (2007). Effect of ferrous ion on the biological activity in a UASB reactor: Mathomatical modeling and verification. Biotechnol Bioeng. 96(5):853-61.

[69] Vallero MVG, Sipma J, Annachhatre A, Lens PNL, Hulshoff, Pol LW. (2003). Biotechnological treatment of sulfur-containing wastewaters. In: Fingerman M, Nagabhushanam R, editors. Recent advances in marine biotechnology: bioremediation, vol. 8. Enfield, NH, USA: Science Publishers. p. 233-68.

[70] van der Zee FP, Villaverde S, Garcia PA, Polanco FFdz. (2007). Sulfide removal by moderate oxygenation of anaerobic sludge environments. Bioresour Technol; 98:518-24.

[71] World Health Organization (1989). Health Guidelines for the Use of Waste water in Agriculture and Aquaculture. Technical Report Series- 778, Geneva: WHO.

[72] Walia R. (2007). Polishing of effluent from UASB reactor: ORP as a monitoring parameter, PhD thesis, Indian Institute of Technology, Roorkee, India.

[73] YingYu A, Yang FL, Bucciali B, Wong FS. (2009). Municipal wastewater treatment using a UASB coupled with cross-flow membrane filtration. J Environ Eng; 135(2):86-91.

[74] YingYu A, Bing W, Wong FS, Yang F. (2010). Post-treatment of up-flow anaerobic sludge blanket effluent by combining the membrane filtration process: fouling control by intermittent permeation and air sparging. Water Environ J; 24: 32-8.

[75] N. Matsuura, M. Hatamoto, H. Sumino, K. Syutsubo, T. Yamaguchi and A. Ohashi. (2010). Closed DHS system to prevent dissolved methane emissions as greenhouse gas in anaerobic wastewater treatment by its recovery and biological oxidation. Water Science & Technology—Wat Sci Technol , Vol 61, No 9, pp 2407-2415 .

Chapter 10

Biodegradation in Animal Manure Management

Matthieu Girard, Joahnn H. Palacios, Martin Belzile,
Stéphane Godbout and Frédéric Pelletier

Additional information is available at the end of the chapter

1. Introduction

Typical manure management strategies from intensive livestock feeding operations in Canada include the pre-storage of manure inside the animal buildings, long-term storage at the farm and finally field application of manure as fertilizer. Different biodegradation phenomena can occur in each of these steps, but naturally occurring biodegradation can cause harmful emissions. However, when used properly, biodegradation can also be beneficial and reduce pollution from animal wastes. This chapter will describe in detail the different processes involved in the biodegradation of manure, the emissions that are produced as well as how biodegradation can be used to treat both the manure and the emissions from manure management. The phenomena and systems described here can be applied to most livestock feeding operations (dairy and beef cattle, poultry, egg production, hog, etc.), but the specific examples and results will be provided for the swine industry.

2. Phenomena

2.1. Manure composition

Manure from animal husbandry contains a wide range of compounds that can be used by microorganisms for energy requirements or anabolic processes. In general, manure contains organic matter, nitrogen, phosphorous and potassium as well as numerous micronutrients (sulphur, copper and zinc for example). The specific concentrations of these components may vary according to several factors: building and storage management as well as the genetics of the animals, their growth stage and their diet. For example, experience has shown that hog diets supplemented with phytase had the effect of, among others, reducing the release of phosphorus in manure. Manure composition may also vary with water dilution when using water-saving drinkers in the building or a roof structure to cover the manure storage pit for example.

According to results from the analysis of various types of manure carried out by the Research and Development Institute for the Agri-Environment (IRDA), the typical swine manure composition for maternity, nursery and growing-finishing stages can be represented by the average values found in Table 1 [1]. The main difference between growth stages is related to the dry matter content. Indeed, manure produced by grower-finisher pigs is generally dryer and therefore more concentrated in nutrients.

Parameter	Unit	Growth stage		
		Maternity	Nursery	Growing-finishing
Dry matter	%	1.8	2.7	4.7
Total nitrogen	%	0.2	0.3	0.6
Ammonium nitrogen	mg/kg	1488	1545	2846
Phosphorus	mg/kg	593	762	1690
Potassium	mg/kg	1049	1964	3405
Copper	mg/kg	8.9	29.6	49.9
Zinc	mg/kg	41.0	208.7	82.9
Manganese	mg/kg	10.3	14.9	29.85
Calcium	mg/kg	697	701	1700
Magnesium	mg/kg	213	311	674

Table 1. Typical composition of swine manure for each growth stage (adapted from [1])

To provide a full description of manure composition, values for organic matter must also be considered. For swine manure, values of 19 to 51 gO_2/L as COD (chemical oxygen demand) are normally encountered [2,3]. The organic matter content depends essentially on the type of feed, manure management and manure age. After excretion, manure decomposes naturally; suspended solids contained in the manure are hydrolyzed into dissolved elements and biodegradation by microorganisms occurs. This decomposition of manure may be favored by appropriate conditions which depend on the proportion of elements contained in the slurry, the amount of oxygen, the pH and the temperature.

2.2. Aerobic biodegradation

Different microorganisms can grow by using various compounds found in manure both in the presence and in the absence of oxygen. The microorganisms can also be classified according to the compounds consumed.

Organic matter represents an important fraction of swine manure and includes many compounds that can be separated in four fractions: readily biodegradable (S_S), slowly biodegradable (X_S), inert soluble (S_I) and inert particulate (X_I). When considering biodegradation, only the biodegradable fractions (S_S and X_S) are taken into account. The S_S fraction is usually in soluble form and is composed of relatively small molecules such as volatile fatty acids (acetic, butyric and valeric acids), monosaccharides (sugar) and alcohols [3]. On the other hand, the X_S fraction is usually found as particles and is composed of high molecular weight organic polymers or dead biomass. This fraction of the organic matter cannot be directly assimilated by microorganisms and must first be hydrolyzed to S_S. The distribution of the organic matter can be quite variable among the different fractions and depends on many factors such as the type of feed and the manure storage time. For the S_S fraction, values from the literature vary from 8 to 30% of the total COD, from 30 to 60% for the X_S and from 10 to 60% for the inert fractions (S_I and X_I) [2-4]. Various types of microorganisms can degrade organic matter: bacteria, protozoa and fungi. As shown in equation 1, these microorganisms degrade the organic matter and release carbon dioxide (CO_2), water and biomass:

$$\text{Organic matter} + O_2 + \text{Nutrients} \rightarrow \text{Biomass} + CO_2 + H_2O \tag{1}$$

Nitrogen in manure can be found as ammonium (NH_4^+), trapped in organic molecules or as urea. Both the organic nitrogen and the urea must be broken down into NH_4^+ by microorganisms before they can be accessible. The aerobic oxidation of NH_4^+, nitrification, follows two distinct steps: the transformation of NH_4^+ into nitrite (NO_2^-) by *Nitroso* bacteria (*Nitrosomonas* for example) followed by the oxidation to nitrate (NO_3^-) performed by *Nitro* bacteria (*Nitrobacter* for example) [5]. The relative kinetics between the two steps are controlled by the temperature and will determine which compound is accumulated (NO_2^- or NO_3^-). The two separate steps of the nitrification reaction are presented in equations 2 and 3 while the combined reaction is given in equation 4 [6]:

$$NH_4^+ + 3/2\ O_2 \rightarrow NO_2^- + 2H_2 + H_2O \tag{2}$$

$$NO_2^- + 1/2\ O_2 \rightarrow NO_3^- \tag{3}$$

$$NH_4^+ + 2\ O_2 \rightarrow NO_3^- + 2H_2 + H_2O \tag{4}$$

Regarding the other compounds in manure (phosphorous, potassium and heavy metals), they can used by microorganisms for different microbiological processes or to synthesise certain compounds such as DNA and amino acids.

2.3. Anaerobic biodegradation

Without oxygen present, (e.g. when all the dissolved oxygen has been exhausted by aerobic respiration) several compounds are released by the anaerobic metabolism of microorganisms still utilizing nutrients in manure. By a complex series of reactions, the anaerobic biodegradation of manure produces different gases, mainly methane (CH_4), hydrogen sulfide (H_2S), ammonia (NH_3) and CO_2, as well as many intermediate compounds; the most noteworthy are volatile fatty acids and other odorous molecules. A study from the North Carolina State University identified a total of 331 compounds that cause odours from manure [7].

Biological decomposition during storage or during anaerobic digestion contributes to the transfer of nutrients, especially nitrogen and phosphorus, between different fractions and chemical forms in manure [8]. For nitrogen, anaerobic digestion can break down organic nitrogen and produce NH_4^+ and NH_3. If oxidised nitrogen compounds (NO_2^- or NO_3^-) are present, heterotrophic microorganisms can use these compounds as an electron acceptor and produce nitrogen gas (N_2). This process is called denitrification and requires a source of easily biodegradable organic carbon. It can also produce nitrous oxide (N_2O), a potent greenhouse gas, as a by-product. For phosphorus, anaerobic digestion contributes to moving some of the dissolved portion into the bodies of bacteria that carry out the anaerobic digestion process. All of the phosphorous present in the manure will still be present in the digester sludge [9]. Anaerobic digestion may also change the pH and the chemical form of salts and metals, such as iron, calcium and magnesium, which may affect the amount of suspended phosphates as a result of precipitation processes [8].

There is a huge interest in controlling anaerobic biodegradation for bioenergy production purposes. In fact, the anaerobic digestion of manure in an airtight container, under certain conditions, will form biogas, an energy source composed of a mixture of CH_4, CO_2 and trace amounts of other gases. Anaerobic digestion is a multi-stage process (Figure 1). Communities of hydrolytic bacteria break down complex organic matter from manure to simpler compounds (sugars, amino acids and fatty acids). Then, acid forming bacteria convert the simple compounds to alcohols and carbon acids (volatile fatty acids), as well as hydrogen, CO_2, NH_3, NH_4^+ and H_2S [10]. An amount of acetic acid is also produced at this stage, which along with hydrogen, can be used directly by methanogens. Other molecules, such as volatile fatty acids must first be catabolised to produce acetic acid, as well as CO_2 and H_2 that can be directly used by methanogens.

3. Biodegradation as a source of emissions

3.1. Buildings

Biodegradation processes begin before manure is even expelled from the animals. Intestinal microorganisms anaerobically degrade nutrients as they pass through an animal's digestive system. Once expelled, manure comes in contact with oxygen and aerobic microorganisms can become dominant after several hours. However, most animal housing systems use some sort of pre-storage for the manure inside the barn where anaerobic conditions can once again take over. Emissions from biodegradation of manure inside animal buildings are therefore a mix of anaerobic and aerobic products and are removed by the ventilation air. The emission rates

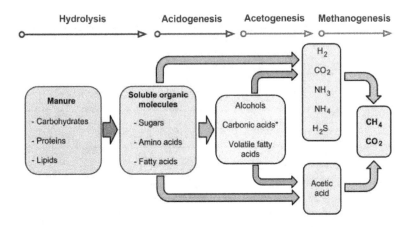

Figure 1. Process stages of anaerobic digestion (Modified from [10])

are affected by many factors such as ventilation flow rate, temperature, manure separation systems and animal activity among others.

Emissions from pig barns include a number of gases (CO_2, CH_4, and N_2O), dust particles (inhalable and breathable), bioaerosols (bacteria, viruses, endotoxins and fungi) and several other volatile compounds such as NH_3 and H_2S. In addition, an increasing importance is given to the odour nuisance associated with swine production. Thus, research in this area has become more important in recent years.

A baseline emission scenario of swine buildings was defined by [11] based on an inventory of gas, odour and dust emission data (Table 2). The resulting scenario provides a good overview of the magnitude of the emissions that are produced in swine production systems for the different growth stages (maternity, nursery and growing-finishing).

Odours, consisting of a complex mixture of several chemical compounds, are one of the major concerns in the emissions from the swine sector. Odours are expelled from barns by the ventilation system at 2.5 to 51.6 EOU/s/pig (EOU: European Odour Unit), depending on the growth stage. According to data, the nursery stage tends to emit fewer odours than the other stages. The use of odour reduction technologies in animal buildings, such as air cleaning technologies, could reduce the level of nuisance. In fact, downwind odours from confined feeding operations are considered to be a nuisance that may lead to a reduced quality of life by nearby residents.

NH_3 is produced by the degradation of urea in the urine on floors or still stored in the building. In a swine barn, average emissions range from 0.33 to 14 gNH_3/d/pig depending on the growth stage (Table 2). The rate of NH_3 emissions from buildings, storage structures and land spreading is favored when the liquid and solid fractions of the manure are not separated and

Parameter	Unit	Growth stage		
		Maternity	Nursery	Growing-finishing
		Europe		
Odours	EOU/s/pig	21.2 (16.3)	10 .69 (8.05)	13.75 (8.23)
NH_3	g/d/pig	14.2 (2.9)	0.94 (0.85)	7.75 (4.95)
CH_4	g/d/pig	57.7	10.7	12.42 (10.41)
CO_2	kg/d/pig	-	-	2.25 (087)
N_2O	g/d/pig	-	-	2.72 (3.26)
$PM_{4,d}$	mg/h/pig		0.4	6.9
PM_{10}	mg/h/pig	8.2 (2.55)	-	4.71 (2.50)
PM_{total}	mg/h/pig	50	-	-
		Canada		
Odours	EOU/s/pig	51.56 (53.45)	2.5 (0.69)	7.82 (8.19)
NH_3	g/d/pig	-	0.33 (0.14)	-
CH_4	g/d/pig	59.9 (45.13)	1.45 (2.37)	3.78 (3.76)
CO_2	kg/d/pig	5,29 (2,26)	0,55 (0,003)	1,71 (1,21)
N_2O	g/d/pig	0,0	0,007 (0,005)	0,04 (0,04)
$PM_{2.5}$	mg/h/pig	-	-	-
PM_{10}	mg/h/pig	-	-	-
PM_{total}	mg/h/pig	-	-	63 (4,12)

Confidence intervals in parentheses.

EOU : European odour unit

$PM_{2.5}$, PM_{10} and PM_{total}: particle matter smaller than 2.5, 10 and 100 micrometers respectively.

Table 2. Emissions scenario from swine buildings for different growth stages (adapted from [11])

when the manure: contains nitrogen from undigested food, has a high pH and has a high temperature. Moreover, a high contact area between the air and the manure as well as a high air movement at the surface increase NH_3 emissions.

In animal production, CH_4 comes from two sources, enteric fermentation in ruminants (cellulose digestion) and the decomposition of manure under anaerobic conditions. In the case of pig production, only the second source applies. At the building, where waste is handled in solid form in aerobic conditions, the production of CH_4 is minimal. Under anaerobic conditions, the production of CH_4 varies with the temperature and the composition of the manure. The emissions inventoried in Table 2 suggest that CH_4 is emitted from swine barns at 1.45 to 57.7 g/d/pig, with higher values at the maternity stage. The concern of CH_4 emissions from animal production systems is due to its high potential as a greenhouse gas and by the large quantities produced.

CO_2, produced by the metabolism of animals, is the most prominent gas in animal housing. Almost all CO_2 (96%) is produced by the respiration of animals; the rest comes from the decomposition of manure [12] and the combustion gases from heating systems. [13] showed that CO_2 emissions from pig manure can be estimated by multiplying the CH_4 emissions by a factor of 1.42 kg CO_2 per kilogram of CH_4.

N_2O emissions are not as high as the other gases expelled from animal buildings. For instance, according to Table 2, the maximal average N_2O emission was 2.7 g/d/pig. However, N_2O is also a major greenhouse gas and air pollutant. Considered over a 100-year period, it has 298 times more impact on climate change than CO_2 [14]. The formation of N_2O occurs during the processes of nitrification and denitrification over the course of manure management. In fact, it is during denitrification that N_2O is emitted, but to do so, nitrification must first take place. Under anoxic conditions, there is not enough oxygen for microorganisms who will take the oxygen they need from NO_3^-. Thus, the NO_3^- is then reduced to N_2. However, when the reaction is not complete (e.g. due to process kinetics or the sudden presence of dissolved oxygen), N_2O is emitted. It should be noted that this cannot occur under complete anaerobic conditions, since these microorganisms cannot survive.

The Environmental Protection Agency of the United States [15] defines particulate matter as a complex mixture of extremely small particles and liquid droplets. Moreover, according to [16], suspended particles in livestock buildings differ from other types of particles for three reasons: their concentration is usually 10 to 100 times greater than other indoor environments, they are vectors of odours and gases and they are biologically active, usually containing a wide variety of bacteria and microorganisms. The dust concentration in the air of buildings depends on several factors, such as relative humidity, temperature, level of animal activity, type and mode of feeding and presence of litter. Such particles have a significant impact on the health and well-being of both workers and animals. The consequences are mainly related to respiratory problems. However, in the inventory performed by [11] (Table 2), particular matter (PM) is the least documented parameter.

3.2. Storage

During manure storage, aerobic conditions are present at the surface of the manure, but after a shallow depth, anaerobic conditions prevail. Emissions from manure storage generally represent the intermediate and end products of anaerobic digestion: NH_3, CH_4, CO_2, H_2S and odours. The composition of the manure as well as the storage and weather conditions (temperature, pH, precipitation and wind) can affect the biodegradation of manure and will dictate the emission rates for each compound.

Typical gas emissions at the surface of manure tanks have been measured in the past using a special device. A sampling chamber, developed at IRDA, floats on the storage tank and takes gas samples in a confined space, swept by an airflow of 100 L/min. The gas concentrations are measured at the outlet of the chamber and the increase of the concentration compared to the ambient air is attributed to the emitting surface. A photo of the sample chamber floating on a manure storage tank is shown in Figure 2.

Figure 2. IRDA sampling device on a manure storage tank

Various research projects have been carried out using this instrument. Typical values for CH_4, CO_2 and N_2O are found in Table 3. Annual emissions represent the summation of the emissions over one year while daily averaged values represent this value distributed on a daily basis and the maximal daily values is the highest emission over one day during the year.

Parameter	Units	Gas		
		CH_4	CO_2	NH_3
Annual emissions	g/year/m²	7 940	19 096	530
Daily mean values	g/day/m²	22	52	1.5
Maximal daily values	g/day/m²	134	2662	6.0

Table 3. Gas concentrations from swine manure storage tanks

Generally, no N_2O is produced during swine manure storage [17] since anaerobic conditions prevail and the NH_4^+ cannot be oxidized to NO_3^-. N_2O is essentially generated once the slurry has been applied to agricultural land as a fertilizer where both aerobic and anoxic conditions can exist [18].

3.3. Fields

Measuring field emissions is a complex area of research and finding representative data for the phenomena which occur after manure application represents a challenge. That is why this section contains only an overview of data measured by different research groups.

Once manure has been applied to agricultural land as a fertilizer, microorganisms continue to degrade it and release additional compounds. Anaerobic conditions can remain for a short time after manure has been applied; therefore NH_3, CH_4, H_2S and odours are emitted directly following application. Other compounds, such as N_2O, require the combination of aerobic and anaerobic processes and can be produced for a long time after manure application. Certain field emissions (NH_3 and odours mainly) can be reduced by quickly incorporating manure into soil after application. Factors that influence these reactions are soil pH, exchange capacity and weather condition. It is therefore difficult to present typical values of gas emissions. An example of NH_3 emissions is however presented in Figure 3; emissions are presented on a daily basis following manure application.

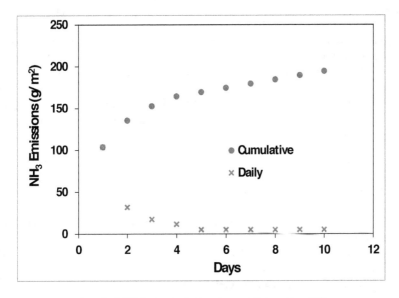

Figure 3. Gases emissions from the field fallowing application of manure (Adapted from [19])

Regarding the production of N_2O, emission values vary greatly. For example, cumulative values measured by [20] for a clay soil cultivated with silage corn varied between 0.255 and 0.873 g/m^2 for the entire growing season (May to October). Although it is only one example, it demonstrates the variation in the measurement of N_2O whereas several factors such as soil type and culture remained constant.

4. Biodegradation as treatment

4.1. Manure

Natural biodegradation phenomena in animal manure can cause harmful emissions, but certain biological processes, whether anaerobic or aerobic, can be used to treat manure. Aerobic biological processes for manure treatment can be relatively simple as in short-term manure aeration, which can remove up to 90% of the biodegradable organic compounds. This process can also significantly reduce odours (up to 96% as evaluated with volatile fatty acids) during manure storage for up to 190 days [21]. Biological processes using suspended biomass, such as aerated lagoons and activated sludge reactors developed for wastewater treatment, can be applied to treat manure [22]. Bioreactors using biomass fixed on a porous filter material can also be used to treat manure, but the solids must be removed prior to treatment in order to avoid clogging problems. The manure can be supplied to these systems from the bottom to obtain a submerged upflow system or from the top and trickle down the filter bed.

Laboratory-scale tests using upflow aerated biological filters showed good results for manure treatment: removal efficiencies of 88% for biodegradable organic matter and 94% for NH_4^+ with two 1.5 m^3 biofilters treating 8 m^3/d of flushed swine manure [23]. In another study, part of the effluent was recirculated to an anoxic reactor at the beginning of the process for complete nitrogen removal [24]. The bioreactor removed 72% of the organic matter as COD, 94% of the NH_4^+ as well as achieving a denitrification rate of 92%. This type of upflow biofilter is available commercially under the name Ekokan® Biofiltration Treatment System and removes between 90 and 98% of the NH_4^+ and between 40 and 70% of the biodegradable organic carbon from swine slurry pre-treated to remove solids [25]. The main disadvantage with this type of system is that the filter bed must be regularly backwashed to remove excess biomass.

In a trickling biofilter, the manure is supplied at the top and flows down through the filter bed. Trickling biofilters have been used for almost 100 years for wastewater treatment [5], but they have only recently been applied to manure treatment. Trickling biofilters can be quite simple consisting of a pile of filter material with passive aeration. However, the performance can be limited; results from preliminary tests showed removal efficiencies up to 56% for biodegradable organic matter and NH_4^+ [26]. Researchers in Québec (Canada) developed a highly engineered system using an enclosed biofilter with forced aeration, the Biosor® biofilter [27]. This type of system can provide a better performance with removal efficiencies up to 99% for the biodegradable organic matter and above 95% for the NH_4^+ [28,29]. Furthermore, a study on the nitrogen elimination mechanisms demonstrated that trickling biofilters can achieve simultaneous nitrification and denitrification which transforms the NH_4^+ directly to N_2 in one system [30]. By means of a mass balance, it was shown that 30% of the nitrogen was eliminated as N_2 and 10% as N_2O. For swine manure, loading rates between 0.017 and 0.035 m^3/m^2/d are generally recommended to avoid clogging problems [29,31]. Due to the high concentrations of nutrients in manure, these values are up to 2000 times lower than the loading rates recommended for wastewater treatment.

While biofiltration can be used to treat liquid manure, composting is a biological system that can treat solid materials to produce a biologically stable product rich in humic compounds [22]. In addition, composting can reach relatively high temperatures (40-60°C) which can reduce pathogenic microorganisms by up to 92% and improve the sanitary quality of the fertilizer produced [32]. Since swine manure is generally managed in liquid form, bulking agents must be added. These additives generally have a high carbon content, such as straw or sawdust, in order to improve the carbon to nitrogen mass ratio of the composting mix. A mass ratio between 25 and 30 is optimal, but values between 15 and 20 can be used to reduce bulking agent requirements. However, this increases reaction time by 30% [33,34]. A major disadvantage of composting is the loss of nitrogen, 10% as NH_3 and 3% as N_2O on average according to [35], which reduces the quality of the fertilizer. Since N_2O is a powerful greenhouse gas, emissions are particularly troubling. By adding nitrite-oxidizing bacteria, [36] were able to reduce N_2O emissions by 80%.

Aerobic bioreactors are usually operated at ambient temperatures with mesophilic microorganisms to maintain low operating costs, but reactors using a thermophilic biomass at temperatures between 50 and 75°C offer interesting advantages. This type of system reduces the quantity of pathogenic microorganisms to improve the sanitary quality of the manure. Furthermore, since nitrification ceases above 40°C, the nitrogen in manure is retained as NH_4^+ [37].

As previously described, anaerobic digestion can be used to treat manure and produce biogas for heat or energy requirements. This process also produces a good quality liquid fertilizer since nitrogen is mainly retained in the liquid. However, the nitrogen in the digestate (liquid effluent of the digester) is mainly NH_4^+ [38] and steps must be taken to reduce NH_3 losses (cover for storage reservoir, incorporation of digestate in soil after spreading, etc.). Manure can be fed to a digester either in batch, steady-state or semi-continuous modes. Batch systems are the simplest systems technically, but biogas production is irregular over time and the reaction rate is temperature dependant. Continuous systems are more complex, but provide consistent biogas production. Several parameters must be controlled for proper digester operation [39]:

• pH (between 6.8 and 7.4)

• dry matter content (maximum value of 14% for proper operation)

• NH_4^+ concentration (must be kept below 3 g/L to avoid inhibiting microorganisms)

• carbon to nitrogen ratio (helps optimise process and reduce reaction time).

A disadvantage specific to anaerobic digesters is the H_2S produced as a by-product which must generally be removed from the biogas to avoid deteriorating equipment.

4.2. Gaseous emissions

Biological treatment systems can also be used to treat gaseous emissions (NH_3, H_2S, greenhouse gases and odours) from manure management in order to improve air quality. Biological treatment of air is based on the capacity of microorganisms to transform organic and inorganic pollutants into non-toxic, odour free compounds.

Biological air treatment units (bioreactors) are an established technology, but research is ongoing for new media and reactor designs, microbial structure analysis and modeling of gas compounds removal [40]. Bioreactors can be used for reducing toxic VOC (volatile organic compound) emissions from industrial sources, but in agricultural applications, pollutant concentrations are lower and bioreactors must be simple, easy to operate and maintain and must meet investment and operating costs below those of the industrial sector.

[40-46] conducted detailed analyses from literature reviews providing interesting information on biological methods for air treatment. Classification of biological reactors, analysis of the mechanisms involved in biological treatment, design of bioreactors and performance analysis.

The basic mechanisms for the biodegradation of pollutants are the same for all biological treatment systems: the contaminant is absorbed from the gas phase (contaminated air) to a liquid phase where biological degradation is initiated. For organic contaminants, oxidation reactions (and sometimes reductions) transform the contaminant in a mixture of CO_2, water vapour and biomass. The air pollutants (organic or inorganic) are used as a source of energy and/or carbon for the development of the microbial population. There are three types of bioreactors with different configurations, which can be used to achieve the transfer between the gas and the liquid phases and promote the microbial metabolic reactions: biofilters, biotrickling filters and bioscrubbers. Such equipment differ by the nature of the microbiological phase (microorganisms attached to the filter bed or suspended in the liquid) and by the circulation mode of the liquid (stationary or flowing) (Table 4).

Reactor	Microorganisms	Liquid phase
Biofilter	Fixed	Stationary
Biotrickling filter	Fixed	Flowing
Bioscrubber	Suspended	Flowing

Table 4. Classification of biological reactors for air treatment [47].

4.3. Biofilters

Biofiltration is the oldest and most widespread biotechnology for the treatment of gaseous emissions. The contaminated gases flow through a humid porous material, usually made of organic waste, where microorganisms capable of degrading the pollutants are present [48]. The microorganisms will grow attached to the material, thereby forming a wet biofilm wherein the air pollutants are absorbed and then degraded by the microorganisms. A liquid nutrient solution can be sprayed periodically over the filter bed to maintain proper moisture levels and to supplement certain nutrients if necessary. The moisture content of the filtration equipment and the maintenance of the biofilm are essential elements for maintaining the performance of this biological reactor. If a biofilter is not irrigated, moisture should be controlled by the humidity of the air fed to the device. This type of control on the moisture content of the filter media is not always effective and the variations in the humidity and temperature of the incoming air can affect the performance of the biofilter [41]. The filter material can also lose its porosity with time; it can even become clogged by excess biomass growth.

The most common type of biofilter is the open biofilter (Figure 4). This equipment can be exposed to atmospheric conditions and can be installed at ground level. Moreover, it usually uses packing materials readily available and affordable (soil or compost for example). The usual height of the filter bed of an open biofilter is between 1.0 and 1.5 m. Open systems are ideal for applications where space is not a constraint and they are known to be the least expensive solutions for odour control [41].

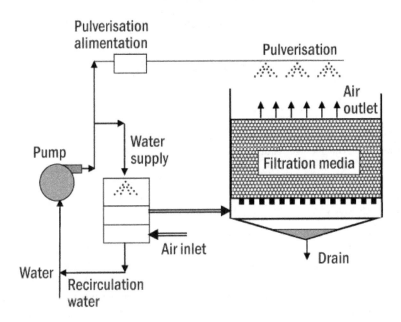

Figure 4. Diagram of an open biofilter system (Adapted from [42])

Closed biofilters (Figure 5) are generally more complex and may have either a circular or rectangular section. These air treatment systems allow a better control of some operating parameters (temperature, moisture, nutrients and pH) while being less sensitive to atmospheric conditions. The filter bed used in closed biofilters generally has a height that varies between 1.0 and 1.5 m and is composed of organic and/or inorganic materials. An air plenum at the inlet and the outlet of the biofilter is generally used for uniform air distribution. For most applications with a closed biofilter, downwards air circulation is more efficient than upward air flow due to a better control of filter bed moisture [47].

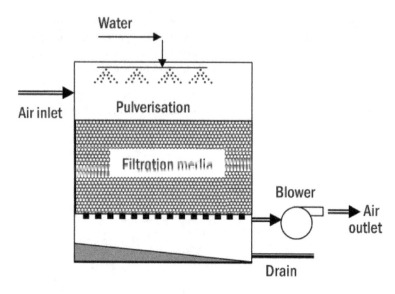

Figure 5. Diagram of a closed biofilter system (adapted from [41])

A study by [49] recommends maintaining the moisture content between 35 and 65% in the filter material. The average reductions of H_2S for low, medium and high relative humidity were 3, 72 and 87 %, respectively. Under the same conditions, the odour and NH_3 reductions were 42, 69 and 79% and 6, 49, and 81%, respectively. The optimal ratio of compost to wood chips recommended by the study for the treatment of air coming from swine buildings is 30% compost and 70% wood chips (on a weight basis).

[50] studied a pilot-scale biofilter treating swine ventilation air to determine the optimum operating conditions. The filter bed had a height of 0.5 m and was built using wood chips of at least 20 mm. The moisture content of the filter bed varied between 64 and 69%. Preliminary tests showed that the installation of a mechanical filter at the air inlet of the biofilter can reduce over 99% of airborne particles with an odour reduction of 19%. During the experiment, the system achieved a removal efficiency between 73 and 87% for NH_3. When the load of NH_3 was increased from 967 to 2 057 mg/h with a maximum volumetric load of 1 898 m³ $_{air}$/m³ $_{filter}$/h, the removal efficiency was reduced by 19%. The study recommended wood chips over 20 mm for biofilters that are used to treat air emitted from swine production facilities. The maximum recommended volumetric load is 1 350 m³ $_{air}$/m³ $_{filter}$/h in order to ensure an odour removal efficiency greater than 90%. In summer operating conditions, the size of the biofilter was 0.148 m²/pig. An efficient humidification system (humidifier at the air inlet and a spraying device above the bedding material) and an adapted air distribution system are determining factors for the design and the operation of treatment systems for high air flow rates.

In another study, [51] compared the effectiveness of two pilot-scale biofilters for the treatment of air from pig barns. The first biofilter used wood chips over 20 mm and the second one used wood chips with dimensions between 10 and 16 mm. The humidity in the filter bed was maintained at 69% and the volumetric load varied between 769 and 1847 $m^3_{air}/m^3_{filter}/h$ for a trial period of 63 days. Both biofilters reduced the odour in the range of 88 to 95%. The reduction of NH_3 was in the range of 64 to 92% for the first biofilter and 69 to 93% for the second. H_2S was reduced by 9 to 66% for the first biofilter while the results for the second ranged from an increase of 147% to a decrease of 51%. The pH was maintained between 6 and 8. Investigations show that there is a risk of forming anaerobic zones in the filter bed (second biofilter) which can release reduced sulphur compounds. The study concluded that biofiltration is an interesting technology for the removal of odour and NH_3 from the air emitted from swine production facilities.

[52] attempted to combine a strategy of minimum ventilation and biofiltration. A minimum air flow rate of 75 $m^3/h/pig$, corresponding to the conditions of summer nights, was established as a reference. The tests were carried out with a biofilter using wood chips with a filter bed height of 27 cm and an area of 80 m^2. The results showed an average removal efficiency of 44% for ammonia, 58% for H_2S and 54% for odours. The results are quite modest, but are partially due to reduced volumes of the treated air.

[53] studied the efficiency of biofilters in reducing NH_3 emitted from livestock buildings. The aim of the research was to test a filter bed composed of non-expensive organic and inorganic materials in combination with a diverse microbial population (multiculture). The tests were conducted on a bench-scale device with a closed-type reactor with a height of 0.5 m. The packing media was composed of peat (91% organic), vermiculite and perlite (ratio 3:1:1). In the second series of tests, the filter material was made from peat and polystyrene (3:1 ratio). The results of the study showed that the removal efficiency of NH_3 can be very high (99 to 100%) under conditions where the inlet concentration is 200 ppmv and flow rates are between 0.03 to 0.06 m^3/h.

A study on a pilot-scale plant by [54] demonstrated the performance of biofilters for odour reduction using different filter materials such as sand, soil, bark and wood mixtures. The reduction of odours analyzed by olfactometry was between 29 and 99.9% with odour concentrations at the inlet ranging between 143 100 and 890 000 OU/m^3. The study highlighted the presence of leachate resulting from wetting of the filter bed. This fluid plays a very important role in maintaining moisture, but it may also have other effects on the quality of the filter bed, such as washing, accumulation of large amounts of pollutants, interference with the airflow, formation of preferential paths, formation of anaerobic zones and release of NH_3 and H_2S. The study demonstrated the need for further research to clarify these aspects that have a direct influence on the performance and longevity of the biofiltration system.

In response to the questions raised regarding the accumulation of nitrogen compounds in the filter bed due to high inlet concentrations of NH_3, a study by Japanese researchers [55] tested the use of a new bacterium (Vibrio alginolyticus), which is able to effectively degrade high concentrations of NH_3. The study demonstrated the feasibility of using this marine bacterium for concentrations of ammonia between 120 and 2 000 ppmv with removal efficiencies greater than 85% for more than 60 days of operation.

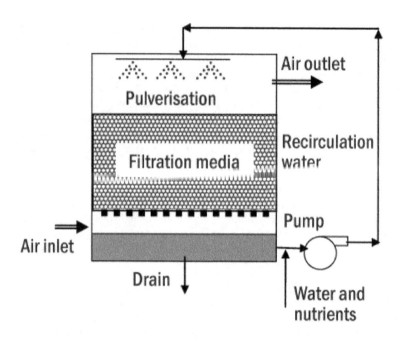

Figure 6. Diagram of a biotrickling filter (adapted from [42])

4.4. Biotrickling filters

Contrary to biofilters, biotrickling filters generally use an inorganic packing material with the liquid solution continuously recirculating over the filter bed (Figure 6). This technology offers many advantages: an easy control of key operating parameters (such as temperature, pH, nutrient supply and concentration of toxic compounds), low pressure drops and reduced space requirements by allowing high flow rates.

Air treatment by biotrickling filters is a relatively new technology and the majority of experimental results are from tests carried out with pilot-scale plants [44]. Various filter media, such as lava rock, random plastic packing, structured blocks of plastic and polyurethane foam have been used. The high porosity of these materials provides a minimal pressure drop on the airflow; higher airflow rates are therefore achievable. One of the main characteristics of biotrickling filters is the continuous flow of the liquid on the filter bed. It is therefore possible to improve process control by the addition of nutrients, adjustment of the pH and the temperature or by removing toxic by-products. For example, in the case of odour reduction and H₂S removal, production of sulfuric acid and reduction of pH and/or accumulation of sodium sulphate are the key controlling parameters. Biotriclickling filters also have other advantages over the other biological treatments in controlling air pollutants [56]: the height of the filter bed, the longer life of the filter media (above 10 years), the ease of control and the ability to treat air containing dust and grease.

The examples cited by [44] show that these reactors have good removal efficiencies for high concentrations of H$_2$S at low residence times (EBRT). Biotrickling filters seem a good option for the treatment of gases with high concentrations of H$_2$S and possibly for other sulfur compounds. Experiments on industrial applications have shown the potential of biofilters and biotrickling filters for the simultaneous removal of odour, H$_2$S and VOCs. From a total of eight cases of biotrickling filters used for the removal of H$_2$S and for inlet concentrations of 1 to 1000 mg/m^3, the reduction efficiencies varied from 95 to 99%. For the reduction of odours, the efficiencies obtained were 65 to 99%.

Biotrickling filters should be inoculated with a variety of microorganisms, since inorganic filter beds generally do not contain bacteria. The addition of nutrients can also become a tool for optimizing the performance of the reactor. The nutrient requirements depend on the type of pollutant to be treated, its concentration, the pollutant loading and the operating conditions of the reactor. However, excess nutrients can lead to an overproduction of biomass and eventually clog the reactor.

A cross-flow biotrickling filter was developed at the IRDA for the treatment of swine ventilation air. Pilot-scale tests were carried out with 6 units treating air from chambers housing 4 grower-finisher pigs. After a start-up phase between 9 and 20 days, the system was able to reduce emissions of NH$_3$ and odours by up to 68 and 82% respectively. Different operating conditions were tested (air residence time, liquid flow rate and type of packing material), but very little effect was observed since the system was probably over-sized. The biotrickling filters had no significant effect on CO$_2$ and CH$_4$ emissions.

4.5. Bioscrubbers

In a bioscrubber, each step of the treatment process is separate: the pollutants are first transferred to a liquid solution in an absorption unit and then the washing liquid is regenerated in a biological reactor which generally resembles an activated sludge reactor (Figure 7). Similar to the biotrickling filter, the operating conditions can easily be controlled in a bioscrubber and it is possible to optimize both the absorption and biodegradation of pollutants. [43] reported several types of absorbers such as the packed tower, the wet cyclone, the spray tower and the venturi scrubber.

The flow of air and water can either be counter-current, co-current or cross-current. The air velocity can vary between 1.5 m/s and 20 m/s in a spray tower, it can reach 25 m/s for a wet cyclone and between 40 and 50 m/s for a venturi. Bioscrubbers have the same space, flexibility and control than biotrickling filter, but they are only suitable for the treatment of highly water soluble compounds (Henry's coefficient below 0.01). Bioscrubbers have thereby increased the scope of application for the biological treatment of waste gases. The greatest advantage of bioscrubbers compared to biofilters and biotrickling filters is the ability to produce and maintain large amount of active microbial mass in smaller units. On the other hand, [57] considers that bioscrubbers and biotrickling filters present greater construction and operation complexity.

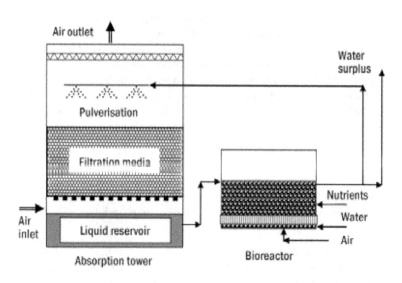

Figure 7. Diagram of a bioscrubber (adapted from [42])

[43] concluded that bioscrubbers offer the greatest efficiencies for the removal of H_2S, NH_3 and organic sulphur compounds. The performance analysis of bioscrubbers used in the industry for H_2S removal showed efficiencies over 98% for low, medium and high inlet concentrations (between 0 and 75 mg/m^3, 2 000 mg/m^3 and between 10 000 and 15 000 mg/m^3 respectively). The results cited for the reduction of odours show an efficiency of 80% for the reduction of organic sulfides with inlet concentrations between 4 000 and 22 000 OU.

There are many technologies available for the reduction of air contaminants and odours using biological reactions. Several systems have configurations that are similar to each other though different in terms of operating conditions. The difficulty of using one of the existing technologies for the treatment of air comes from the specific constraints of each application. Livestock buildings in general are characterized by a very large number of parameters that influence the application of air treatment.

Biofiltration has been the subject of many scientific publications and several units are currently installed throughout the world. Its efficiency has been demonstrated for reducing odours and to a lesser extent, for the reduction of NH_3 and H_2S emitted from barns. However, despite the advantage of being simple, the use of biofilters in livestock buildings is limited by several problems such as the accumulation of pollutants, the potential for clogging, the high pressure losses and the relatively rapid degradation of the filter bed. On the other hand, even if biotrickling filters and bioscrubbers have better features compared to biofiltration (e.g. rapid response to changes in operating conditions and longer life time), experimental research on these two technologies has just started and only specific solutions for specific applications and partial results are available. There are not many experimental studies on full-scale systems.

[45] highlighted that the technologies with the greatest potential for reducing the contaminants emitted from livestock buildings should, in all likelihood, come from the combination of different treatment systems. A combination of an air scrubber and a biotrickling filter or a combination of a biofilter and a biotrickling filter could provide greater capabilities than each technology used individually. However, there is no information in the literature on the efficiency of these different combinations of technologies.

5. Conclusion

This chapter outlined the importance of biodegradation associated with modern manure management practices. Depending on the type of management strategy and the production stage, both aerobic and anaerobic conditions can prevail, providing a wide range of emissions. The second part of the chapter explored the different technologies where biodegradation can be used to treat both the manure and the gaseous emissions. Biological treatment systems generally provide good removal efficiencies at relatively low costs and are well adapted to the agricultural sector.

Author details

Matthieu Girard*, Joahnn H. Palacios, Martin Belzile, Stéphane Godbout and Frédéric Pelletier

*Address all correspondence to: matthieu.girard@irda.qc.ca

Research and Development Institute for the Agri-Environment (IRDA), Québec (Qc), Canada

References

[1] Godbout, S. and Trudelle, M. 2002. Évaluation des performances techniques des séparateurs mécaniques à lisier et de leur rapport efficacité/coût. Final Report, IRDA, 84 pages.

[2] Andreottola, G., Bortone, G., and Tilche, A. 1997. Experimental validation of a simulation and design model for nitrogen removal in sequencing batch reactor. Water, Science and Technology, 35(1): 113-120.

[3] Boursier, H., Béline, F., and Paul, E. 2005. Piggery wastewater characterisation for biological nitrogen removal process design. Bioresource Technology, 96: 351-358.

[4] Aubry, G. 2008. Étude du comportement de l'azote dans un biofiltre à lit ruisselant traitant du lisier de porc. Ph.D. thesis, Department of Civil Engineering, Laval University, Quebec City, Qc. Canada.

[5] Metcalf and Eddy. 2003. Wastewater engineering: treatment and reuse - 4th edition. McGraw-Hill, New-York.

[6] Henze, M., Harremoës, P., la Cour Jansen, J., and Arvin, E. 2002. Wastewater treatment: biological and chemical processes – 3rd edition. Springer, New York, United States.

[7] Schiffman, S.S., J.L. Bennett and J.H. Raymer. 2001. Quantification of odors and odorants from swine operations in North Carolina. Agric. Forest Meteor. 108(3): 213–240.

[8] Møller, H.B., S.G. Sommer and B.K. Ahring. 2002. Separation e•ciency and particle size distribution in relation to manure type and storage conditions. Bioresource Technology 85, 189–196.

[9] Natural Resources Conservation Service. 2007. Manure chemistry - Nitrogen, Phosphorus, & Carbon. Manure management information sheet, number 7.

[10] Hamilton, D. W. 2009. Anaerobic Digestion of Animal Manures: Understanding the Basic Processes. Oklahoma Cooperative Extension Service. BAE-1747.

[11] Godbout, S., L. Hamelin, J. H. Palacios, F. Pelletier, S.P. Lemay and F. Pouliot. 2011. État de référence des émissions gazeuses et odorantes provenant des bâtiments porcins québécois. Final report. IRDA. 53 pages.

[12] Marquis, A. and P. Marchal. 1998. Qualité de l'atmosphère à proximité des bâtiments d'élevage. Cahiers d'études et de recherches francophones – Agricultures. 7(5): 377-385.

[13] Hamelin, L., M. Wesnæs, H. Wenzel and B.M. Petersen. 2010. Life cycle assessment of slurry management technologies II – emphasis on biogas production. Danish Ministry of the Environment, Environmental Protection Agency. http://www2.mst.dk/udgiv/publications /2010/978-87-92668-03-5/pdf/978-87-92668-04-2.pdf. Accessed January 30, 2013.

[14] IPCC. 2007. Fourth Assessment Report (AR4) by Working Group 1 (WG1), Chapter 2 "Changes in Atmospheric Constituents and in Radiative Forcing".

[15] EPA. Environmental Protection Agency. 2010. Particulate matter. http://www.epa.gov/pm/. Accessed January 30, 2013.

[16] Cambra-López, M., A.J.A. Aarnink, Y. Zhao, S. Calvet and A.G. Torres. 2010. Airborne particulate matter from livestock production systems: A review of an air pollution problem. Environmental Pollution, 158: 1–17.

[17] Chadwick, D.R., Sneath, R.W., Phillips, V.R., and Pain, B.F. 1999. A UK inventory of nitrous oxide emissions from farmed livestock. Atmospheric Environment, 33: 3345-3354.

[18] Velthof, G.L., Nelemans, J.A., Oenema, O., and Kuikman, P.J. 2005. Gaseous nitrogen and carbon losses from pig manure derived from different diets. Journal of Environmental Quality, 34: 698-706.

[19] Chantigny, M.H., MacDonald, J.D., Beaupré, C., Rochette, P., Angers, D.A., Massé, D.I., and Parent, L.-É. 2009. Ammonia volatilization following surface application of raw and treated liquid swine manure. Nutrient Cycling in Agroecosystems, 85(3), p. 275-286.

[20] Perron, M-H. 2010. Disponibilité de l'azote de biosolides de traitement de lisier de porc et de deux boues de papetière et émissions de N2O consécutives à leur épandage au champ. Master thesis. Laval University, Québec, Canada. 138 pages.

[21] Zhang, Z., and Zhu, J. 2006. Characteristics of solids, BOD5 and VFAs in liquid swine manure treated by short-term low-intensity aeration for long-term storage. Bioresource Technology, 97: 140-149.

[22] BAPE. 2003. L'état de la situation de la production porcine au Québec. Rapport d'enquête et d'audience publique du Bureau d'audiences publiques sur l'environnement, ISBN: 2-550-41393-8.

[23] Westerman, P.W., Bicudo, J.R., and Kantardjieff, A. 2000. Upflow biological aerated filters for the treatment of flushed swine manure. Bioresource Technology, 74: 181-190.

[24] Lanoue, M. 1998. Traitement du lisier de porc par procédé de biofiltration aérobie. M.Sc. thesis, Department of Chemical Engineering, Université de Sherbrooke, Sherbrooke, Qc. Canada.

[25] Westerman, P.W. and Arogo, J. 2004. Ekokan biofiltration technology performance verification. Available from the North Carolina State University's College of Agriculture and Life Sciences: http://www.cals.ncsu.cdu/waste_mgt/smithfield_ projects/ phase1report04/A.6EKOKAN%20final%20.pdf. [Accessed in January 2013].

[26] Sommer, S.G., Mathanpaal, G., and Dass, G.T. 2005. A simple biofilter for treatment of pig slurry in Malaysia. Environmental Technology, 26: 303-312.

[27] Buelna, G. 2000. Biofilter for purification of waste waters and method therefore. United States Patent Number 6,100,081.

[28] Dubé, R. 1997. Traitement du lisier de porc par biofiltration sur milieu organique : influence de l'aération. M.Sc. thesis, Department of Civil Engineering, Laval University, Quebec, Qc. Canada.

[29] Aubry, G., Lessard, P., Gilbert, Y., Le Bihan, Y., and Buelna, G. 2006. Nitrogen behaviour in a trickling biofilter treating pig manure, Water Science and Technology, IWA Conference BIOFILMS Systems VI, Amsterdam.

[30] Garzón-Zúñiga, M., Lessard, P., Aubry, G., and Buelna, G. 2005. Nitrogen elimination mechanisms in an organic media aerated biofilter treating pig manure. Environmental Technology, 26: 361-371.

[31] Garzón-Zúñiga, M., Lessard, P., Aubry, G., and Buelna, G. 2007. Aeration effect on the efficiency of swine manure treatment in a trickling filter packed with organic materials. Water, Science and Technology, 55(10): 135-143.

[32] Ros, M., Garcia, C., and Hernández, T. 2005. A full-scale study of treatment of pig slurry by composting: kinetic changes in chemical and microbial properties. Waste Management, 26: 1108-1118.

[33] Huang, G.F., Wong, J.W.C., Wu, Q.T. and Nagar, B.B. 2004, Effect of C/N on composting of pig manure with sawdust. Waste Management, 24: 805–813.

[34] Zhu, N. 2007. Effect of low initial C/N ratio on aerobic composting of swine manure with rice straw. Bioresource Technology, 98: 9–13.

[35] Hassouna, M., Espagnol, S., Robin, P., Paillat J-M., Levasseur, P. and Li, Y. 2008. Monitoring NH3, N2O, CO2 and CH4 emissions during pig solid manure storage - effect of turning. Compost Science & Utilization. 16(4): 267-274.

[36] Fukumoto, Y., Suzuki, K., Osada, T., Kuroda, K., Hanajima, D., Yasuda, T., and Haga, K. 2006. Reduction of nitrous oxide emission from pig manure composting by addition of nitrite-oxidizing bacteria. Environmental Science and Technology, 40: 6787-6791.

[37] Juteau, P. 2006. Review of the use of aerobic thermophilic bioprocesses for the treatment of swine waste, Livestock Science, 102: 187– 196.

[38] Ortenblad, H. 2000. The use of digested slurry within agriculture. Available from: http://homepage2.nifty.com/biogas/cnt/refdoc/whrefdoc/d9manu.pdf [accessed in January 2013.]

[39] Ricard, M.-A., V. Drolet, A. Coulibaly, C.B. Laflamme et al. 2010. Développer un cadre d'analyse et identifier l'intérêt technico-économique de produire du biogaz à la ferme dans un contexte québécois. Report produced by the Centre de développement du porc du Québec (CDPQ), ISBN 978-2-922276-35-0, 110 pages.

[40] Deshusses, M., Z. Shareefdeen. 2005. Modeling of biofilter and biotrickling filters for odour and VOC control applications. Biotechnology for odour and air pollution control. ed. Z. Shareefdeen and A. Singh. Springer. Verlag, Berlin, Heidelberg.

[41] Devinny, J.S., M.A. Deshusses, T.S. Webster. 1999. Biofiltration for air pollution control. Lewis Publishers. Washington, DC, USA.

[42] Revah, S. and J.M. Morgan-Sagastume. 2005. Methods of odor and VOC control. Biotechnology for odour and air pollution control. ed. Z. Shareefdeen and A. Singh. Springer. Verlag, Berlin, Heidelberg.

[43] Singh, A., Z. Shareefdeen and O.P. Ward. 2005. Bioscrubber technology. Biotechnology for odour and air pollution control. ed. Z. Shareefdeen and A. Singh. Springer. Verlag, Berlin, Heidelberg.

[44] Iranpour, R., H.H.J. Cox, M.A. Deshusses and E.D. Schroeder. 2005. Literature review of air pollution control biofilters and biotrickling filters for odor and volatile organic compound removal. Environmental Progress. 24 (3).

[45] Lemay S.P., M. Martel, M. Belzile, D. Zegan, J. Feddes, S. Godbout, F. Pelletier. 2009. A systematic literature review to identify an air contaminant removal technology for swine barn exhaust air. Written for presentation at the CSBE/SCGAB 2009 Annual Conference Rodd's Brudenell River Resort, Prince Edward Island, Canada. 12-15 July 2009.

[46] Godbout S., S.P. Lemay, F. Pelletier, M. Belzile, J.P. Larouche, L.D. Tamini, J. H. Palacios and D. Zegan. 2010. Réduction des émissions gazeuses et odorantes aux bâtiments porcins : techniques simples et efficaces applicables à la ferme. Final Report. IRDA. 143p.

[47] Godbout, S., F. Pelletier, J.P. Larouche, M. Belzile, J.J.R. Feddes, S. Fournel, S.P. Lemay and J.H. Palacios. 2012. Greenhouse Gas Emissions from Non-Cattle Confinement Buildings: Monitoring, Emission Factors and Mitigation. Greenhouse Gases - Emission, Measurement and Management, Guoxiang Liu (Ed.), ISBN: 978-953-51-0323-3, InTech, DOI: 10.5772/31948.

[48] Delhoménie, M.-C, and Heitz, M. 2005. Biofiltration of air: a review. Critical Reviews in Biotechnology, 25(1-2): 53-72.

[49] Nicolai, R.E. and Janni, K.A. 2001. Biofilter media mixture ratio of wood chips and compost treating swine odors. Dept of Biosystems and Agricultural Engineering, University of Minnesota, USA.

[50] Sheridan, B. A., T. P. Curran and V. A. Dodd. 2002. Assessment of the influence of media particle size on the biofiltration of odorous exhaust ventilation air from a piggery facility. Bioresource Technology. 84: 129-143.

[51] Sheridan, B. A., T. Curran, V. Dodd and J. Colligan. 2002. Biofiltration of odour and ammonia from a pig unit - A pilot-scale study. Biosystems Engineering. 82 (4): 441-453.

[52] Hoff, S. J. and J. D. Harmon. 2006. Biofiltration of the critical minimum ventilation exhaust air. Workshop on Agricultural Air Quality. Washington, D.C., USA.

[53] Kalingan, A. E., C. M. Liao, J. W. Chen and S. C. Chen. 2004. Microbial Degradation of Livestock -Generated Ammonia Using Biofilters at Typical Ambient Temperatures. Journal of Environmental Science and Health. B 39 (1): 185-198.

[54] Luo, J. 2001. A pilot-scale study on biofilters for controlling animal rendering process odours. Water Science and Technology. 44 (9): 277-285.

[55] Kim, N. J., Y. Sugano, M. Hirai and M. Shoda. 2000. Removal of a high load of ammonia gas by a marine bacterium, Vibrio Alginolyticus. Journal of Bioscience and Bioengineering. 90 (4): 410-415.

[56] Deshusses, M. A. and D. Gabriel. 2005. Biotrickling filter technology. Biotechnology for odour and air pollution control. ed. Z. Shareefdeen and A. Singh. Springer. Verlag, Berlin, Heidelberg.

[57] Kraakman, A. 2005. Biotrickling and Bioscrubber Applications to control odour and air pollutants. Biotechnology for odour and air pollution control. ed. Z. Shareefdeen and A. Singh. Springer. Verlag, Berlin, Heidelberg.

Permissions

The contributors of this book come from diverse backgrounds, making this book a truly international effort. This book will bring forth new frontiers with its revolutionizing research information and detailed analysis of the nascent developments around the world.

We would like to thank Rolando Chamy and Francisca Rosenkranz, for lending their expertise to make the book truly unique. They have played a crucial role in the development of this book. Without their invaluable contribution this book wouldn't have been possible. They have made vital efforts to compile up to date information on the varied aspects of this subject to make this book a valuable addition to the collection of many professionals and students.

This book was conceptualized with the vision of imparting up-to-date information and advanced data in this field. To ensure the same, a matchless editorial board was set up. Every individual on the board went through rigorous rounds of assessment to prove their worth. After which they invested a large part of their time researching and compiling the most relevant data for our readers. Conferences and sessions were held from time to time between the editorial board and the contributing authors to present the data in the most comprehensible form. The editorial team has worked tirelessly to provide valuable and valid information to help people across the globe.

Every chapter published in this book has been scrutinized by our experts. Their significance has been extensively debated. The topics covered herein carry significant findings which will fuel the growth of the discipline. They may even be implemented as practical applications or may be referred to as a beginning point for another development. Chapters in this book were first published by InTech; hereby published with permission under the Creative Commons Attribution License or equivalent.

The editorial board has been involved in producing this book since its inception. They have spent rigorous hours researching and exploring the diverse topics which have resulted in the successful publishing of this book. They have passed on their knowledge of decades through this book. To expedite this challenging task, the publisher supported the team at every step. A small team of assistant editors was also appointed to further simplify the editing procedure and attain best results for the readers.

Our editorial team has been hand-picked from every corner of the world. Their multi-ethnicity adds dynamic inputs to the discussions which result in innovative outcomes. These outcomes are then further discussed with the researchers and contributors who give their valuable feedback and opinion regarding the same. The feedback is then collaborated with the researches and they are edited in a comprehensive manner to aid the understanding of the subject.

Apart from the editorial board, the designing team has also invested a significant amount of their time in understanding the subject and creating the most relevant covers. They scrutinized every image to scout for the most suitable representation of the subject and create an appropriate cover for the book.

The publishing team has been involved in this book since its early stages. They were actively engaged in every process, be it collecting the data, connecting with the contributors or procuring relevant information. The team has been an ardent support to the editorial, designing and production team. Their endless efforts to recruit the best for this project, has resulted in the accomplishment of this book. They are a veteran in the field of academics and their pool of knowledge is as vast as their experience in printing. Their expertise and guidance has proved useful at every step. Their uncompromising quality standards have made this book an exceptional effort. Their encouragement from time to time has been an inspiration for everyone.

The publisher and the editorial board hope that this book will prove to be a valuable piece of knowledge for researchers, students, practitioners and scholars across the globe.

List of Contributors

Edixon Gutiérrez
Centro de Investigación del Agua. Facultad de Ingeniería, Universidad del Zulia, Maracaibo, estado Zulia, Venezuela

Yaxcelys Caldera
Laboratorio de Investigaciones Ambientales, Núcleo Costa Oriental del Lago, Universidad del Zulia Cabimas, estado Zulia, Venezuela

Mehdi Hassanshahian
Department of Biology - Faculty of Science- Shahid Bahonar University of Kerman - Kerman, Iran

Simone Cappello
Istituto per l'Ambiente Marino Costiero (IAMC) - C.N.R. U.O.S. di Messina, Italy
Istituto Sperimentale Talassografico (IST) di Messina, Italy

Isabel Natalia Sierra-Garcia and Valéria Maia de Oliveira
Microbial Resources Division, Research Center for Chemistry, Biology and Agriculture (CPQBA), University of Campinas, Campinas, Sao Paulo, Brazil

Beatriz Pérez-Armendáriz, Amparo Mauricio-Gutiérrez and Angélica Santieste-ban-López
Universidad Popular Autónoma del Estado de Puebla, Interdisciplinary Center for Postgraduate Studies, Research and Consulting, Santiago. CP, Puebla, Mexico

Teresita Jiménez-Salgado and Armando Tapia-Hernández
Benemérita Universidad Autónoma de Puebla. ICUAP, Centro de Investigaciones en Ciencias Microbiológicas, Laboratorio de Microbiología del Suelo. Edificio J 1er. Piso, C.U., Puebla, Mexico

Magdalena Urbaniak
European Regional Centre for Ecohydrology under the auspices of UNESCO, Łódź, Poland
University of Łódź, Department of Applied Ecology, Łódź, Poland

Álvaro Torres and David Jeison
Chemical Engineering Department, Universidad de La Frontera, Temuco, Chile

Fernando G. Fermoso, Bárbara Rincón and Rafael Borja
Instituto de la Grasa (CSIC). Avenida Padre García Tejero, Sevilla, Spain

Jan Bartacek
Department of Water Technology and Environmental Engineering, Institute of Chemical Technology Prague, Prague, Czech Republic

Mohamed Abdallah and Kevin Kennedy
Department of Civil Engineering, University of Ottawa, Ottawa, Canada

Zeynep Cetecioglu, Samet Azman, Nazli Gokcek and Nese Coskun
Istanbul Technical University, Environmental Engineering Department, Istanbul, Turkey

Bahar Ince
Boguzk I University, Institute of Environmental Sciences, Istanbul, Turkey

Abid Ali Khan
Department of Civil Engineering, IIT Roorkee, India
Royal HaskoningDHV, India

Absar Ahmad Kazmi
Department of Civil Engineering, IIT Roorkee, India

Rubia Zahid Gaur
Water & Sanitation Specialist, Plan Environ, H-273, GK1, New Delhi, India

Beni Lew
The Volcani Center, Institute of Agriculture Engineering, Bet Dagan, Israel
Department of Civil Engineering, Ariel University Center of Judea and Samaria, Ariel, Israel

Matthieu Girard, Joahnn H. Palacios, Martin Belzile, Stéphane Godbout and Frédéric Pelletier
Research and Development Institute for the Agri-Environment (IRDA), Québec (Qc), Canada

Printed in the USA
CPSIA information can be obtained
at www.ICGtesting.com
JSHW011450221024
72173JS00004B/1016